< QUICK >
< CONCISE >
< PRACTICAL >

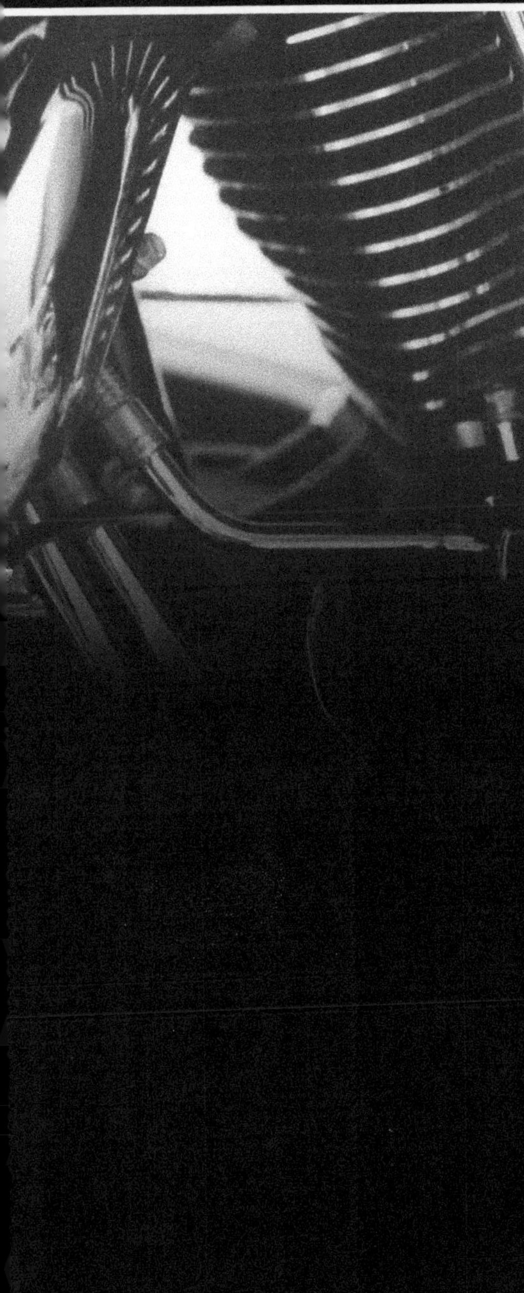

Microsoft .NET Compact Framework Kick
up to speed on how to use Smart Devi
Visual Studio 7.1 to write, deploy, and
that target the .NET Compact Framew

If you have a Windows CE device and you wonder how to get the .NET Compact Framework running on it, this book can help. It details how to get around the most common obstacles developers face when trying to target Windows CE devices.

Who Should Read This Book

This book is aimed at developers already familiar with C# or Visual Basic .NET. It does not waste your time introducing elements of language syntax or programming constructs. If you are an experienced .NET Framework developer, this book can help you understand the functionality available to you on the .NET Compact Framework.

All of the topics covered in this book are treated with enough detail so that a developer who knows nothing about the topic can master it by the end of the chapter. Each chapter is filled with code snippets and full sample applications to drive home the chapter's major points. All code snippets and sample applications are presented in both in C# and Visual Basic.

Microsoft .NET Compact Framework Kick Start covers the following topics:

- Setting up the development environment
- Targeting nonstandard Windows CE devices
- Understanding the connectivity layer between Visual Studio and your device
- Deploying the .NET Compact Framework manually
- Designing .NET Compact Framework applications by using the Visual Studio.NET Form Designer
- Optimizing your code for the Just-In-Time compilers used by the .NET Compact Framework
- Avoiding code pitching
- Using the .NET Compact Framework performance counters
- Packaging a .NET Compact Framework application
- Building a application CAB file by using the CAB Wizard
- Distributing a .NET Compact Framework application
- DiffGram support on the .NET Compact Framework
- Investigating the XML Web service architecture
- Creating a Web service client

Erik Rubin
Ronnie Yates

Microsoft® .NET Compact Framework

KICK START

800 East 96th Street, Indianapolis, Indiana 46240

Microsoft .NET Compact Framework Kick Start

International Standard Book Number: 0-672-32570-5

Library of Congress Catalog Card Number: 203093139

Printed in the United States of America

First Printing: September 2003

This product is printed digitally on demand.

Trademarks

All terms mentioned in this book that are known to be trademarks or service marks have been appropriately capitalized. Sams Publishing cannot attest to the accuracy of this information. Use of a term in this book should not be regarded as affecting the validity of any trademark or service mark.

Warning and Disclaimer

Every effort has been made to make this book as complete and as accurate as possible, but no warranty or fitness is implied. The information provided is on an "as is" basis. The authors and the publisher shall have neither liability nor responsibility to any person or entity with respect to any loss or damages arising from the information contained in this book.

Bulk Sales

Sams Publishing offers excellent discounts on this book when ordered in quantity for bulk purchases or special sales. For more information, please contact

U.S. Corporate and Government Sales
1-800-382-3419
corpsales@pearsontechgroup.com

For sales outside of the U.S., please contact

International Sales
1-317-428-3341
international@pearsontechgroup.com

Associate Publisher
Michael Stephens

Acquisitions Editor
Neil Rowe

Development Editor
Mark Renfrow

Managing Editor
Charlotte Clapp

Project Editor
Matthew Purcell

Copy Editor
Publication Services, Inc.

Indexer
Publication Services, Inc.

Proofreader
Publication Services, Inc.

Technical Editors
Narayana Rao
Surapaneni
Reena Pramodh
Dhananjay Katre
Prasad Naik
from Patni Computers
Systems Ltd., India

Team Coordinator
Cindy Teeters

Interior Designer
Gary Adair

Cover Designer
Gary Adair

Page Layout
Publication Services, Inc.

Graphics
Tammy Graham
Laura Robbins

Contents at a Glance

Table of Contents

About the Authors

Erik Rubin has been with Microsoft since 2001. While there, he worked on portions of the .NET Compact Framework and currently is a Software Design Engineer on the Smart Device Extensions team.

Erik has been involved with software development for all of his professional life. Before coming to Microsoft, he worked as an embedded software developer and became interested with managed code while at Virginia Tech. When he is not working with computers, he and his wife spend time with their three dogs and daughter.

Erik received his bachelor of science degree in electrical engineering from the Pennsylvania State University. He holds two master of science degrees in computer science, from West Chester University and Virginia Polytechnic Institute and State University.

Ronnie Yates is a Software Design Engineer on the .NET Compact Framework development team. He has been involved with the .NET Compact Framework from its origin. He works on the XML, ADO.NET, and Web Services libraries.

Ronnie received his bachelor of science degree from Loyola University in New Orleans, LA. He has worked at Microsoft since his graduation in December 1999.

Dedication

This book is dedicated to Adalia Mae. —Erik Rubin

This book is dedicated to the four special women in my life: my darling wife Moya, my loving mother Dianne, my beautiful sister Veronica, and my wonderful mother-in-law Montez. —Ronnie Yates

Acknowledgments

I first want to thank my coauthor, Ronnie Yates, without whose hard work this book would not have been possible. I also appreciate all of the hard work that the editorial staff at SAMS Publishing has contributed to this book.

I wish to thank Microsoft's Seth Demsey, David Rasmussen, and Bruce Hamilton for providing the `ControlInvoker`. They created a very elegant solution to a problem many developers will face and were very helpful and unselfish in sharing it. Additionally, I want to thank Ximing Zhou and the rest of the Smart Device Extensions team for their support and willingness to answer questions.

Finally, I want to thank my wife, Charsa, who was so helpful and supportive while I was writing this book. She made a difficult undertaking much easier to bear.

—Erik Rubin

I want to thank Neil Rowe, the acquisitions editor, and Jan Fisher, the project manager, for all of their hard work. I also want to thank the development editor, Mark Renfrow, the sponsoring editor, Peter Nelson, the project editor, Matt Purcell, and the copyeditors, Eric Tucker, Jessica Barringer, and Alysia Cooley. Also, thanks to the technical editors who kept me in line.

I want to thank all my teammates at Microsoft for answering any questions I asked and for not kicking me out of their offices even when I got a bit annoying. Huge thanks goes to my coauthor, Erik Rubin, for getting me involved in this project.

Finally, I want to thank my wife, Moya, for all her support and encouragement while I was writing this book.

—Ronnie Yates

We Want to Hear from You!

As the reader of this book, *you* are our most important critic and commentator. We value your opinion and want to know what we're doing right, what we could do better, what areas you'd like to see us publish in, and any other words of wisdom you're willing to pass our way.

As an associate publisher for Sams Publishing, I welcome your comments. You can email or write me directly to let me know what you did or didn't like about this book—as well as what we can do to make our books better.

Please note that I cannot help you with technical problems related to the topic of this book. We do have a User Services group, however, where I will forward specific technical questions related to the book.

When you write, please be sure to include this book's title and author as well as your name, email address, and phone number. I will carefully review your comments and share them with the author and editors who worked on the book.

Email: `feedback@samspublishing.com`

Mail: Associate Publisher
Sams Publishing
800 East 96th Street
Indianapolis, IN 46240 USA

For more information about this book or another Sams Publishing title, visit our Web site at `www.samspublishing.com`. Type the ISBN (0672325705) or the title of a book in the Search field to find the page you're looking for.

Introduction

Welcome to the .NET Compact Framework, the fast, trimmed-down version of the Microsoft .NET Framework tailored especially for devices. The .NET Compact Framework is Microsoft's next-generation development framework for easily creating complex applications for the Pocket PC and Windows CE platforms. Using the .NET Compact Framework, developers can target Windows CE using C# or Visual Basic. For the first time, developers need not become experts in writing for Windows CE in C/C++, nor need they settle for the previous, watered-down Visual Basic dialect, Embedded Visual Basic.

This book will bring you up to speed on how to use Smart Device Extensions in Visual Studio 7.1 to write, deploy, and test applications that target the .NET Compact Framework.

If you have a Windows CE device and you wonder how to get the .NET Compact Framework running on it, this book can help. It details how to get around the most common obstacles developers face when trying to target Windows CE devices.

This book is aimed at developers already familiar with either C# or Visual Basic .NET. It does not spend any time introducing elements of language syntax or programming constructs. If you are an experienced .NET Framework developer, this book can help you understand what functionality is available to you on the .NET Compact Framework.

All of the topics covered in this book are treated with enough detail that a developer who knows nothing about the topic can master it by the end of the chapter. Each chapter is filled with code snippets and full sample applications to drive home the chapter's major points. All code snippets and sample applications are presented in both in C# and Visual Basic. Some sample listings include input data or special steps needed to use the code. Such information is always included after the C# and Visual Basic code listings.

What's in This Book

This book covers a full spectrum of topics, teaching readers how to solve a wide variety of problems by using the .NET Compact Framework, such as these:

- Setting up the development environment
- Targeting nonstandard Windows CE devices
- Understanding the connectivity layer between Visual Studio and your device
- Deploying the .NET Compact Framework manually
- Understanding the Common Language Runtime for the .NET Compact Framework

- Optimizing your code for the Just-In-Time compilers used by the .NET Compact Framework
- Avoiding code pitching
- Designing .NET Compact Framework applications by using the Visual Studio.NET Form Designer
- Designing applications for different target platforms
- Working with the .NET Compact Framework UI controls
- Multithreading on the .NET Compact Framework
- Using `Timers` on the .NET Compact Framework
- Waiting for a thread to finish
- Invoking user interface elements from outside threads
- Coordinating threads with a `Mutex`
- Managing threads with a thread pool
- Atomic variable updates
- Controlling access to data objects with a `Monitor`
- Connecting with sockets to a remote party by using TCP/IP
- Accepting socket connections from a remote party by using TCP/IP
- Reading and writing data by using a connected socket
- Communicating with the UDP protocol and multicasting
- Serializing objects
- Interacting with HTTP and HTTPS servers
- Communicating with another device by using the infrared port
- Creating and populating a `DataSet`
- Extracting and altering data in a `DataSet`
- Adding constraints to a `DataSet`
- Setting up parent-child relationships in a `DataSet`
- Defining columns with an expression
- Setting up an autoincremented column
- Creating sorted views of data with a `DataView`

- Binding a `DataView` to a `DataGrid`
- Creating a Microsoft SQL Server CE database
- Interacting with a Microsoft SQL Server CE database by using the SQL CE managed providers
- Using the `DataSet` to manage a Microsoft SQL Server CE database
- Loading and saving a `DataSet` as XML by using files and streams
- Using XML schema to describe a `DataSet`
- Using the Visual Studio XML editor to alter the contents of a `DataSet`
- Specifying how the `DataSet` treats schema information
- DiffGram support on the .NET Compact Framework
- Investigating the XML Web service architecture
- Creating a Web service client
- Consuming a Web service that exposes a `DataSet`
- Reading XML documents with the `XmlTextReader`
- Writing XML documents with the `XmlTextWriter`
- Parsing XML documents with the `XmlDocument`
- Using reflection to dynamically load .NET assemblies
- Dynamically discovering type information by using reflection
- Instantiating objects by using reflection
- Invoking object members by using reflection
- Encrypting and decrypting data by using a password or a session key
- Creating, storing, loading, and sharing session keys
- Computing a hash
- Measuring the performance of a .NET Compact Framework application
- Using the .NET Compact Framework performance counters
- Packaging a .NET Compact Framework application
- Building an application CAB file by using the CAB Wizard
- Distributing a .NET Compact Framework application
- Targeting the SmartPhone with the .NET Compact Framework

The sample applications for this book are available for download from the Sams Web site at `www.samspublishing.com`. All of the sample applications for this book are written in both C# and Visual Basic.

To download the code on the Sams Web site at `www.samspublishing.com`, enter this book's ISBN (without the hyphens) in the Search box and click Search. When the book's title is displayed, click the title to go to a page where you can download the code.

Conventions Used in This Book

The following typographic conventions are used in this book:

- Code lines, commands, statements, variables, and code-related terms appear in a `monospace` typeface.
- You'll see sidebars throughout the book, which are meant to give you something more, such as a little more insight or some new technique. Here's what a sidebar looks like:

SHARING ENCRYPTED TEXT

Although password-based encryption and decryption is convenient, the resulting encrypted text is not necessarily portable to other devices. If your application calls for encrypting and decrypting data that can be securely exchanged between two devices, see the section titled "Encrypting and Decrypting by Using a Handle to a Session Key."

What You'll Need

To use this book, you will need Visual Studio .NET Enterprise Developer or Enterprise Architect editions, version 7.1. The Professional edition does not include the Smart Device Extensions through which you can develop applications that target the .NET Compact Framework.

Smart Device Extensions and the .NET Compact Framework enable you to target Pocket PC and Windows CE devices. It is strongly recommended that you have a device like the one you plan to support with your projects so that you can test your work. If you do not have a compatible device with which to test, you can use one of the device emulators included with the Smart Device Extensions.

Chapter 1, "Setting Up Your Development Environment," describes specifically what Pocket PC and Windows CE devices are supported by the Smart Device Extensions. Also, Chapter 1 holds information about what specifications your desktop computer needs to use the Smart Device Extensions.

Online Resources

There are several places on the Internet where you can get help with the .NET Compact Framework and participate in discussions about it. There is an official Usenet group at `http://microsoft.public.dotnet.framework.compactframework`. Also, you can find useful tweaks and tools for Smart Device Extensions at `http://smartdevices.microsoftdev.com`.

You are now ready to start learning about how to use the .NET Compact Framework to create compelling applications for Pocket PC and Windows CE devices. I hope you have fun and learn a lot about this exciting platform.

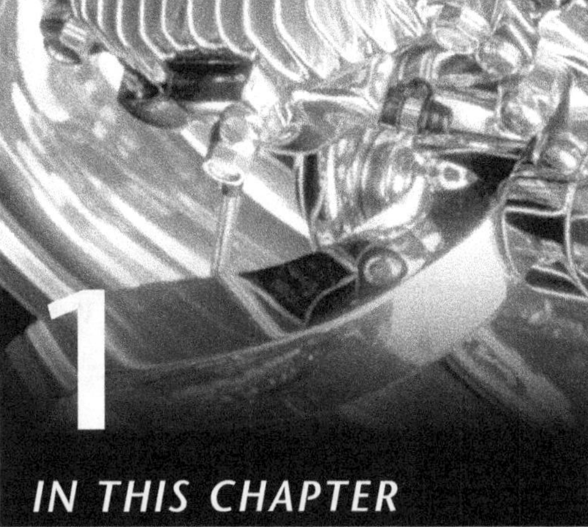

1

Setting Up Your Development Environment

IN THIS CHAPTER

Smart Device Extensions and the .NET Compact Framework

System Requirements

Smart Devices Extensions is the integrated development environment (IDE) through which developers target the .NET Compact Framework. It is included with Visual Studio .NET version 7.1 or later, Enterprise Developer and Enterprise Architect Editions. To run Visual Studio .NET 7.1 Enterprise Developer or Enterprise Architect, you need a system with the minimum specifications outlined in Table 1.1.

In addition, you need to have a supported device with which to run your programs. The .NET Compact Framework is compatible with every device certified to run the Pocket PC operating system. To target the SmartPhone platform, developers must first update Visual Studio with the SmartPhone add-on pack, available from Microsoft. Details about the SmartPhone add-on pack are discussed in Chapter 16, "Deploying a .NET Application."

The Development Process Using Smart Device Extensions

The most straightforward way to develop for the .NET Compact Framework is to use the Smart Device Extensions in Visual Studio .NET 7.1. Simply an

TABLE 1.1
System Requirements for Visual Studio .NET 2003

AREA	REQUIREMENT
Operating system and RAM	Windows 2000 Professional; 96MB RAM, 128MB recommended Windows 2000 Server; 192MB RAM, 256MB recommended Windows XP Professional; 192MB RAM, 256MB recommended Windows XP Home; 96MB RAM, 128MB recommended Windows .NET Server 2003; 192MB RAM, 256MB recommended
Hard disk space	At least 900MB on system drive (where the `\WINNT` or `\WINDOWS` directory resides) and approximately 4.1GB free on installation target drive
Processor speed	Minimum Pentium II 450MHz or equivalent; Pentium III 600MHz or greater recommended
Device connectivity	ActiveSync 3.5 or greater

extension for Visual Studio 7.1, Smart Device Extensions introduces new project types that allow you to target Windows CE devices supporting the .NET Compact Framework, such as the Pocket PC. This means using Smart Device Extensions makes developing applications for Windows CE as easy as developing applications for Windows 2000 or XP. If you have ever developed an application by using previous versions of Visual Basic, then you will be instantly comfortable with the Smart Device Extensions for Visual Studio. Smart Device Extensions is smart enough to compile your application so that it specifically targets the .NET Compact Framework. When you run your code for the first time on a new device, Smart Device Extensions chooses the correct flavor of the .NET Compact Framework for your device and installs it for you automatically.

Accessing the Smart Device Extensions occurs automatically when you create a new application for a Pocket PC or Windows CE device or when you open a previously created Pocket PC or Windows CE application. A project that targets a Pocket PC or Windows CE device is called a Smart Device Application.

USING WINDOWS XP HOME EDITION

Although you can install Visual Studio .NET on a system running Windows XP Home Edition, you will not be able to develop programs that host Web Services without adding a Web server, such as IIS, to your system.

IF YOUR DEVICE ISN'T LISTED IN TABLE 1.2

It is also possible to develop for a wide range of Windows CE devices that are not officially supported, if you are willing to explore the work-arounds needed to get managed code to run on these unsupported devices. The "Targeting Nonstandard Devices" section will discusses how to get your managed code to run on an unsupported device. This section gets technical, and if you have a supported device, you can safely skip it.

Creating a New Application for Pocket PC Devices

The easiest way to introduce the Smart Device Extensions is by example. We'll create a simple "Hello World" style application by using both Visual Basic .NET and C# in the following tutorials. The steps in both tutorials are nearly identical, so to keep the tutorials compact, we'll look at the process of creating a Visual Basic project in great detail, then at just what to do differently to create a C# project.

In this first set of steps, we'll work with Visual Basic:

1. When Visual Studio .NET first launches, it displays the Start Page, as shown in Figure 1.2. To create a new project, click the button labeled New Project near the bottom of the screen;

IF YOU HAVEN'T USED VISUAL STUDIO BEFORE

It is assumed that you have worked inside the Visual Studio IDE and are familiar with deploying and debugging your applications. If you are not familiar with the Visual Studio IDE, you should carefully follow the examples later in this chapter and consider an introductory book on Visual Studio .NET. Using the Smart Device Extensions is so similar to standard desktop development in Visual Studio .NET, you will be able to apply what you learn in an introductory book to the Smart Device Extensions.

INSTALL SMART DEVICE PROGRAMMABILITY

When you first install Visual Studio .NET 7.1, be sure to select installation of Smart Device Programmability, as shown in Figure 1.1.

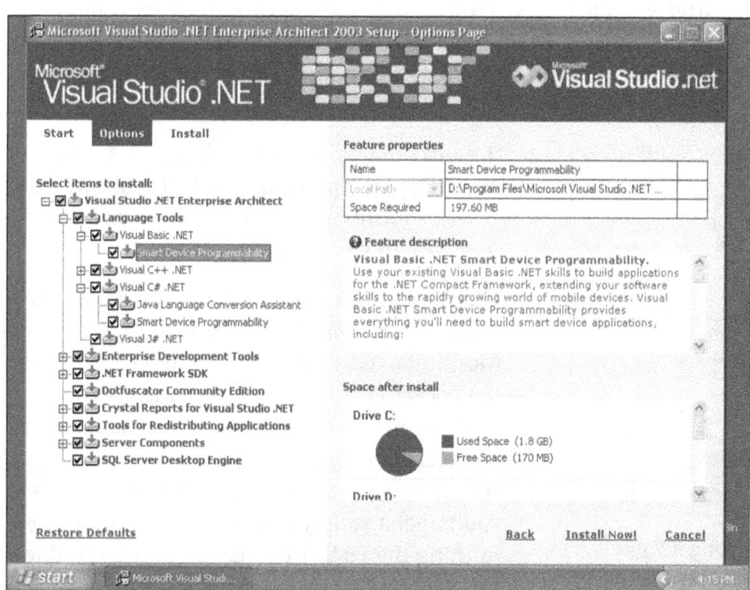

FIGURE 1.1 Be sure to select Smart Device Programmability when installing Visual Studio.

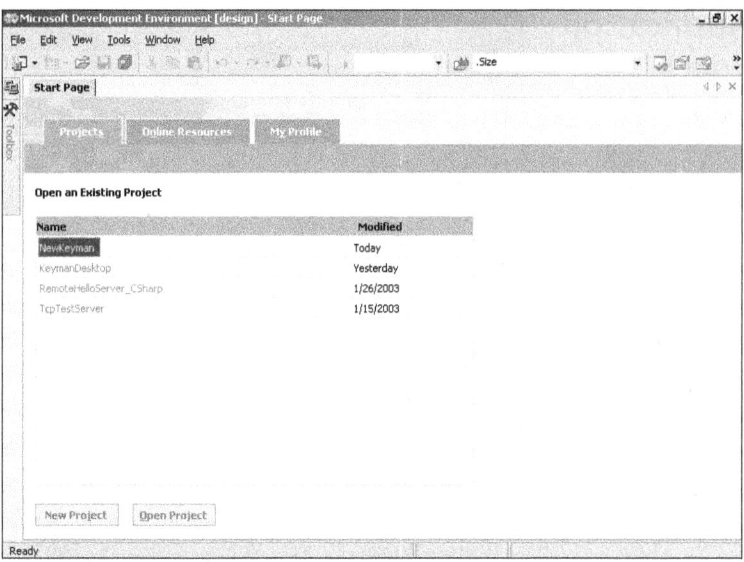

FIGURE 1.2 The Start Page is shown when Visual Studio first launches.

use the pull-down menu by selecting File, New, Project; or use the key combination Ctrl+Shift+N.

2. (For Visual Basic only) After you click New Project, a dialog appears showing the types of projects available. Select the Visual Basic Projects folder and the Smart Device Application template, as shown in Figure 1.3. Provide a name and location for the new application and then press the OK button.

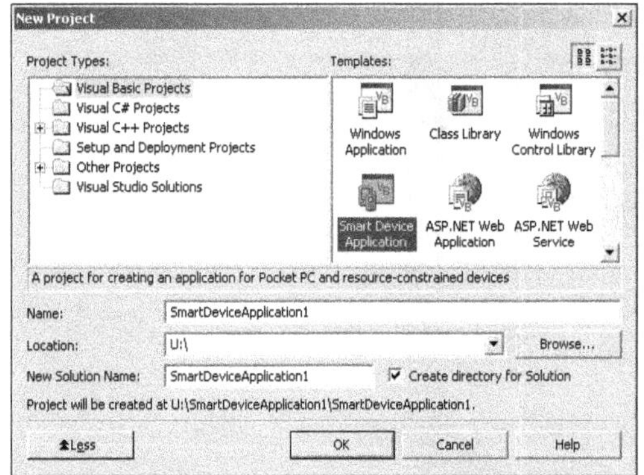

FIGURE 1.3 The dialog for creating a Visual Basic Smart Device Application has you select a folder and a template.

3. The next dialog, shown in Figure 1.4, the Smart Device Application Wizard, allows you to select more options for the new project. They are divided into two categories.

   "What platform do you want to target?" This category lets you select what kind of device you want to target. The Pocket PC selection targets the Pocket PC platform, which means your application will run on all devices that support the Pocket PC operating system, including SmartPhones. The Windows CE selection lets you target a variety of devices other than those running the Pocket PC version of the Windows CE operating system. For the tutorial, we'll target the Pocket PC platform.

"What project type do you want to create?" The choices here are Windows Application, Class Library, Non-graphical Application, and Empty Project. We'll choose Windows Application. This type of project sets up the main form automatically and provides the designer environment that makes it easy to add more controls.

> **WORKING WITH WINDOWS CE DEVICES**
>
> The first release of Smart Devices Extensions is aimed specifically at Pocket PC devices because they are so abundant. The Windows CE selection lets intrepid developers target a wider variety of devices, but there are pitfalls, which we discuss in the "Targeting Nonstandard Devices" section.

The Class Library project is used to create a dynamic link library (DLL) for the .NET Compact Framework. The Non-graphical Application lets users create a console mode application, which is useful on Windows CE devices that offer a command prompt. The Non-graphical Application project sets up a minimal amount of startup code so that users can drop in their own code without setting up a main class. Finally, the Empty Project creates a blank source code file into which developers must enter all of the code to set up the user interface by hand.

4. After we make the selections as shown in Figure 1.4 and click OK, Visual Studio *automatically* activates the Smart Device Extensions and brings up the Forms designer, shown in Figure 1.5. The Forms designer is almost exactly the same as the designer used in desktop projects, so we won't dwell on it here.

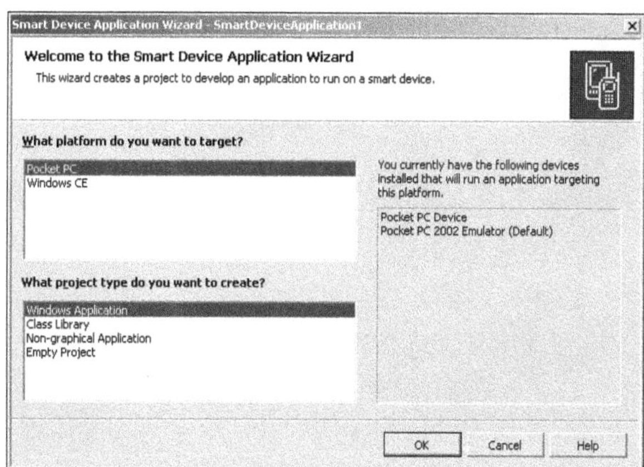

FIGURE 1.4 Select a target platform and an application template.

5. To the left of the Forms designer is a small button labeled Toolbox. Clicking this button brings up the Toolbox, as shown in Figure 1.6.

6. Every item in the Toolbox is a control that is available to developers for the .NET Compact Framework. Most of the controls are related to Winforms, and they are discussed in detail in Chapter 3, "Designing GUI Applications with Windows Forms." For our tutorial we will select a TextBox, drag it onto the form, and then select a button and drag *it* onto the form. Figure 1.7 shows the result.

7. Now double-click Button1, and the IDE brings up the code editor with the cursor blinking in the `Button1_Click` method. Whatever code is entered there is executed when Button1 is clicked. We now enter some Visual Basic code, as shown in Figure 1.8.

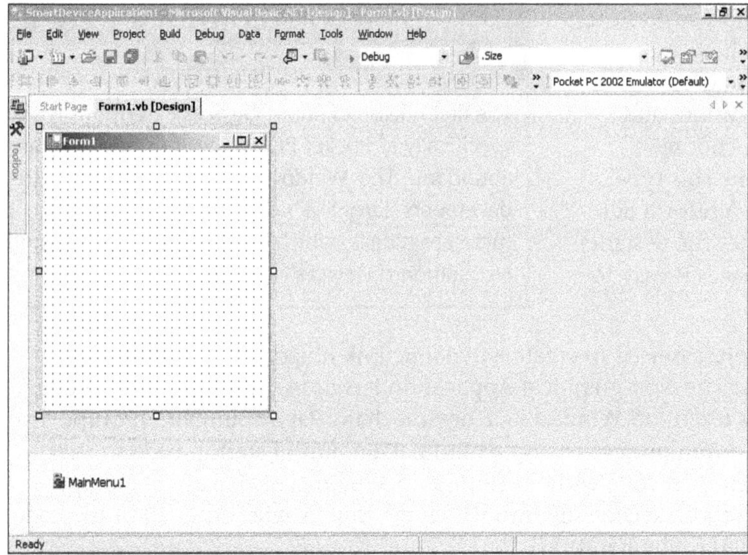

FIGURE 1.5 The Forms designer appears after a project is created.

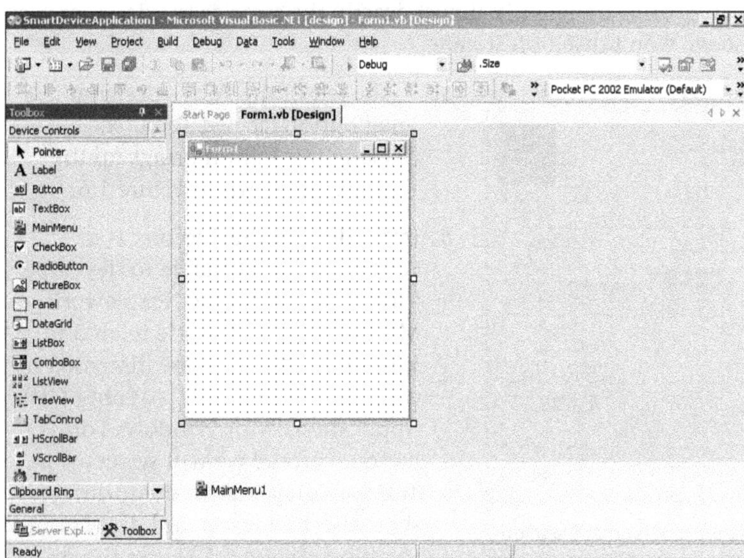

FIGURE 1.6 The Toolbox for Smart Device Application projects is very similar to that for desktop applications.

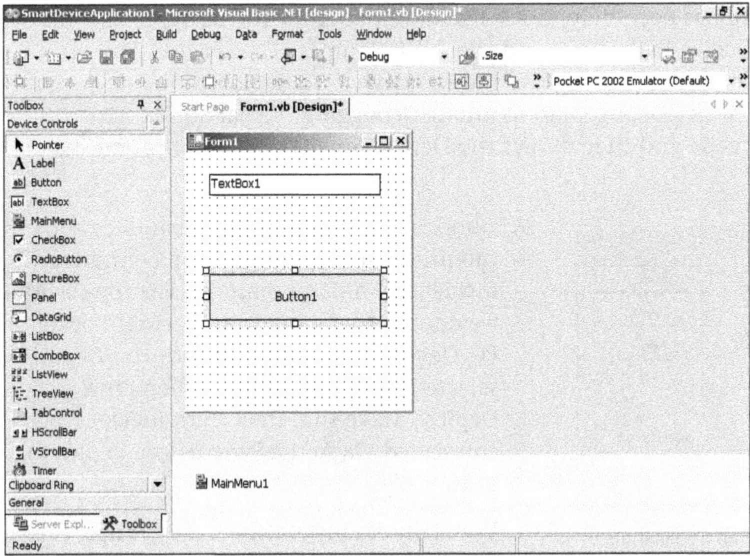

FIGURE 1.7 After both controls are on the form, Form1 in the HelloWorld project looks like this.

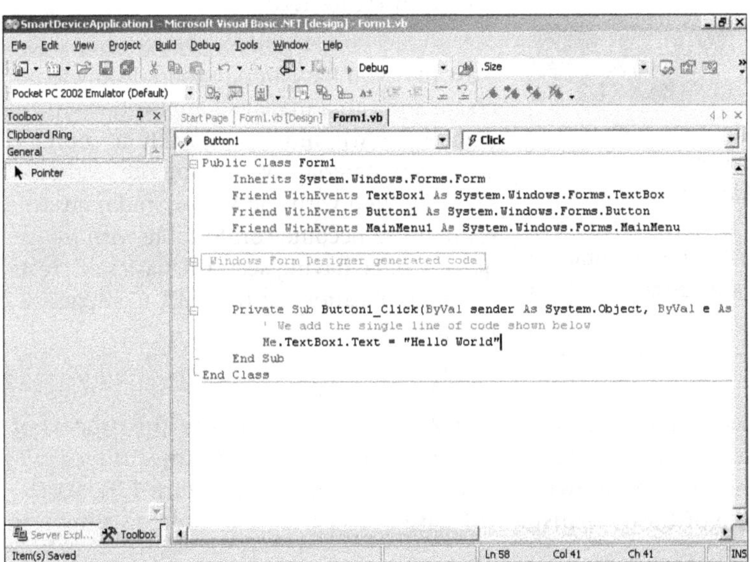

FIGURE 1.8 Visual Studio displays the code view after the button in the Form editor is clicked.

8. So far, the experience has been very much like developing for the desktop, and that is the point of Smart Device Extensions and the .NET Compact Framework. The project is now ready to be compiled and deployed to the device. To deploy to the device and run the application, select Debug, Start Without Debugging. Visual Studio first compiles the source code and then shows the Deploy SmartDeviceApplication dialog, as in Figure 1.9.

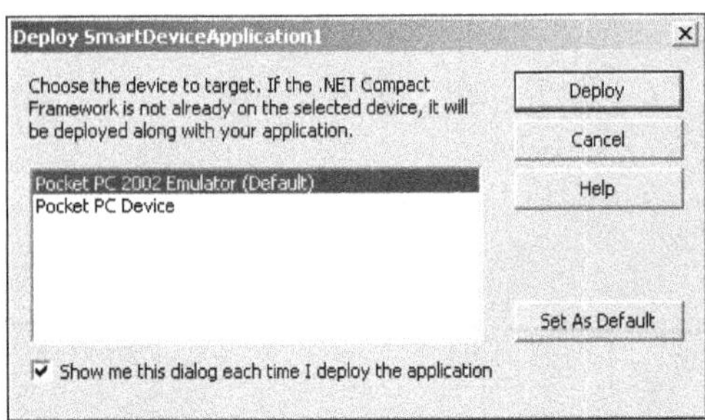

FIGURE 1.9 Before deploying an application to a device, Visual Studio shows the Deployment dialog.

9. Choosing to deploy to the emulator runs the program on your desktop computer inside an emulator window. For the tutorial we choose to deploy to the actual Pocket PC Device, so we select it (not the Default selected in Figure 1.9) and then click Deploy. Make sure that your device is connected via ActiveSync before deploying to it.

10. Visual Studio installs the .NET Compact Framework and launches your application on the device. If you click the button, you will see the message "Hello World," as shown in Figure 1.10.

THE .NET COMPACT FRAMEWORK IS AUTOMATICALLY INSTALLED

When a project is deployed to the device, the .NET Compact Framework is automatically installed to the device if it is not already present. Large-scale deployment of a completed application to customers in a marketplace or an enterprise environment entails special considerations, which are discussed in Chapter 16, "Deploying a .NET Application."

Now that you have briefly toured how to create and run a managed application on devices, we will move on quickly to more advanced material. We assume the reader is savvy enough with the Visual Studio environment to become comfortable with it quickly once having seen the basics of creating an application for the .NET Compact Framework.

HelloWorld for C#

Compared to the Visual Basic sequence, only steps 2 and 6 are different in the process of building a HelloWorld example for C#. For step 2 we must choose a C# project for Windows instead of Visual Basic, as shown in Figure 1.11 (compare to Figure 1.3). For step 6 we must insert a line of C# code instead of a line of Visual Basic code, as shown in Figure 1.12.

The rest of the steps are the same because regardless of the language used, the applications target the .NET Compact Framework. The controls available in the ToolBox are virtually unchanged, since they are part of the Winforms implementation in the .NET Compact Framework.

FIGURE 1.10 The device emulator runs the HelloWorld application.

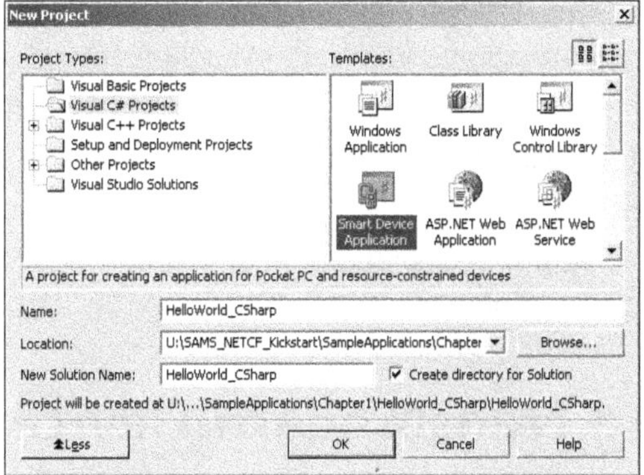

FIGURE 1.11 This figure shows what step 2 of the tutorial looks like for a Smart Device Application written in C#.

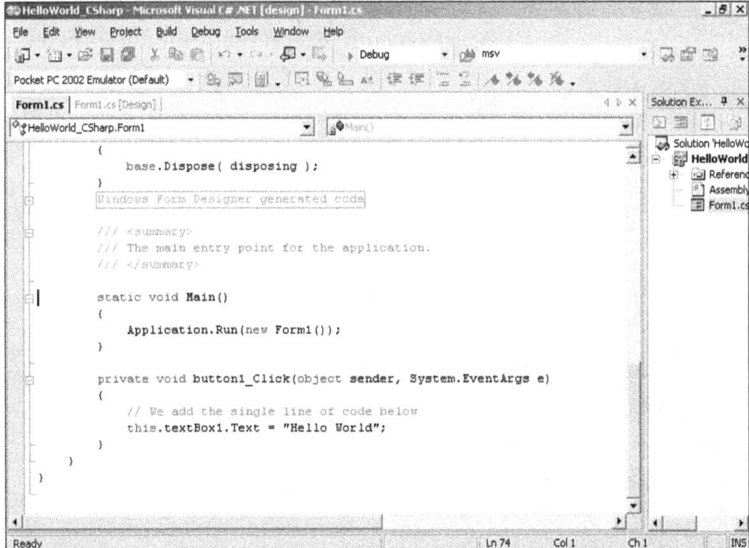

FIGURE 1.12 This figure displays the C# code to insert for step 6 of the tutorial.

The completed project files for the tutorials are on the companion CD-ROM in the folders `SampleApplications\Chapter1\HelloWorld_VB` and `SampleApplications\Chapter1\HelloWorld_CSharp`.

Opening an Existing Smart Device Extensions Project

Opening an existing Smart Device Extensions project is no different from opening any other kind of project within Visual Studio .NET. Simply select File, Open, Project and select a project; otherwise, choose a recent project from the Start Menu.

Targeting Nonstandard Devices

The Hello World tutorials show just how easy it is to develop managed code that runs on standard supported devices, such as the Pocket PC. The .NET Compact Framework can also run on a wide variety of hardware that runs Windows CE. Table 1.2 shows what processors are compatible with the .NET Compact Framework and what operating systems are supported for specific processors.

> **.NET COMPACT FRAMEWORK IS IN ROM FOR POCKET PC 2003**
>
> All Pocket PC devices running Pocket PC version 2003 or later have the .NET Compact Framework in ROM. If you are unable to deploy or debug your application on these devices, this section teaches how Smart Device Extensions communicates with devices for debugging and deployment and discusses work-arounds to a variety of problems.

The .NET Compact Framework binaries are stored as a CAB file on your desktop computer. There is a unique CAB file for each operating system and processor type that the .NET Compact Framework supports. Smart Device Extensions pushes the correct CAB file to your device when it detects that your device does not have the Compact Framework installed. In this section we discuss in detail how this process works and how to deploy the CAB files by hand if you are unable to get deployment to work automatically.

TABLE 1.2
Processors and Operating Systems Supported by the .NET Compact Framework

CPU NAME	SUPPORTED OS VERSIONS
Intel ARM 4	Pocket PC 2000, 2002, 2003, and WinCE 4.1 or greater
Intel ARM 4i	Pocket PC 2000, 2002, 2003, and WinCE 4.1 or greater
Hitachi SH3	Pocket PC 2000, 2002, 2003, and WinCE 4.1 or greater
Hitachi SH4	Pocket PC 2003 and WinCE 4.1 or greater
Intel 80x86	Pocket PC 2000, 2002, 2003, and WinCE 4.1 or greater
MIPS 16	Pocket PC 2000, 2002, 2003, and WinCE 4.1 or greater
MIPS II	Pocket PC 2000, 2002, 2003, and WinCE 4.1 or greater
MIPS IV	Pocket PC 2000, 2002, 2003, and WinCE 4.1 or greater

Table 1.2 demonstrates that the .NET Compact Framework runs on a wide variety of hardware. The trick is getting the IDE to allow you to deploy and debug to that hardware. There are three levels of support for nonstandard devices:

Full deploy and debug support This level of support means that the IDE can deploy binaries to the device and debug code running on the device.

Deploy-only support This means that the IDE can deploy only binaries to the device but cannot debug code running on the device. Nonstandard devices will rarely support ConmanClient enough to allow deployment without being able to run the debug host.

Target-only support This level of support means that you can develop your application in Visual Studio, but you must install the Compact Framework onto the device manually and also copy your device binaries manually. It is the "last resort" for developing applications to a nonstandard device.

The overall strategy for targeting a nonstandard device is to attempt to achieve full deploy and debug support, then fall back to deploy only, and settle for target only as a last resort. To achieve the first two levels, we need to examine carefully how Visual Studio gets binaries to a device and communicates with the device-side debugger host. To work with Target Only support, see the section entitled "Installing the .NET Compact Framework Manually."

How Visual Studio Communicates with Devices

If you have a device that you believe fulfills the requirements for running the .NET Compact Framework, but you are unable to use the Smart Device Extensions to debug or deploy to the device, there are a variety of tricks for getting successful debug and deployment sections. In order to fully understand how to use these tricks, we need a strong understanding of how Smart Device Extensions communicates with the device.

Conman

Conman is the connectivity technology for Smart Device Extensions. It is through Conman that connections are made between the desktop and the device for deploying files and communicating debugging messages.

The role of Conman is to abstract the physical connection between a desktop running Smart Device Extensions and the Windows CE device with which it is communicating. For example, the communication might occur through a TCP/IP connection, an ActiveSync connection, or when the remote device is an emulator, through Direct Memory Access. Regardless of the underlying connection, the deployment and debugger code in Smart Device Extensions knows only that it is communicating with a Conman connection.

The specific underlying connections are called *transports*. The code for each transport is held in a separate DLL. Two examples of transports are the TCP transports and the Direct Memory Access transports. Table 1.3 shows the name of all of the transport DLLs available on the desktop and the device.

TABLE 1.3
Conman Transport DLLs

TRANSPORT DLL NAMES	PLATFORM	DESCRIPTION
`cmtnpt_TCPConnect.dll` and `TCPTransportConnect.dll`	Desktop	Used to connect to a device with a network card or ActiveSync connection
`cmtnpt_TCPAccept.dll`	Device; binaries for each CPU and OS build type are in the Target directory	Accepts incoming TCP or ActiveSync connections from the desktop
`cmtnpt_Emulator.dll`	Desktop	Connects to an emulator
`cmtnpt_Emulator.dll`	Device; binaries for each CPU and OS build type are in the Target directory	Allows `ConmanClient.exe` running on emulator to accept desktop connection
`IRDATransportAccept.dll`	Desktop	Used to connect to device via infrared; works for deployment but not debugging
`Cptnpt_IRDAConnect.dll`	Device; binaries for each CPU and OS build type are in the `Target` directory	Used to connect to device via infrared; works for deployment but not debugging

WHEN COMMUNICATION WITH CONMAN FAILS

Communication occurs through Conman transports anytime a project is deployed or debugged. Deployment uses a single instance of a Conman transport. Debugging opens two additional transports used for communicating debug information between the device and the desktop. Whenever either of these two features is not working correctly, a likely cause of the problem is that Conman is not being initialized on the device correctly. If you cannot find these DLLs on your desktop computer, then your Visual Studio installation is damaged, and deployment and debugging will fail. Missing DLLs can indicate that you did not select to install Smart Device Extensions as shown in Figure 1.1.

The desktop side gains access to the Conman transports when the Visual Studio executable, `devenv.exe`, loads and initializes the transports because the user tried to debug or deploy to the device. On the device an executable called `ConmanClient.exe` runs. When it is first launched, `ConmanClient.exe` loads a transport and listens until a desktop transport connects to it. At that time internal protocols determine whether the connection is for deploying files, launching an executable, or passing debugging information, and the transport connection is handed off to the correct party.

Initializing Conman on the Device

One of the major problem areas with nonstandard devices is getting `ConmanClient.exe` launched on them. If it is not successfully launched, then no connection will ever succeed for deployment or debugging.

When the Smart Device Extensions tries to deploy or debug to the device, it uses the transport that is specified in the `Device Tools` folder in the Visual Studio options dialog. If you are having trouble connecting to a device, you can specify a specific transport that might work for you. To set the transport used to connect with the device, follow these steps:

1. Select Tools, Options from within Visual Studio.
2. Double-click the `Device Tools` folder and select the `Devices` subcategory. Figure 1.13 shows the resulting Options dialog.

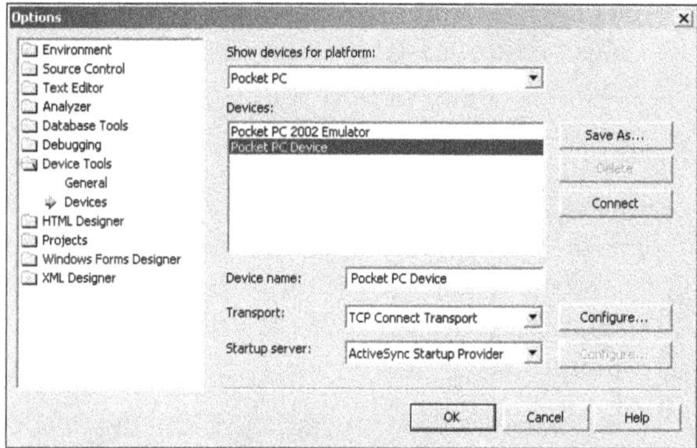

FIGURE 1.13 Use the Device connection Options dialog to choose what kind of device to connect to.

3. Choose either Pocket PC or Windows CE as the platform, depending on whether you are trying to connect to a Pocket PC device.
4. Choose what kind of device you are deploying to. In Figure 1.13 the choices are Emulator or Pocket PC Device. Note that for the emulator there is only one Transport choice available.
5. Select the transport to use. For the Pocket PC device, there are two choices: TCP Connect or IRDA. The TCP Connect Transport means that the desktop device will connect to `ConmanClient.exe` on the device by making a TCP connection to the device. The IRDA Tranport uses the device's IRDA transport to communicate. This is especially useful if your computer is a laptop with an IRDA port.
6. If you choose the TCP Connect Transport, then you can alter the behavior of that transport by clicking the Configure... button shown in Figure 1.13. Clicking Configure... yields the dialog box seen in Figure 1.14.

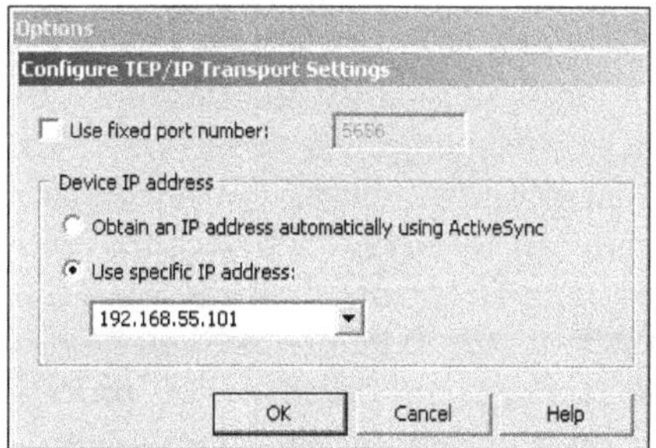

FIGURE 1.14 The TCP Connect Transport lets you choose settings related to connecting to your device via TCP.

7. The new dialog lets you choose the IP address of your device, if you know it. If your device is connected by ActiveSync, then Visual Studio can discover the address automatically. You can also choose to use a specific port number other than the default value of 5656. To use the non-default port number, you must configure `ConmanClient.exe` by hand on the device. This procedure is specified later in this chapter.

When the user tries to deploy or debug on a device, Smart Device Extensions attempts to connect to the device, assuming that ConmanClient.exe is running on the device and following the parameters specified in the configuration dialog from Figure 1.14. If there is no response, then Smart Device Extensions installs a new copy of ConmanClient.exe and the associated DLLs by using ActiveSync. There is a prebuilt binary of ConmanClient.exe for devices running Pocket PC and devices running Windows CE version 4.1 or greater. Also, there are binaries for each of the possible processor types: SH3, ARM, MIPS, and X86. Smart Device Extensions determines which specific build of ConmanClient to send to the device by querying ActiveSync about the operating system and processor type of the target device. There is a version of ConmanClient.exe and the transport libraries for every processor and operating system version that the .NET Compact Framework supports.

BE CAREFUL WHEN DEFINING PARAMETERS FOR CONMANCLIENT.EXE

PITFALL—These configurations give users a lot of freedom in specifying how the desktop connects to the device. If you are trying to connect to a Windows CE device (not a Pocket PC), specifying the details of how the desktop connects to ConmanClient.exe on the device is often helpful. However, be careful to match your choices with the choices you make for ConmanClient.exe.

IF YOU ARE USING A DEVICE WITH WINDOWS CE 4.1 OR HIGHER

There is a bug in ActiveSync that causes it to return improper processor-type information on non–Pocket PC devices running Windows CE 4.1 or greater. This creates a problem for users working with such devices. Because of the inaccurate information from ActiveSync, Smart Device Extensions sends the wrong build of ConmanClient.exe to the device, and it does not launch successfully. Thus, deploy and debug always fails. There is a utility pack available on the Microsoft Smart Devices developer Community Web site at: http://smartdevices.microsoftdev.com/Downloads/default.aspx called Windows CE Utilities for Visual Studio .NET 2003 Add-on Pack. The Add-on Pack includes the CPU Picker, which is useful for working around the ActiveSync bug. CPU Picker is discussed in great detail later in this chapter.

If Visual Studio is installed in the default location of C:\Program Files, then the Conman binaries are located in the directory ConnectionManager\Target\wce300 for Pocket PC 2000 and 2002 devices, and in ConnectionManager\Target\wce400 for Windows CE 4.1 or greater and Pocket PC 2003 devices. Each of these directories holds subdirectories for each supported CPU.

At the same time that Smart Device Extensions sends ConmanClient.exe to a device, it creates a security relationship with the device by exchanging a shared key. This is called *prepping* the device. It ensures that no desktop can connect to a device that has not been prepped by that desktop. Since the device must be physically connected to the device with ActiveSync to be prepped, an implicit trust relationship exists between a desktop and a device it has prepped. This process prevents rogue hackers from connecting to your device from the Internet, and it also prevents hackers from creating devices that can accept connections from your desktop.

After prepping, Smart Device Extensions uses ActiveSync to launch the `ConmanClient.exe` program and attempts to connect to it again using the transport specified in the transport preferences. If the connection fails again, then Smart Device Extensions gives up and returns an error message to the user.

THE REMOTE PROCESS VIEWER

Many developers interested in deploying to nonstandard devices also use Microsoft Embedded Visual Tools version 3.0 or 4.0. These products include the Remote Process Viewer, which shows which executables are running on the device. If `ConManClient.exe` is running on your device, you will see an entry for it by using the Remote Process Viewer, as shown in Figure 1.15.

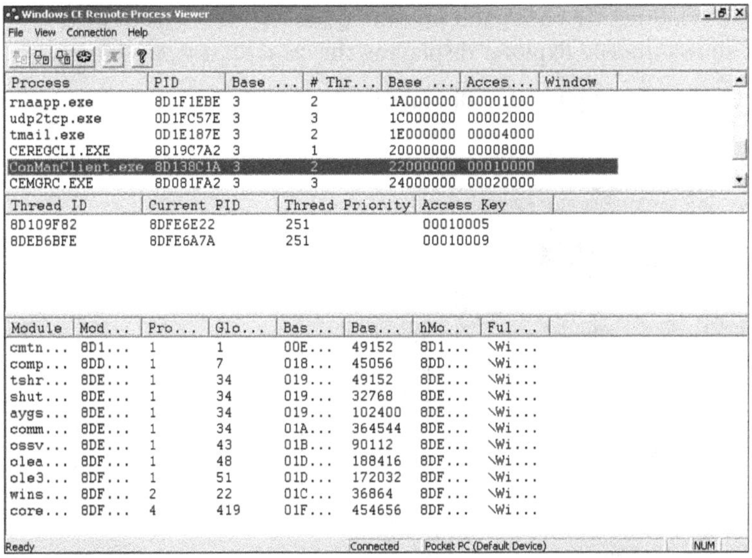

Windows CE Remote Process Viewer

File View Connection Help

Process	PID	Base ...	# Thr...	Base ...	Acces...	Window
rnaapp.exe	8D1F1EBE	3	2	1A000000	00001000	
udp2tcp.exe	0D1FC57E	3	3	1C000000	00002000	
tmail.exe	0D1E187E	3	2	1E000000	00004000	
CEREGCLI.EXE	8D19C7A2	3	1	20000000	00008000	
ConManClient.exe	8D138C1A	3	2	22000000	00010000	
CEMGRC.EXE	8D081FA2	3	3	24000000	00020000	

Thread ID	Current PID	Thread Priority	Access Key
8D109F82	8DFE6E22	251	00010005
8DEB6BFE	8DFE6A7A	251	00010009

Module	Mod...	Pro...	Glo...	Bas...	Bas...	hMo...	Ful...
cmtn...	8D1...	1	1	00E...	49152	8D1...	\Wi...
comp...	8DD...	1	7	018...	45056	8DD...	\Wi...
tshr...	8DE...	1	34	019...	49152	8DE...	\Wi...
shut...	8DE...	1	34	019...	32768	8DE...	\Wi...
aygs...	8DE...	1	34	019...	102400	8DE...	\Wi...
comm...	8DE...	1	34	01A...	364544	8DE...	\Wi...
ossv...	8DE...	1	43	01B...	90112	8DE...	\Wi...
olea...	8DF...	1	48	01D...	188416	8DF...	\Wi...
ole3...	8DF...	1	51	01D...	172032	8DF...	\Wi...
wins...	8DF...	2	22	01C...	36864	8DF...	\Wi...
core...	8DF...	4	419	01F...	454656	8DF...	\Wi...

Ready Connected Pocket PC (Default Device) NUM

FIGURE 1.15 The Remote Process Viewer can show when `ConmanClient.exe` is running on a device.

Launching `ConmanClient.exe` Manually

If you have a Windows CE device without ActiveSync, you must launch `ConmanClient.exe` yourself. There are two ways to do it. The most convenient and most strongly recommended way is to use the Smart Device Authentication Utility, because it preps the device and launches ConmanClient in one easy step.

IF YOU RESET YOUR DEVICE

If you reset your device, `ConmanClient.exe` is stopped. You must connect to your desktop computer through ActiveSync the next time you try to deploy or debug to your device. If your proprietary device does not have an ActiveSync connection, see the "Devices without ActiveSync" section.

However, when the Smart Device Authentication Utility launches `ConmanClient.exe`, it sets `ConmanClient.exe` to listen to the default TCP port, 5656, for the TCP transport. If you are running software that conflicts with this port, then you can launch `ConmanClient.exe` manually and specify the port by following these steps:

1. Connectivity will fail if the device is not prepped. Use the Smart Device Authentication Utility, which copies `ConmanClient.exe` and the transport libraries to the device, preps the device, and launches `ConmanClient.exe` configured to use the TCP Accept transport on port 5656.
2. Warm-reboot the device. This stops `ConmanClient.exe` but retains the prep information.
3. Launch ConmanClient on the device. The easist way to do this is to use the Windows CE File Explorer or the Windows CE command prompt. `ConmanClient.exe` is located in `\Windows`. Figure 1.16 shows the File Explorer displaying the `ConmanClient.exe` binary.
4. When ConmanClient is launched by hand, it displays the interface shown in Figure 1.17.

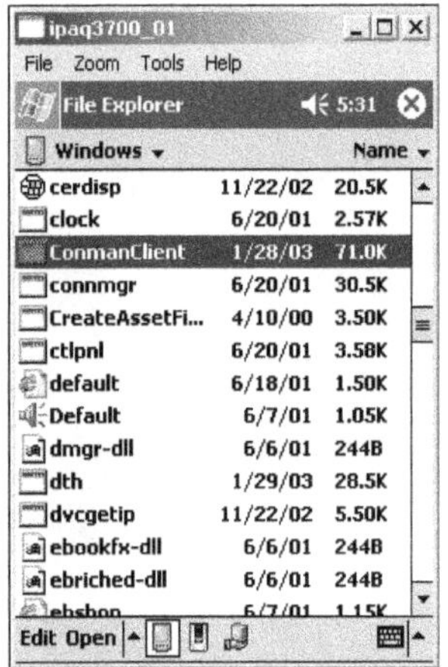

FIGURE 1.16 You can use the Windows CE File Explorer to launch with `ConmanClient.exe`.

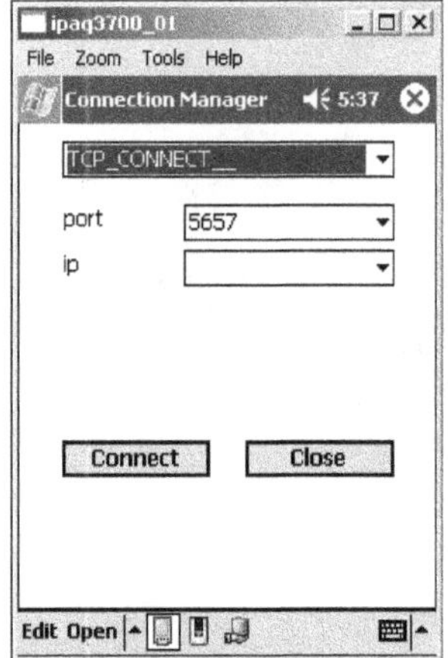

FIGURE 1.17 `ConmanClient.exe` has a user interface with which you can alter connection settings.

5. In the top menu, pull down the selection TCP_ACCEPT__. Although it is technically possible to choose TCP_CONNECT__, this choice is worthless because after the Beta 1 release of Visual Studio .NET 2003, the desktop has no way to accept a connection.
6. Choose the port on which the device should listen and click the Connect button. `ConmanClient.exe` is now listening with the TCP Accept transport on the specified port, and the device is prepped because of the Smart Device Authentication Utility.

Non–Pocket PC Devices Running Windows CE 4.1 and Greater

Users who try to deploy and debug to devices running Windows CE 4.1 and greater and who also have an ActiveSync connection to their devices often run into problems due to a bug in ActiveSync. ActiveSync reports improper CPU information for the device, causing Smart Device Extensions to attempt to launch a version of `ConmanClient.exe` that will not run on the device. To solve this problem, Microsoft has provided a utility called CPU Picker. Using CPU Picker, you can determine exactly what kind of CPU your device has and specifically start the correct version of `ConmanClient.exe`.

Devices using Windows CE 4.0 sometimes work with the .NET Compact Framework, but it is not an officially supported platform.

When to Use CPU Picker

CPU Picker might help when the following are true:

- You have a device using Windows CE 4.1 or greater, and it is connected to your desktop computer through ActiveSync.
- The device exhibits one of the following behaviors:

  Attempts to deploy to the device fail after the initial connection is established.

  The CAB file that Smart Device Extensions pushes to the device are copied to the device, but they do not explode successfully.

  CAB files are dropped onto the root directory of the device and deleted after they are successfully exploded. If you see one of the CAB files shown in Table 1.4 on your device and it is the wrong CAB file for your device CPU, then Smart Device Extensions is unable to tell which CPU your device uses. This is a condition that CPU Picker can remedy.

The .NET Compact Framework CAB Files

The .NET Compact Framework is delivered to your device through CAB files. Smart Device Extensions uses ActiveSync to determine what kind of CPU your device has and copies the needed CAB files to the device for you. Because there are so many supported-device CPU types, there is a wide variety of CAB files. However, there are only six basic types of CAB files, and their names adhere to the following conventions:

`netcf.core.ppc3.{CPU}` This kind of cab file holds the central .NET Compact Framework binaries compiled specifically for the Pocket PC 2000 and 2002 operating systems. The "`{CPU}`" is a placeholder for the extension indicating what kind of CPU the CAB works with. For example, `netcf.core.ppc3.arm` is for the ARM processor.

`netcf.all.wce4.{CPU}` Like `netcf.core.ppc3.{CPU}`, except for Windows CE version 4.1 or higher.

`sqlce.ppc.{CPU}` This kind of cab file holds the SQL CE engine. It is deployed to your device if your project requires the SQL CE engine. The "`{CPU}`" is a placeholder for the extension indicating what kind of CPU the CAB works with. For example, `sqlce.ppc3.arm` is for the ARM processor.

`sqlce.wce4.{CPU}` Like `sqlce.ppc.{CPU}`, except for Windows CE version 4.1 or higher.

`sql.ppc.{CPU}` This kind of cab file holds the code that supports the SQLClient namespace, which is used for desktop SQL Server connectivity. It is deployed to your device if your project requires it. The "`{CPU}`" is a placeholder for the extension indicating what kind of CPU the CAB works with. For example, `sql.ppc3.arm` is for the ARM processor.

`sql.wce4.{CPU}` Like `sql.ppc.{CPU}`, except for Windows CE version 4.1 or higher.

The CAB files come in sets of three: one for the .NET Compact Framework binaries, one for the SQL CE engine, and one for desktop SQL connectivity. Each set of three CAB files corresponds to a specific CPU and operating system. Furthermore, each set of three is held in its own special directory. Table 1.4 shows which directories hold sets of CAB files for specific operating systems and CPU types.

TABLE 1.4

CAB Files for Various CPUs

Note: These directories are relative to the Visual Studio subdirectory `\CompactFrameworkSDK\v1.0.5000\WINDOWS CE\`.

CAB PATH AND NAME	CPU	OS
`wce300\arm\`	ARM	Pocket PC 2000 Pocket PC 2002
`wce300\armv4\`	ARMv4	Pocket PC 2003 Win CE 4.1 or higher
`wce300\mips\`	MIPS	Pocket PC 2000 Pocket PC 2002
`wce300\sh3\`	SH3	Pocket PC 2000 Pocket PC 2002
`wce300\wce420x86\`	80x86	Pocket PC 2000 Pocket PC 2002
`wce300\x86\`	80x86	Pocket PC 2000 Pocket PC 2002
`wce400\armv4\`	ARMv4	Pocket PC 2003 Win CE 4.1 or higher
`wce400\armv4t\`	ARMv4t	Pocket PC 2003 Win CE 4.1 or higher
`wce400\mips16\`	MIPS16	Pocket PC 2003 Win CE 4.1 or higher
`wce400\mips16\`	MIPS16	Pocket PC 2003 Win CE 4.1 or higher

TABLE 1.4
Continued

CAB PATH AND NAME	CPU	OS
`wce400\mipsii\`	MIPS II	Pocket PC 2003 Win CE 4.1 or higher
`wce400\mipsii_fp\`	MIPS II FP	Pocket PC 2003 Win CE 4.1 or higher
`wce400\mipsiv\`	MIPS IV	Pocket PC 2003 Win CE 4.1 or higher
`wce400\mipsiv_fp\`	MIPS IV FP	Pocket PC 2003 Win CE 4.1 or higher
`wce400\sh3\`	SH3	Pocket PC 2003 Win CE 4.1 or higher
`wce400\sh4\`	SH4	Pocket PC 2003 Win CE 4.1 or higher
`wce400\x86\`	80x86	Pocket PC 2003 Win CE 4.1 or higher

Installing CPU Picker

CPU picker is available as an MSI file on the GOTDOTNET Web site at `www.gotdotnet.com`. As of this writing, Visual Studio–related utilities were located by following these links: .NET Downloads, Visual Studio .NET, Tools and Utilities.

Before installing CPU Picker, exit all instances of Visual Studio .NET and click the MSI file to start installation. After installing the CPU Picker, start Visual Studio. During startup, Visual Studio brings up the CPU Picker interface, as shown in Figure 1.18. Select one of the CPU architecture types shown in the list to match your device hardware and click OK.

After you click OK, Visual Studio asks you to confirm your choice. After Visual Studio confirms your CPU selection, restart Visual Studio to make the changes take effect. The next time you deploy or debug against the device, Visual Studio will use your CPU choice instead of trying to detect the CPU type using ActiveSync.

IF YOU CHOOSE THE WRONG CPU WITH CPU PICKER

If you make a bad CPU selection, you can change it at any time by quitting Visual Studio and restarting it from the command line by typing **devenv.exe/setup**. Typically, Visual Studio is installed into the `C:\Program Files` folder, and `devenv` is in the folder `\Program Files\Microsoft Visual Studio .NET 2003\Common7\IDE\devenv.exe`. If you installed Visual Studio in a custom location, simply look in the `Common7\IDE` subdirectory wherever you installed Visual Studio.

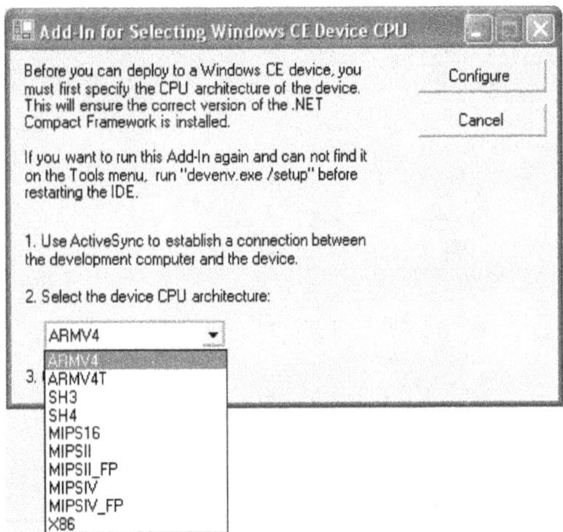

FIGURE 1.18 You can manually select the device CPU type with the CPU Picker.

Devices without ActiveSync

The primary target audience for the first release of the .NET Compact Framework is the Pocket PC and the SmartPhone, two classes of devices that support ActiveSync connections. The previous discussions show that Smart Device Extensions rely on ActiveSync to copy `ConmanClient.exe` to a device and to prep the device. This means developers who want to target Windows CE devices without ActiveSync connections will not have any way to deploy or debug against their devices. In recognition of this problem, Microsoft created the Smart Device Authentication Utility.

Smart Device Authentication Utility

The Smart Device Authentication Utility allows users to prep their devices and launch `ConmanClient.exe` by using a TCP connection instead of a ActiveSync connection. This is the utility that lets developers target devices without ActiveSync. It is available at the Microsoft Smart Devices Developer Community Web site at `http://smartdevices.microsoftdev.com/Downloads/default.aspx`. As of this writing, the Smart Device Authentication Utility is part of the Windows CE Utilities for Visual Studio .NET 2003 Add-on Pack. There is a link on the Web page to download this product.

The Smart Device Authentication Utility comes in two parts—the desktop- and device-side executables. The desktop side is composed of the `SDAuthUtil.exe` executable. The device side is a single client executable called `SDAClient.exe`.

To use the Smart Device Authentication Utility, follow these steps:

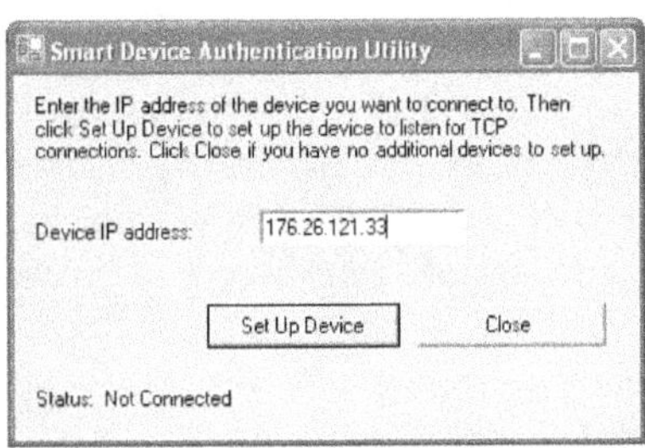

FIGURE 1.19 Users interact with the Smart Device Authentication Utility mainly through the Desktop Client user interface.

1. Make sure your desktop has Visual Studio 2003 installed and launch the `SDAuthUtil.exe` program. Note that `SDAuthUtil.exe` requires version 1.1 of the .NET Framework. This version of the Framework is automatically installed when Visual Studio .NET 2003 is installed. When `SDAuthUtil.exe` is launched, it shows a simple user interface, as shown in Figure 1.19.
2. Copy the correct version of `SDClient.exe` to the device and launch it. You can use any convenient means to get the executable to the device, such as emailing it to yourself and checking your email with the device. Some

devices support FTP from the Windows CE command line. If you are a developer using a nonstandard device, we assume that you know enough about your device to get the `SDClient.exe` program to your device.

There are separate builds of `SDClient.exe` available, each for a different operating system version and processor. Each version of `SDClient.exe` is in a separate directory, as shown in Table 1.5. Use the table to determine which directory holds the version of `SDClient.exe` that will run on your device.

TABLE 1.5
SDClient Device Builds

DIRECTORY NAME	CPU	OS VERSION
wince3\arm	ARM	Pocket PC 2002, Pocket PC 2003
wince3\mips\	MIPS	Pocket PC 2002, Pocket PC 2003
wince3\sh3\	SH3	Pocket PC 2002, Pocket PC 2003
wince3\x86	80x86	Pocket PC 2002, Pocket PC 2003
wince4\armv4\	ARMv4	Win CE 4.1 or greater
wince4\armv4t\	ARM v4t	Win CE 4.1 or greater
wince4\mips16	MIPS 16	Win CE 4.1 or greater
wce400\mipsii\	MIPS II	Win CE 4.1 or greater
wince4\mipsii_fp	MIPS II FP	Win CE 4.1 or greater
wince4\mipsiv\	MIPS IV	Win CE 4.1 or greater
wince4\mipsiv_fp\	MIPS IV FP	Win CE 4.1 or greater
wince4\sh3\	SH3	Win CE 4.1 or greater
wince4\sh4\	SH4	Win CE 4.1 or greater
wince4\x86	80x86	Win CE 4.1 or greater

3. When `SDClient.exe` launches on the device, a simple user interface shows up. Note that `SDClient.exe` displays all of the IP addresses for the device. If `SDClient.exe` does not launch, it is possible that you chose the wrong build of `SDClient.exe`. It is especially easy to make this mistake because of all of the different versions of the ARM processor and MIPS processor. Look at Table 1.5 and try another build that looks like it might run on your hardware. For example, if MIPS II did not work, MIPS IV may. You might have to try several different builds before you find one that runs on your device.

4. Enter the device IP address that you discovered in step 3 into `SDAuthUtil.exe` on the desktop, as shown in Figure 1.19. In this example we set up the desktop to try connecting to a device whose IP Address is `172.26.121.33`.

5. Click the Connect button on `SDAuthUtil.exe` (on the device).

6. Click the Launch Conman on Device button on the desktop. The Smart Device Authentication Utility determines what version of `ConmanClient.exe` to install based on

what build of `SDClient.exe` is running on the device. It installs ConmanClient, preps the device, and displays a message of success. If there are more than one IP address for the device, you may have to try more than one before you get a successful connection.

7. After you have successfully prepped your device and launched Conman, you can quit by clicking the Close button. This causes both the desktop- and device-side programs to quit. After a successful interaction with the Smart Device Authentication Utility, your device is running the correct version of `ConmanClient.exe`, and it has been prepped with the desktop.

8. Attempt to deploy and/or debug against the device. *Do not reboot the device, or you will have to run the Smart Device Authentication Utility all over again.*

Installing the .NET Compact Framework Manually

If nothing seems to get debugging and deployment to work, some devices can still run the .NET Compact Framework if it is installed onto the device manually. Once the .NET Compact Framework is successfully installed to a device, managed applications can be copied over to the device and launched by hand. While this is not an ideal scenario, it helps make some nonstandard devices viable for use with managed code. If you are forced into this situation, you can still debug your code using the Pocket PC or Windows CE emulator and deploy your application and the .NET Compact Framework manually.

The .NET Compact Framework is stored in CAB files on your desktop computer. There are three CAB files that must be copied to the device. One holds the .NET Compact Framework runtime and class libraries. Another holds the SQL CE database engine, and the third holds the .DLLs that implement the SQLClient namespace that is needed to connect to a desktop SQL Server (see Chapter 7, "Programming with Microsoft SQL Server CE"). These CAB files are built for Pocket PC operating systems and Windows CE 4.1 or greater. There is one set of CAB files for each combination of operating system version and processor type. Table 1.4 shows the names of the CAB files and their contents according to the processor and operating system in the device.

The CAB files are all in the Visual Studio .NET subdirectory `\CompactFrameworkSDK\v1.0.5000\WINDOWS`

RISKS ASSOCIATED WITH SMART DEVICE AUTHENTICATION UTILITY

Using the Smart Device Authentication Utility carries a small risk because it uses a TCP/IP connection to transmit the shared key that is created when prepping the device. When the shared key is transmitted from the desktop to the device, it is encrypted using the device's exchange key. This means that even if a hacker accessed the shared key as it traveled between the desktop and the device, it would be useless to him. Thus, using the utility is still a reasonably safe thing to do, but nothing beats the security of using an ActiveSync connection with a USB cable, since the data transmitted on the USB cable cannot go out to the internet.

`CE\`. You can use the Windows Explorer Search function to search for the CAB files directory.

To manually install your application and the .NET Compact Framework to your device, follow these steps:

1. Locate the directory holding the CAB files appropriate for your device.
2. Copy the two CAB files to the root directory of your device by using ActiveSync or any other convenient means.
3. Click the .NET Compact Framework CAB file to explode it first, and then click the SQL CE CAB file to explode it second.
4. Copy all of the files associated with your application to your device in any directory you wish.

.NET COMPACT FRAMEWORK APPLICATIONS ARE PORTABLE

Although you must choose the correct build of the .NET Compact Framework to install manually on your nonstandard device, you do not have to make a special build of your application for a specific device. This is because your application's executable is managed code that is JITed specifically for the processor that is running the application. Chapter 2, "Introducing the .NET Compact Framework," discusses the JIT engine in greater detail.

You can tell if the CAB files exploded successfully by looking in the device's Windows directory. The files listed in Table 1.6 are dropped there if the .NET Compact Framework installed successfully.

TABLE 1.6

Files in the `\Windows` Directory for the .NET Compact Framework

Note: The filenames may differ slightly if you installed a non-English localized version of the .NET Compact Framework.

GAC_Microsoft.VisualBasic_v7_0_5000_0_cneutral_1.dll
GAC_Microsoft.WindowsCE.Forms_v7_0_5000_0_cneutral_1.dll
GAC_Mscorlib_v7_0_5000_0_cneutral_1.dll
GAC_System.Data_v7_0_5000_0_cneutral_1.dll
GAC_System.Drawing_v7_0_5000_0_cneutral_1.dll
GAC_System.Net.IrDA_v7_0_5000_0_cneutral_1.dll
GAC_System.SR_v7_0_5000_0_cneutral_1.dll
GAC_System.Web.Services_v7_0_5000_0_cneutral_1.dll
GAC_System.Windows.DataGrid_v7_0_5000_0_cneutral_1.dll
GAC_System.Windows_Forms_v7_0_5000_0_cneutral_1.dll
GAC_System.XML_v7_0_5000_0_cneutral_1.dll
GAC_System_v7_0_5000_0_cneutral_1.dll

Devices with Unusual Memory Maps

Some proprietary devices have memory maps that mix flash memory with standard static memory. The result is that some entries that are written to the file system remain even after the device is cold-rebooted. The only way to reset the file system to a "pristine" state is to re-flash the device, which means downloading the original operating system image by using a special tool that overwrites the flash memory.

This scenario can become a field support nightmare because unsuspecting users can alter the root filesystem in such a way as to disable the device completely, unless a tech support specialist comes to re-flash the device. This alteration is especially problematic if the memory map is set up so that the device registry is stored on flash memory. Users can do things to damage the registry that causes the device to malfunction, and there is nothing that can be done without re-flashing the device.

Holding the system registry in writable flash memory causes a very specific problem when working with Smart Device Extensions. Recall that the process of prepping a device means that a private key is exchanged between the desktop and the device. To ensure safety, the private key is encrypted using the device's *key exchange key* before it is transmitted between the desktop and the device.

The key exchange key is created when it is needed the first time, and then it is stored in the device registry. If the device is cold-rebooted but the registry is not cleared, then the old key exchange key remains. However, the old key exchange key is incompatible with the newly cold-rebooted device, and so the device now has no way to exchange private keys with other parties. Thus, prepping will fail, and trying to deploy or debug to the device will never work again until the device is re-flashed.

DEVICES WITH REGISTRY INFORMATION IN FLASH

A device with a memory map in this state is likely to exhibit problems other than those related to the .NET Compact Framework. It is not a good design. However, this discussion is included to give readers every chance of getting deployment and debugging to work on their nonstandard devices.

If you have a device with which you could deploy and debug until it was cold-rebooted, and even the Smart Device Authentication Utility does not seem to help after a cold reboot, it is possible that the registry is surviving cold reboots. To check for this condition, follow these steps:

1. Look for the default key exchange key in the device registry. The easiest way to do this is to use the device registry editor that is included with embedded Visual C++ version 3.0 or 4.0. The default exchange key is stored under the key `HKEY_CURRENT_USER\Comm\Security\Crypto\UserKeys\*Default*`. Figure 1.20 shows the device registry editor as it examines the default key exchange key.

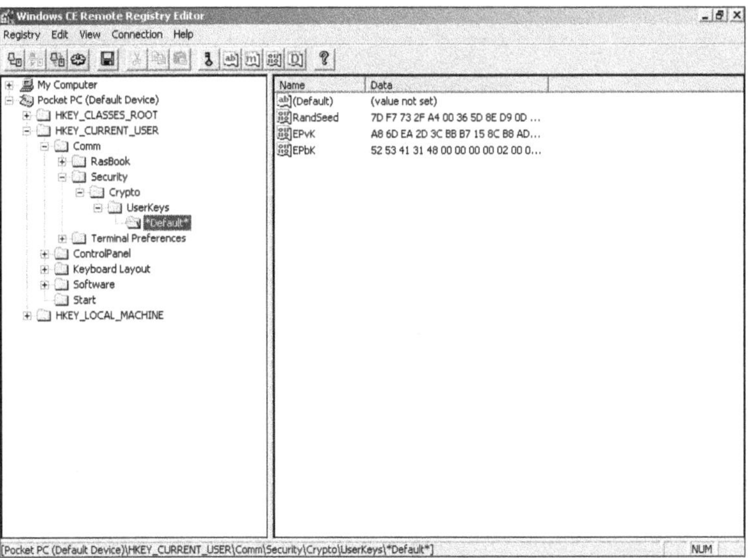

FIGURE 1.20 You can use the Device registry editor, shipped with Embedded Visual Tools 3.0 and 4.0, to display the Default exchange key.

2. Write down the value for the key exchange key and cold-reboot the device.
3. The default key exchange key entry should be gone. If it is not, then the registry is surviving cold reboots and will cause the problems just described.
4. To make deployment possible again after a cold reboot, delete the default key exchange key after a cold reboot by using the device registry editor.

Connectivity Summary and Troubleshooter

Table 1.7 lists the most common problems users experience when connecting to their devices and possible solutions. The table is most effective when the reader has read the previous sections of this chapter and has an understanding of how device connectivity works.

TABLE 1.7

Common Connectivity Problems and Solutions

PROBLEM	SOLUTION
Cannot deploy or debug against CE device running ActiveSync and Windows CE 4.1 or greater.	1. Use the CPU Picker utility. 2. Use the Smart Device Authentication Utility to prep the device and launch the correct version of `ConmanClient.exe`. 3. Install the .NET Compact Framework onto the device manually.

TABLE 1.7
Continued

PROBLEM	SOLUTION
Device has TCP/IP connectivity but not ActiveSync.	1. Use the Smart Device Authentication Utility. 2. Install the .NET Compact Framework onto the device manually.
Deployment and debugging to a nonstandard device without ActiveSync worked until device was rebooted.	`ConmanClient.exe` is stopped. Relaunch it using the Smart Device Authenticaiton Utility.
Nonstandard device was set up successfully by using the Smart Device Authentication Utility until device was cold-rebooted; subsequent attempts to use the Smart Device Authentication Utility fail and deployment and debugging fail.	Device has a nonstandard memory map. See discussion on such devices in previous section.
Deployment and debugging to a Pocket PC device work until Pocket PC device is rebooted.	Device must be connected through ActiveSync so that Smart Device Extensions can restart `ConmanClient.exe`.

In Brief

- Smart Device Extensions is a part of Visual Studio 7.1 that lets developers target the .NET Compact Framework for PocketPC and Windows CE devices.
- Creating and deploying applications for PocketPC devices is very easy and usually works "out of the box." However, deploying and debugging applications on Windows CE devices can be more challenging.
- Conman is the connectivity technology by which Smart Device Extensions communicates with devices. Understanding Conman makes it possible to effectively troubleshoot connectivity issues.
- CPU Picker lets users dictate to Conman exactly which CPU architecture their device uses, which can solve problems caused when Conman cannot correctly determine the architecture for itself.
- The Smart Device Authentication Utility lets developers set up the security infrastructure needed to make Conman connections to devices. This utility makes it possible to set up such an infrastructure on devices that do not support ActiveSync.
- If Conman cannot be set up successfully, then it is still possible to install the .NET Compact Framework manually.

Introducing the .NET Compact Framework

IN THIS CHAPTER

Why the .NET Compact Framework?

In Chapter 1, "Setting Up Your Development Environment," we discussed Smart Device Extensions for Visual Studio 7.1 in enough depth to create a very simple application and run it on a device. We also gained an understanding of what hardware is supported and how to deal with nonstandard devices. Now we examine the internals of the .NET Compact Framework as a platform.

Traditional compiled programming languages, such as C, C++, Pascal, and so on, use a compiler to translate source code directly into machine code understandable by the processor of the computer for which the code was compiled. Most of the time, developers were left responsible for managing low-level details, such as memory allocations and interacting with external code libraries. This includes DLL files on Windows systems. Programs could interact with the host operating system by directly calling into the operating system's API or by using one of many function libraries, such as MFC or Borland's OWL.

Portability between platforms by using native code is often difficult because the underlying operating system APIs are usually radically different. Even worse, interacting with components written in different programming languages on the same operating system is problematic because the languages use different binary representations for data structures. In short, interoperability was difficult among three major axes: across operating systems, across different

CPU types running the same operating system, and across different languages on the same machine.

Until Visual Studio 7.1, developers for Pocket PC and Windows CE platforms who wanted real programming power were stuck in the compartmentalized world of traditional, or native, programming languages. With the introduction of the .NET Compact Framework, all of the problems previously described and more are solved.

Visual Studio 7.1 with Smart Device Extensions offers these important features to help devlopers write applications for Windows CE and Pocket PC devices more quickly and easily than ever before:

- Two languages, C# and Visual Basic.NET, are supported by Smart Device Extensions out-of-the-box. Interoperability between these two languages is easy because both compile to the same bytecode format and provide access to data objects within the .NET Compact Framework.
- Any language that targets the .NET Compact Framework can exchange objects from the .NET Compact Framework with another language that supports the .NET Compact Framework.
- The compiled binaries for managed applications are the same, regardless of the underlying processor architecture. Developers can build an application just once and deploy it to a wide variety of Pocket PC and Windows CE hardware.
- The .NET Compact Framework is a rich class library that provides a uniform API that developers can target from C#, Visual Basic .NET, and future languages supporting the .NET Compact Framework.
- The .NET Compact Framework includes support for developing Windows Forms to rapidly develop compelling user interfaces.
- The .NET Compact Framework includes a wide variety of data access classes, XML manipulation classes, a rich set of fundamental data types, easy-to-use networking support, and more.
- The .NET Compact Framework and the associated Common Language Runtime provide garbage collection, thereby eliminating memory leak and double free bugs from your code automatically. These two bug classes are among the most time consuming, subtle, and difficult to diagnose when using native code.
- The .NET Compact Framework provides a rich threading model without having to understand the complexities of creating threads by directly targeting the operating system.
- Multiple versions of the .NET Compact Framework can operate side by side. Applications target the version that they know they are compatible with. This arrangement eliminates the classic "DLL Hell" problem that occurs when a DLL component is

upgraded and breaks some of the applications that target a previous version of the component.

- The .NET Compact Framework exposes a mechanism for calling into native code to help retain compatibility with legacy code and call the Windows CE operating system directly.

The rest of this book is devoted to exploring all of the power and features of the .NET Compact Framework. Whereas the previous list gives us an idea of what the .NET Compact Framework offers, drilling down to a deeper understanding of how it works helps us become more effective developers. The rest of this chapter fills in the details about how the .NET Compact Framework is designed.

Examining the .NET Compact Framework in Detail

The .NET Compact Framework does not exist in a vacuum. It rests upon three layers of technology: the Common Language Runtime, the Just-In-Time compiler, and the Windows CE Operating System. Figure 2.1 is a block diagram that describes how an application that targets the .NET Compact Framework interacts with the .NET Compact Framework and the layers below.

Figure 2.1 gives us a launching pad for discussing the architecture of the .NET Compact Framework. The discussion starts at the deepest level and works upward.

The first layer is the hardware itself, which is composed minimally of the CPU and main memory. Typical devices include a video adapter with touch screen, network connectivity, sound hardware, and so on.

All of the hardware is controlled by the second layer, the Windows CE operating system. Windows CE provides memory management routines and a program loader. The program loader pushes an executable into memory and launches it. It also manages threads, exposes methods for drawing windows and reacting to GUI events, handles network connectivity, and manages a host of other responsibilities.

The Common Language Runtime

The next layer above the Windows CE operating system is the Common Language Runtime, or CLR. Applications written in traditional languages, such as C or C++, target Windows CE as a platform. They are compiled into code understandable by the CPU on the device, but they rely on Windows CE to load them and provide services, such as drawing windows and reacting to user input.

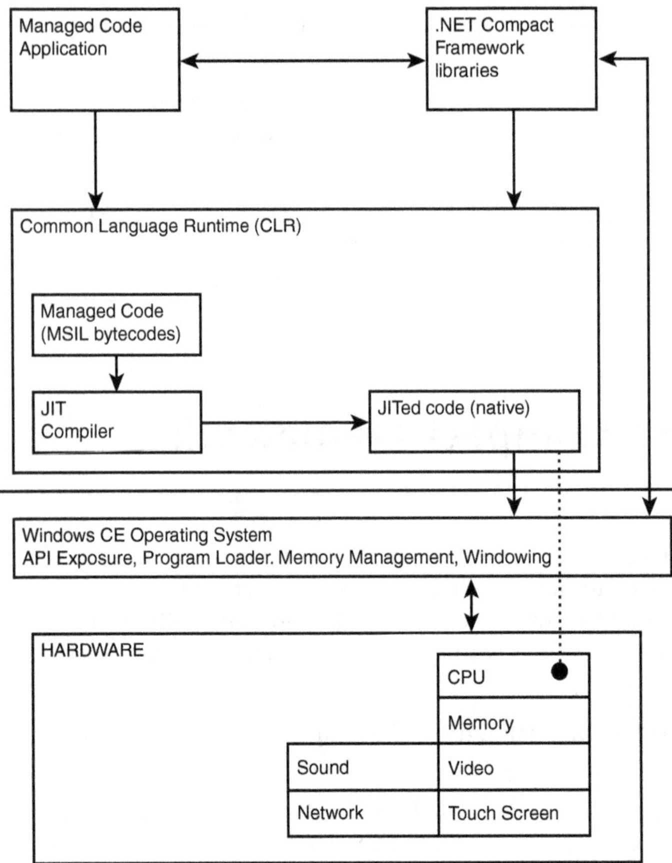

FIGURE 2.1 This block diagram depicts the .NET Compact Framework architecture.

The CLR is the platform that "managed" applications target. Standard "native" applications are compiled to machine code directly understandable by the device's CPU. In contrast, managed applications are compiled into Microsoft Intermediate Language bytecodes, or *IL code,* an intermediate format that loosely resembles machine code for a CPU architecture that does not physically exist.

Code that is compiled to IL code is referred to as *managed code*. Similarly, a *managed executable* is an entire EXE that is composed of IL code, and a *managed library* is a DLL file composed of IL code. A very important fact about managed code is that it cannot be directly executed by an existing CPU. It must be translated into native code first.

The CLR's job is to execute managed code. In order to do so, the managed code must be rendered into a native form understandable by the CPU. The CLR follows these three basic steps to transform managed code into native code and execute it:

1. The IL code must be loaded into memory from a filesystem.
2. Some or all of the IL must be translated into native code so that the CPU can actually execute it. This includes dealing with housekeeping issues, such as dealing with threads, exceptions, argument passing, and so on.
3. If the IL code refers to chunks of code in a DLL, then the DLL must be located and loaded. Then the correct portions of the DLL must be translated into native code and executed.

Transforming Managed Code to Native Code with the Just-In-Time Compiler

The JIT, the Just-In-Time compiler, is responsible for translating managed code into native code so that a managed program can execute. There are two JITs available with the .NET Compact Framework, the "SJIT" and the "IJIT."

IJIT The IJIT compiler is available for every CPU that is supported by the .NET Compact Framework: ARM, MIPS, SHx, and x86. The IJIT compiler is the faster of the two JITs, but it is simpler. Thus, although the IJIT compiler compiles managed code into native code quickly, the resulting native code is not as optimal as that produced by the SJIT compiler.

SJIT The SJIT compiler is available only for ARM processors. The ARM is the most common processor for the Pocket PC platform, and it is thus likely to represent the largest fraction of the .NET Compact Framework customer base. The SJIT compiler is very heavily tuned to take advantage of the ARM processor. Although compiling managed code with the SJIT compiler takes longer than compiling with the IJIT compiler, the resulting native code can run up to twice as quickly.

By default, the IJIT compiler is used only on platforms for which the SJIT compiler is unavailable. This makes sense because if your device is not under memory pressure, then the time spent JITing code with either compiler seems insignificant compared to the time executing code. Thus, you want the SJIT compiler whenever available.

An important part of writing well-performing applications is avoiding excess JIT activity. Both JIT engines compile code only as it as needed on a method-by-method basis. That is, if a class contains 100 methods but only 10 of them are ever executed by an application, then only 10 of them are ever JITed by the CLR. Once a method is JITed, the CLR tries to retain the native code in memory for the life of the application. Thus, the cost of JITing a method is paid only once, regardless of how many times the method is executed.

The CLR retains all JITed code in memory for as long as possible. It will kick out JITed code on a method-by-method basis if the device encounters memory pressure. This is called *code*

CONSIDER YOUR DEVICE AUDIENCE

The discussion comparing the SJIT and the IJIT underscores the fact that you must consider your device audience when writing an application. Will all of your users have ARM-based devices? Can you assume a certain amount of performance of all devices? You can get yourself into trouble if you assume that the performance you see on your ARM-based 400MHz Pocket PC device is the same as what a customer with a 200MHz MIPS-based device will see. It is generally invalid to compare different architectures simply by comparing clock speeds. Even if comparing clock speeds were a good way to extrapolate performance, the 200MHz MIPS device will probably run managed code at less than half the speed of the 400MHz ARM device because it is using the IJIT instead of the SJIT.

pitching. If the JITed code for a method is pitched and then the method is called again, then the method's IL code must be JITed all over again. The CLR pitches code for methods based on how recently the methods were executed; the code that was least recently executed is pitched first.

In extreme situations, applications can end up in a scenario where a method is re-JITed each time it is called in a loop. For very complex applications running under memory pressure, this is a plausible scenario that will horribly impact performance. This problem is analogous to running too many applications on a computer with insufficient memory. Pages of memory are swapped to disk, and if the memory pressure is severe enough, then the computer spends most of its time swapping pages instead of executing program code.

There are some simple steps you can take to avoid this situation. If you are writing a custom application to deploy to your company's workforce, equip the devices that will run the application with enough memory. Also, note that the default behavior when closing an application on the Pocket PC is to retain it in memory. In this state, applications still consume memory. To see what applications you have running on a Pocket PC, follow these steps:

1. Select Start, Settings.
2. At the bottom of the settings window, click the tab labeled System.
3. Click the Memory icon.
4. At the bottom of the window, click the Running Programs tab. You will see a list of programs that are currently running and consuming memory.
5. You can highlight a program and click Stop to terminate it. You can click Stop All to terminate all programs.

There are other tricks and tools in addition to those just listed that can help developers determine whether code pitching is occuring and where in the code it happens. For example, performance counters are very effective tools for determining what causes an application to perform poorly. Chapter 15, "Measuring the Performance of a .NET Compact Framework Application," describes such tools in detail.

Comparing the .NET Compact Framework CLR to the Desktop CLR

The architecture of the CLR for the .NET Compact Framework is markedly different from that of the desktop .NET Framework. For example, the CLR for the .NET Compact Framework is built upon a Platform Abstraction Layer (PAL) that abstracts away differences in hardware from the rest of the CLR. To port the CLR to a new platform, one needs only to change the PAL on which the CLR rests and create a JIT for the target CPU, if it is different from all existing supported CPUs. This approach gives the .NET Compact Framework flexibility and nimbleness in keeping up with evolving mobile hardware. However, there are no publicly available white papers or utilities to allow third parties to port the CLR to a new hardware platform.

Another difference between the CLR for the .NET Compact Framework and the desktop is how JITed code is handled. On the desktop, JITed code is retained even after a managed program exits, speeding up its load the next time. It is only re-JITed if the managed application's IL changes. The CLR for the .NET Compact Framework stores JITed code only for the lifetime of the application. The next time the application is launched, JITing must occur again.

The CLR that supports the .NET Compact Framework is similar to the one supporting the desktop .NET Framework in that it supports the notion of assemblies and application domains. An *assembly* is the atomic unit by which an application is identified. For example, a managed EXE file or a managed DLL file is each an assembly. Each assembly contains metadata inside the binary that describes its structure, and it can contain a signature to allow the CLR to detect if the assembly has been tampered with.

The desktop CLR supports the notion of an assembly that is composed of more than one file. The CLR for the .NET Compact Framework does not support this feature. This can have ramifications for developers who are trying to port managed code from a desktop application to a device. For example, if an assembly for a desktop component is composed of three individual DLLs, then the code must be merged into a single DLL when it is ported to the .NET Compact Framework.

Another concept related to a CLR is that of an *application domain*, which is a reasonably secure container in which each managed program runs. In a modern operating system with protected memory, each stand-alone process is completely insulated from all others. The language used to write the program running in the process may allow developers to use pointers to wreak havoc on the process memory space. However, because the process lives in a virtual address space, it cannot touch the virtual address space of other applications.

Similarly, the CLR enforces insulation between the application domains running within it. This level of insulation is cheaper than having separate processes, because the CLR actually operates as one process. Thus, application domains do live in the same virtual address space. This means it is technically possible for code in one application domain to write into memory being used by another application domain.

The CLR can prevent this behavior in virtually all cases because, with few exceptions, IL code can be guaranteed type safe. There is no possibility of pointer use going awry and accessing unexpected locations if there is no notion of a pointer in the language and if type references are guarded carefully by the CLR.

The C# language allows the direct manipulation of pointers if you embed the pointer activity in an `unsafe` block. One of the reasons the name "unsafe" was chosen as the keyword to delineate such a block is because it makes it possible for the C# code to escape the CLR type checking mechanisms and access unexpected memory locations. There are times when using unsafe code is completely necessary. For example, unsafe code can help when calling into native code with custom marshalling routines (see Chapter 12). Unsafe managed code should be avoided and used only if absolutely necessary.

The .NET Compact Framework Class Libraries

We now have enough of an understanding of the underlying technologies involved with the .NET Compact Framework to very specifically define the .NET Compact Framework itself. The .NET Compact Framework is a set of DLL files containing classes capable of performing all of the useful things that the rest of this book is devoted to. The DLL files are written in mostly managed code, although a handful of them interact through native code directly with the Windows CE operating system.

Out of the box, developers can use Smart Device Extensions in Visual Studio 7.1 to target the .NET Compact Framework by using the C# and Visual BASIC.NET programming languages. Third-party languages that can also target the .NET Compact Framework may become available in the future.

The specific DLLs comprising the .NET Compact Framework and the roles they play are now outlined:

> `Mscorlib.dll` This library comprises one part of the base classes. It holds the fundamental data type classes, such as `Int16`, `Int32`, and so on, and the classes for the following namespaces: `System.Collections`, `System.Diagnostics` (Debugger classes), `System.IO`, `System.Reflection`, `System.Runtime`, `System.Text`, and `System.Threading`. Some of the classes held in this library require calling into native code, but they present to managed code developers the illusion of being managed code only classes.
>
> `System.dll` This library holds classes like `Uri`, `FileUri`, and `UncName`. It also houses `System.Collections.Specialized`, `System.Net`, and `System.Text.RegularExpressions`.
>
> `Microsoft.VisualBasic.dll` This library includes helper methods and objects for Visual Basic programs, such as the `ErrObject` that is used in Visual Basic exception handling. All Visual Basic projects must reference this DLL.
>
> `System.Windows.Forms.dll` This library holds most of the classes in the `System.Windows.Forms` namespace. These classes provide common GUI elements, such as

buttons, list boxes, labels, and so on. Chapter 3, "Designing GUI Applications with Windows Forms," discusses how to use these classes to create professional-looking applications for Windows CE by using the .NET Compact Framework. Note that not all of the WinForms classes available on the desktop are available in `System.Windows.Forms.dll`. Additionally, many of the WinForms classes on the .NET Compact Framework are missing some of the methods and fields available on their desktop counterparts.

`System.Windows.DataGrid.dll` This library holds the `DataGrid` class, a GUI component that makes it easy to show the contents of a database in a structured, tabular format by using only a few lines of code. It is used in conjunction with the `DataView` and the `DataSet` classes, and it is described in detail in Chapter 6, "ADO.NET on the .NET Compact Framework."

`System.Drawing.dll` This library includes the API for low-level drawing methods, such as for drawing lines, circles, squares, images, and so on. Game developers will use this library often.

`Microsoft.WindowsCE.Forms.dll` This library handles implementation of the input panel (SIP) and the `MessageWindow` class, which allows developers to directly handle the messages from the Windows CE message pump.

`System.SR.*.dll` This library holds string resources for exceptions. You can choose not to ship this library with an application to save space on a device. If you make this choice and if an exception is thrown, then there will be no textual information included with the exception.

`System.Data.dll` This library holds the implementation for classes in the `System.Data` namespace. The central class in this namespace is the `DataSet`, which is described in Chapter 6. The `DataSet` is capable of XML serialization, as described in Chapter 8, "XML and the `Dataset`," and it can interact with the SQL CE database engine, as described in Chapter 7, "Programming with Microsoft SQL Server CE."

`System.XML.dll` This library holds the classes in the `System.XML` namespace. Chapter 10, "Manipulating XML with the `XmlTextReader` and the `XmlTextWriter`," covers them in detail. The classes available on the .NET Compact Framework version of `System.XML.dll` are a subset of those available on the desktop.

`System.Web.Services.dll` This library holds classes in the `System.Web.Services` namespace that provide the ability to interact with XML-based Web services. This feature is described in Chapter 9, "Using XML Web Services"

`System.Net.IrDA.dll` This library holds classes in the `System.Net` and `System.Net.Sockets` namespaces that relate to communication by using a device's infrared (IR) port. The central class in this namespace is the `IrDAClient`, which is described in detail in Chapter 5, "Network Connectivity with the .NET Compact Framework."

In Brief

- The .NET Compact Framework empowers developers by letting them write powerful applications for devices in C# and Visual Basic.
- Unlike traditional programming languages, the .NET Compact Framework is founded on "managed code." Using managed code provides many advantages, such as those mentioned in the following items.
- Managed code applications feature garbage collection, eliminating bugs related to erroneous memory allocations and releases.
- Managed code applications target the same rich class framework regardless of the programming language used.
- Managed code programs can pass each other instances of data simply, safely, and easily.
- Managed code is portable. The same binaries execute on devices regardless of the CPU on the device.
- The Common Language Runtime, or CLR, executes managed code. Its duties include loading managed code and compiling it to native executable code that the device CPU understands. The CLR also insulates managed applications from one another, preventing one malicious application from crashing other running managed applications.
- One of the two Just-In-Time (JIT) compilers translates managed code to native code. The SJIT creates faster native code than the IJIT, but it is available only on ARM-based devices.
- If a device comes under memory pressure, it will discard the JITed code, a process called code pitching. The least recently executed code is pitched first.
- Code pitching is undesirable and can cause severe performance problems if code is frequently pitched, only to be JITed soon thereafter.
- The roughly dozen libraries that comprise the .NET Compact Framework are mostly managed code.

Designing GUI Applications with Windows Forms

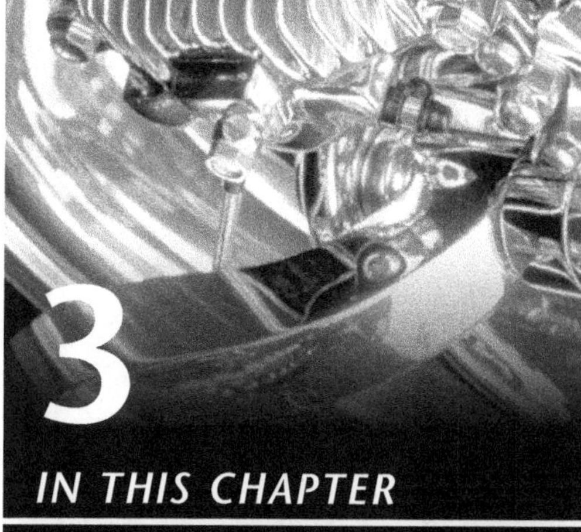

IN THIS CHAPTER

The .NET Compact Framework provides a rich functionality for building graphical user interfaces for your application. However, this functionality is a subset of what the full .NET Framework supplies. Before we instigate the controls that the .NET Compact Framework supports, let's investigate what is missing.

Investigating Unsupported Controls in the .NET Compact Framework

The .NET Compact Framework provides a subset of the controls available on the full .NET Framework.

The following list contains the controls that are not supported on the .NET Compact Framework.

- `CheckedListBox`
- `ColorDialog`
- `ErrorProvider`
- `FontDialog`
- `GroupBox`
- `HelpProvider`
- `LinkLabel`
- `NotificationBubble`

- `NotifyIcon`
- All `Print` controls
- `RichTextBox`
- `Splitter`

Investigating Unsupported `System.Windows.Forms` Functionality in the .NET Compact Framework

In addition to missing controls, the .NET Compact Framework also lacks functionality provided by the remaining controls that are supported. The following list gives the functionality that is missing in the .NET Compact Framework controls.

- `AcceptButton`
- `CancelButton`
- `AutoScroll`
- `Anchor`
- Multiple Document Interface (MDI)
- `KeyPreview`
- `TabIndex`
- `TabStop`
- Drag and drop
- All printing capabilities
- Hosting ActiveX controls

Working with the Visual Studio .NET Form Designer

The Visual Studio .NET Form Designer allows you to design your application's user interface visually by dragging controls onto a design-time representation of your application. Once the controls are on the form, you can visually position the controls, set their properties through the Properties window, and create event handlers for the events the controls fire.

The Form Designer Window

When you create a Smart Device Extension (SDE) project that is a Windows application, Visual Studio .NET will open the project in Designer view. You can also select the Designer option from the View menu to put the project in Designer view. The Form Designer is usually the windows in the middle of the Visual Studio .NET environment that contains the design-time representation of the application's form object. Figure 3.1 shows the Form Designer of a SDE Pocket PC project in Designer view. Notice the `mainMenu1` component at the bottom of the Designer window. This region of the designer is reserved for controls that do not have visual representations, such as the `MainMenu` control, the `ContextMenu` control, the `Timer` control, and so on.

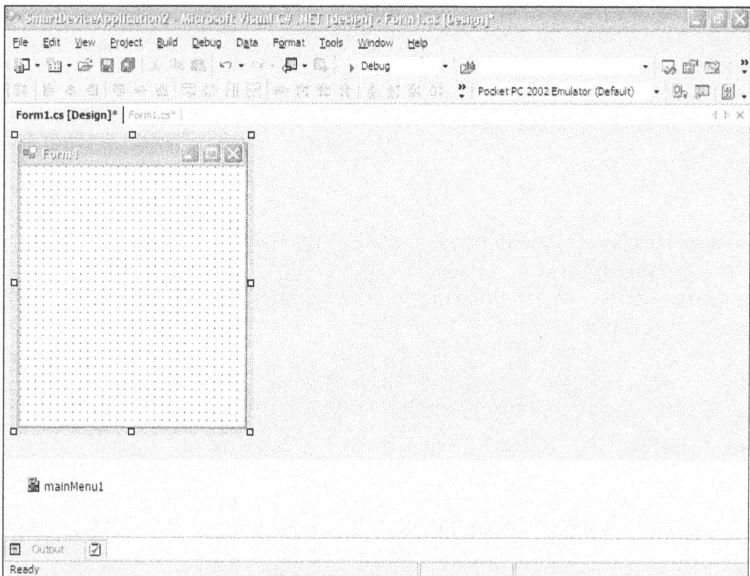

FIGURE 3.1 This SDE Pocket PC is shown in Designer view.

INCREASED RUNTIME FORM CREATION PERFORMANCE BY MODIFYING GENERATED CODE

When the Form Designer is used to build an application, the `InitializeComponent` method contains the code that build an application's user interface. This code has a big impact on performance if your form contains several nested controls. On the .NET Compact Framework it is recommended that windows be created top-down instead of bottom-up. For example, if a panel is on a form and the panel contains several controls, the panel should be added to the form, and then the controls should added to the panel. Unfortunately, the Visual Studio .NET Form Designer will generate code that creates forms from the bottom-up. For example, here is a section of code that was generated by the designer:

```
// This code is generated by the VS.NET Form Designer. It builds
// forms from the bottom-up. Button added to Panel…Panel added to Form
this.panel1.Controls.Add(this.button1);
this.panel1.Location = new System.Drawing.Point(16, 16);
this.panel1.Size = new System.Drawing.Size(208, 168);
//
// button1
//
this.button1.Location = new System.Drawing.Point(8, 16);
this.button1.Size = new System.Drawing.Size(72, 24);
this.button1.Text = "button1";
//
// Form1
//
this.Controls.Add(this.panel1);
this.Menu = this.mainMenu1;
this.Text = "Form1";
```

The following code contains code that will boost performance when it runs on the .NET Compact Framework:

```
// This code is hand written and modifies the code generated by the VS.NET
// designer. It builds forms from the top-down.
// Panel added to Form…Button added to Panel…
this.panel1.Location = new System.Drawing.Point(16, 16);
this.panel1.Size = new System.Drawing.Size(208, 168);
this.Controls.Add(this.panel1);
//
// button1
//
this.button1.Location = new System.Drawing.Point(8, 16);
this.button1.Size = new System.Drawing.Size(72, 24);
this.button1.Text = "button1";
```

SHOP TALK

```
//
// Form1
//
this.panel1.Controls.Add(this.button1);
this.Menu = this.mainMenu1;
this.Text = "Form1";
```

The `ToolBox` Window

The `ToolBox` window contains all of the .NET Compact Framework controls that can be added to an application. Adding a control to an application at design-time is as easy as dragging the control from the `ToolBox` and dropping it on the application's form in the Form Designer window. Figure 3.2 shows the `ToolBox` for an SDE Pocket PC project.

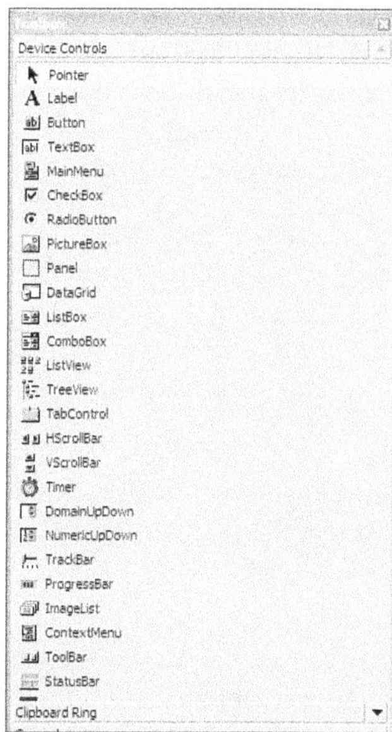

FIGURE 3.2 The `ToolBox` window for an SDE Pocket PC project.

The Properties Window

The Properties window contains all of the public properties of the control currently selected in the Form Designer window. You can change these properties by typing values into the `TextBox` controls next to the property names. If the property has a limited number of values, then a drop-down box is displayed next to the property name that contains the possible values for the property. Finally, if the property's value is a collection of objects or a complex object, there may be an ellipsis located next to the property name. Clicking this ellipsis will display a dialog box that allows you to edit the value of the property further. Figure 3.3 displays the Properties window when a `TextBox` is selected.

Adding Events Handlers

When a control is manipulated at runtime, it will fire an event to notify the application that the control's state has changed. This event notification follows the same event handling convention used throughout the .NET Compact Framework. To handle an event published by a control, you must first create a method containing the code to execute when the event is fired. Then attach the method to the control's published event. This can be done through the Properties window.

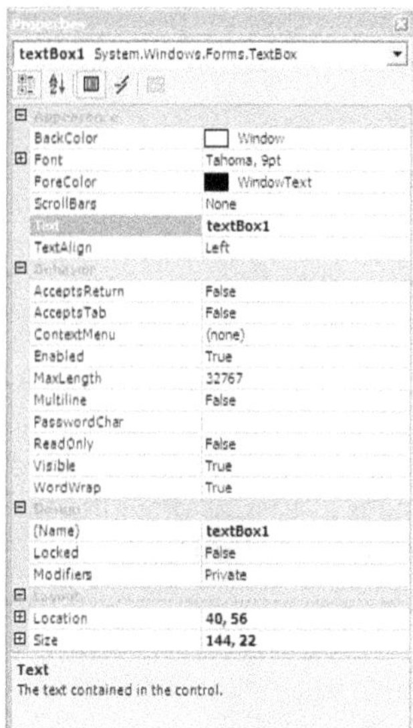

FIGURE 3.3 The Properties window for a `TextBox` control.

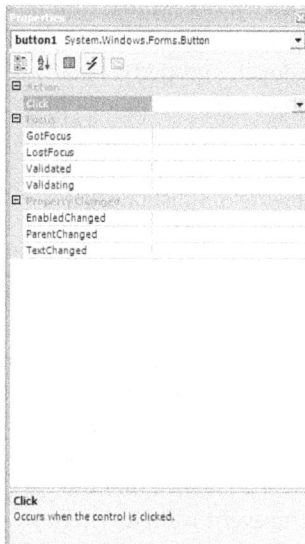

FIGURE 3.4 The Properties window that displays the `Button` control's published events.

The Properties window lists the events that are published by the control. Clicking the Events button at the top of the Properties window will list these events. The Events button is labeled with a lightning bolt. Figure 3.4 shows the Properties window displaying the event published by the `Button` control. (In Visual Basic .NET, events are found in drop-down boxes above the code editor.) You can assign a method as the event handler by clicking the event and selecting a method from the drop-down control next to the event name. (In Visual Basic .NET, selecting the event name from the drop-down box will automatically generate a method to handle the event.)

You can also add an event handle by double-clicking the event name. Doing so will switch the project to code view and will leave the editor inside the event-handler method. Code to wire the event handler to the control is also automatically generated. You can find this code in the `InitializeComponent` method of the application's `Form` class. (In Visual Basic .NET an event handler is wired to its event in its method signature.) The `InitializeComponent` method may be in a collapsed region labeled "Windows Form Designer generated code."

The following code was automatically generated in the `InitializeComponent` method. This code wires a `Click` event handler to the `Click` event of a `Button` control.

```
C#
#region Windows Form Designer generated code
/// <summary>
/// Required method for Designer support; do not modify
/// the contents of this method with the code editor.
/// </summary>
private void InitializeComponent()
{
  this.button1 = new System.Windows.Forms.Button();
  //
  // button1
  //
  this.button1.Location = new System.Drawing.Point(56, 128);
  this.button1.Text = "button1";
  this.button1.Click += new System.EventHandler(this.but-
ton1_Click);
  //
```

```
    // Form1
    //
    this.Controls.Add(this.button1);
    this.Text = "Form1";

}
#endregion

VB
Private Sub Button1_Click(ByVal sender As System.Object, _
        ByVal e As System.EventArgs) Handles Button1.Click
  MessageBox.Show("Hello")
End Sub
```

Understanding the Different Windows Forms Target Platforms

Smart Device Extensions (SDE) projects must target either the Pocket PC OS or the Windows CE .NET OS. These two platforms have different user interface APIs. An SDE project handles this by calling different libraries on each platform.

Understanding Windows CE .NET Projects

Windows CE .NET projects are similar to full .NET Framework Windows application projects. First, the minimize button, maximize button, and close button appear in the control box of the application just as they do in the full .NET Framework `Form` object. These buttons also behave as they would on the desktop. You can remove the control box from the form by setting the `ControlBox` property to false. You can remove the minimize button and maximize button by setting the `MinimizeBox` and `MaximizeBox` properties, respectively, to false.

When a Windows CE .NET application form is created by the Visual Studio.NET designer, its size is set to be 640 × 450. You can modify the `Size` property if this is not appropriate for your application. Although the `Form` class exposes the `FormBorderSytle` property, setting the property to `Sizable` will not affect the window's border. No Windows CE .NET application is resizable. It can only be minimized, maximized to full screen, or sized according to the `Size` property.

Understanding Pocket PC Projects

Pocket PC applications deviate further from full .NET Framework Windows application projects. First, a `MainMenu` object is always added to a Pocket PC application. You can remove this menu, but doing so will cause an exception to be thrown when interacting with the Soft Input Panel (SIP). The SIP is a software implementation of the QWERTY keyboard.

The Visual Studio .NET `ToolBox` window contains an `InputPanel` control. On the Pocket PC this control allows you to interact with the SIP. The `InputPanel` allows you to raise and lower the SIP. The `InputPanel` will also notify your application when the SIP has been enabled. Interestingly, your form must have a `MainMenu` control on it in order for the `InputPanel` control to be added to the form. If there is no `MainMenu` control on the form, then an exception will be thrown at runtime when you attempt to make the `InputPanel` visible.

Pocket PC applications must adhere to certain guidelines and recommendations. First, there must be only one instance of an application running at once. The .NET Compact Framework runtime guarantees this functionality, so there is no need to write code that ensures that only one instance of your application is running.

The second guideline is that once an application is running, the user should not have the ability to close the application. Instead, the application will be deactivated when it is no longer the application in the foreground. The guidelines state that the application should save any data, release any resources, and disconnect any connections (database connections, for example) when it is deactivated. If the application is started again, it will be activated by the OS. When it is activated, the application should reallocate resources and reestablish any connections. The application will be notified when it is activated or deactivated by `Form.Activate` and `Form.Deactivate` events. Unfortunately, the `Activate` and `Deactivate` events are not listed in the Properties Designer, so you have to wire these events manually. The following code demonstrates how to handle these events as well as wire them to the `Form` object.

```
C#
private void InitializeComponent(){
  this.Activated += new EventHandler(Form1_Activated);
  this.Deactivate += new EventHandler(Form1_Deactivate);

  this.mainMenu1 = new System.Windows.Forms.MainMenu();
  this.Menu = this.mainMenu1;
  this.Text = "Form1";
}

private void Form1_Activate(object sender, System.EventArgs e) {
  // Re-connect to databases, re-claim resources, and restore saved data
}
```

```
private void Form1_Deactivate(object sender, System.EventArgs e) {
  // Disconnect from databases, release resources, and save data
}

VB
Private Sub
Form1_Activated(ByVal sender As Object, ByVal e As System.EventArgs)
Handles MyBase.Activated
  ' Re-connect to databases, re-claim resources, and restore saved data
End Sub

Private Sub Form1_Deactivate(ByVal sender As Object, ByVal e As System.EventArgs)
Handles MyBase.Deactivate
  ' Disconnect from databases, release resources, and save data
End Sub
```

If available memory becomes scarce, the Pocket PC OS will start terminating minimized applications. If you have handled the `Deactivate` event correctly, then your application will have saved all data, and the user will not lose any unsaved data. Handle the `Form.Closing` and `Form.Closed` events to perform some task when Pocket PC terminates the application.

Because applications cannot be allowed to terminate themselves, you must not give the user the ability to close the application by clicking the OK button in the application's control box. To remove the OK button set the `MinimizeBox` property to true. This will add the X button to the application's control box. When the X button is clicked, the application is minimized instead of being closed.

Closing a dialog box is handled more traditionally. All dialog boxes should be closable by clicking the OK button (`MinimizeBox` property set to false), and they should also be displayed full screen. If they are not full screen, they must appear in the center of the screen. Use this code to center a dialog box:

```
C#
private void CenterForm() {
  Rectangle screenBounds = Screen.PrimaryScreen.Bounds;
  int x = ( screenBounds.Width - this.Width ) / 2;
  int y = ( screenBounds.Height - this.Height ) / 2;
  this.Location = new Point(x, y);
}

VB
Private Sub CenterForm()
  Dim screenBounds as Rectangle
```

```
    Dim x as Int32
    Dim y as Int32

    screenBounds = Screen.PrimaryScreen.Bounds
    x = ( screenBounds.Width - Me.Width ) / 2
    y = ( screenBounds.Height - Me.Height ) / 2
    Me.Location = new Point(x, y)
End Sub
```

SHOP TALK

ASSOCIATING AN ICON WITH YOUR APPLICATION

An application is never really complete until it has its icon. Associating an icon with an application can be done through Visual Studio .NET by opening the project Property Pages dialog box. Then select the General section under the `Common Properties` folder. Next click the ellipsis next to the `Application Icon` property. This will display the `OpenFileDialog`. Locate the icon in the dialog box. Figure 3.5 shows the project Property Pages dialog box.

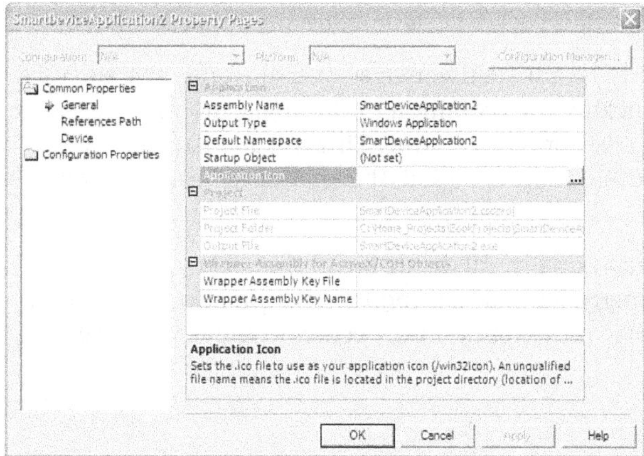

FIGURE 3.5 The project Property Pages dialog box.

Working with the `Form` Control

The `Form` control is the container for an application's entire user interface. The `Form` control is the actual window that contains the application's controls. The `Form` class has several properties that cause the form to act differently depending on the target platform.

Understanding the Effects of the `FormBorderStyle` Property

The `FormBorderSytle` property determines the border style of the form. The default value for this feature is `FormBorderStyle.FixedSingle`.

On the Pocket PC, setting the property to `FormBorderStyle.None` creates a form with no border and no title bar. This type of form can be resized and moved in code but not by the user. Setting the property `FillBorderStyle.FixedSingle` or any other value will create a form that fills the desktop area, and the form will not be movable or resizable.

On the Windows CE .NET, setting the property to `FormBorderStyle.FixedDialog` or `FormBorderStyle.None` will create a form with no border or title bar. The form will be movable and resizable through code only. Setting the property to `FormBorderStyle.FixedSingle` or any other value will create a form sized to the `Size` property with a border and a title bar. The form will only be resizable or moved through code, and the user will be able to move the form.

Using the `ControlBox` Property

The form's `ControlBox` property determines whether the control box is displayed. Setting the `ControlBox` property to true will display the control box. Set the property to false to hide the control box.

Understanding the `MinimizeBox` and `MaximizeBox` Properties

On Pocket PC the control box only contains at most one button, either the minimize button, labeled X, or the close button, labeled OK. On Windows CE .NET the control box can contain the minimize button, the maximize button, and the close button. The visibility of these buttons is controlled by the `MinimizeBox` and `MaximizeBox` properties. Table 3.1 describes the possible values of the `MinimizeBox` and their effects on each target platform. Table 3.2 does the same for the `MaximizeBox` property.

TABLE 3.1

The Possible Values for the `MinimizeBox` Property and Their Effects on Each Target Platform

PROPERTY VALUE	POCKET PC APPLICATION	WINDOWS CE .NET APPLICATION
`True`	X (minimize) button in the title bar	Traditional windows' minimize button in the title bar.
`False`	OK (close) button in the title bar	There is no minimize button in the title bar.

TABLE 3.2
The Possible Values for the `MaximizeBox` Property and Their Effects on Each Target Platform

PROPERTY VALUE	POCKET PC APPLICATION	WINDOWS CE .NET APPLICATION
`True`	No effect	Traditional windows' maximize button in the title bar.
`False`	No effect	There is no maximize button in the title bar.

Understanding the `WindowsState` Property

The `WindowsState` property determines the initial visible state of the window. This property can be set only to `FormWindowState.Normal` or `FormWindowState.Maximized`. Table 3.3 describes each value and its effect on an application running on both target platforms.

TABLE 3.3
The Possible Values for the `WindowState` Property and Their Effects on Each Target Platform

`FormWindowState` MEMBER NAME	POCKET PC APPLICATION	WINDOWS CE .NET APPLICATION
`Normal`	The application will fill the entire desktop area, which is the entire screen area minus the area of the start menu and the area of the main menu bar.	The application is sized according to the `Size` property.
`Maximize`	The application fills the entire screen. This will hide the start menu, but the main menu will still be visible.	The application fills the entire desktop area.

Understanding the `Size` Property

The `Size` property determines the size of the application window. Depending on the value of the `FormBorderStyle` property, the application can either ignore the value of the `Size` property or draw the application at the specified size. On Pocket PC, in order for the `Size` property to be honored, the `FormBorderStyle` must be set to `FormBorderSytle.None`. On Windows CE the `Size` property is always honored.

Setting the Location of the Form by Using the `Location` Property

The `Location` property determines the position of the top-left corner of the form. On Pocket PC the `Location` property has no effect unless the `FormBorderSytle` property is set to

FormBorderSytle.None. On Windows CE the location of the window is always equal to the Location property, unless the application been put in the minimized or maximized state.

Programming the Button Control

The System.Windows.Forms.Button class is the .NET implementation of a button control. When the user clicks the button with the stylus, a Click event is raised. You can handle this event by implementing a System.EventHandler delegate. The code that follows is an implementation of the EventHandler that displays the current time.

```
C#
Private void button_Click(object sender, System.EventArgs e) {
  MessageBox.Show(DateTime.Now.ToShortTimeString(),
                 "The Current Time Is",
                 MessageBoxButtons.OK,
                 MessageBoxIcon.Exclamation,
                 MessageBoxDefaultButton.Button1);
}

VB
Private Sub
Button1_Click(ByVal sender As System.Object, ByVal e As System.EventArgs)
Handles Button1.Click
  MessageBox.Show(DateTime.Now.ToShortTimeString(),
    "The Current Time Is",
    MessageBoxButtons.OK,
    MessageBoxIcon.Exclamation,
    MessageBoxDefaultButton.Button1)
End Sub
```

Figure 3.6 shows the GiveEmTime.exe running on the Pocket PC emulator. The button labeled What is the Time has been clicked, and the current time is being displayed in the dialog box.

TABLE 3.4
The KeyCodes Generated by the Directional Pad on a Pocket PC Device

KeyCode VALUE	ASSOCIATED HARDWARE BUTTON
Keys.Up	The top of the pad was pressed.
Keys.Down	The bottom of the pad was pressed.
Keys.Left	The left side of the pad was pressed.
Keys.Right	The right side of the pad was pressed.
Keys.Return	The center of the pad was pressed.

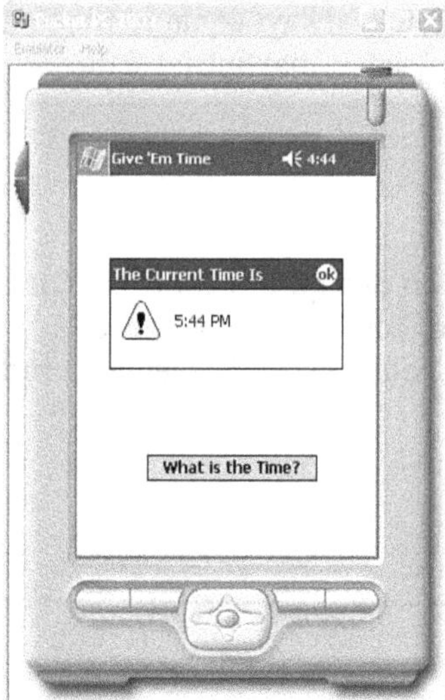

FIGURE 3.6 The `GiveEmTime` application running on the Pocket PC 2002 emulator.

Using the `TextBox` Control

The `TextBox` control is used to accept input from the user. The `TextBox` control supports the `BackColor` and `ForeColor` properties, unlike most other controls in the .NET Compact Framework. The `Click` event is not supported, but the `KeyPress`, `KeyUp`, and `KeyDown` events are. The `PasswordChar` property is supported. No matter what character this property is set to, the asterisks will always be used to mask the input.

Using the `Label` Control

The `Label` control allows you to display text to the user. This is a simple control that does not warrant much explanation. The `Text` property of the control determines what text will be visible to the user. The display text can have a different alignment based on the `TextAlign` property. The possible align values are `TopLeft`, `TopCenter`, and `TopRight`. The `TextChanged` event is fired when the text in a `Label` control changes.

SHOP TALK

HANDLING THE POCKET PC HARDWARE BUTTONS

The Pocket PC has a directional pad that can be used to interact with Pocket PC applications. It is possible to capture the events that this pad generates. When the pad is pressed, a `KeyDown` event is generated by the `Form` object. The event handler receives a `KeyEventArgs` object. The `KeyEventArgs` object exposes the `KeyCode` property. You can check the value of the `KeyCode` property to determine which direction on the pad was pressed. Table 3.4 lists the `Key` enumeration members that can be assigned to the `KeyCode` property and their associated buttons.

Working with `RadioButton` Controls

Radio button controls are commonly presented to give users an array of choices that are mutually exclusive. When a radio button within a group is selected, the others clear automatically. Radio buttons are considered to be in the same group if they are in the same container. An application can have multiple radio button groups by putting radio buttons in different `Panel` controls (see the "Using the `Panel` Control" section for a description of the `Panel` class).

SHOP TALK

HANDLING THE SOFT INPUT PANEL

The InputPanel control corresponds to the SIP on the Pocket PC OS. This control provides the ability to display and hide the SIP. The InputPanel exposes the Enabled property. If the property is set to true, the SIP is displayed; setting the property to false hides the SIP.

Some TextBox controls would be hidden when the SIP appears. The Pocket PC application design guidelines suggest that these controls should be moved so that they remain visible. To help achieve this, the InputPanel control will raise the EnabledChanged event when the Enabled property changes. You should handle this event and move the TextBox above the SIP when the Enabled property is set to true. When the Enabled property is set to false, the TextBox controls should be moved back to their original positions.

The RadioButton class publishes two events that are fired when the checked state of a RadioButton changes: Click and CheckedChanged. The Click event is raised when a user clicks the radio button with the stylus. You can handle this Click event just as you handled the Click event for the Button class (see the "Handling a ToolBar's ButtonClick Event" section). The CheckedChanged event is raised when the RadioButton's checked state changes, either programmatically or graphically.

The Click event will not be raised if the RadioButton's Checked property is changed programmatically. The Arnie.exe application demonstrates how to use a group of RadioButton controls. (You can find the code for this application in this book's Arnie sample program.) Figure 3.7 shows the application running in the Pocket PC emulator. The application is a simple trivia question about the name of the first movie Arnold Schwarzenegger ever starred in.

When a movie is selected, the application traps the RadioButton's CheckedChanged event, and a message box is displayed if the correct RadioButton is checked. The following code demonstrates how to handle the CheckedChanged event for an incorrect answer (I don't want to give away the correct answer).

```
C#
private void radioButton2_CheckedChanged(object sender,
        System.EventArgs e) {
  if(this.radioButton2.Checked)
    MessageBox.Show
      ("Wrong, The Terminator (1984) O.J Simpson almost got the role...",
              "Wrong!");
}

VB
Private Sub
```

```
radioButton2_CheckedChanged(ByVal sender As System.Object,
       ByVal e As System.EventArgs)
Handles radioButton2.CheckedChanged
  If radioButton2.Checked Then
    MessageBox.Show
      ("Wrong, The Terminator (1984) O.J Simpson almost got the role...",
              "Wrong!")
  End If
End Sub
```

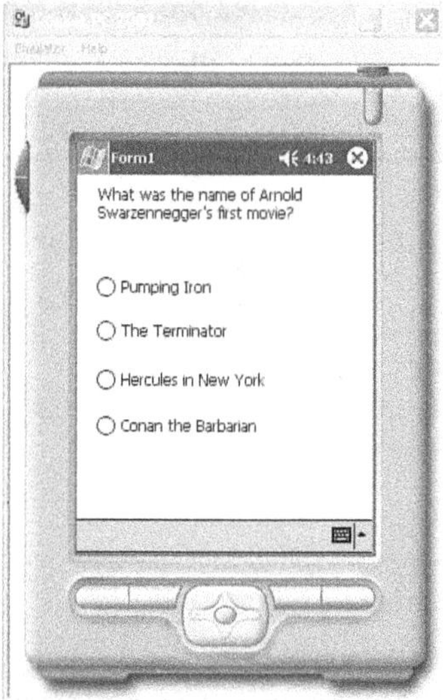

FIGURE 3.7 The Arnie application running on the Pocket PC 2002 emulator.

Using the CheckBox Control

The CheckBox control is similar to the RadioButton control in that it presents a list of choices to the user. The difference is that multiple CheckBox controls can be selected at once, while RadioButton controls are mutually exclusive. The CheckBox control is usually a GUI for a Boolean value or expression.

The CheckBox control provides the CheckState property, which determines whether the CheckBox is checked. The CheckState property is actually an enumeration, the CheckState enumeration. Its members are Unchecked, Checked, and Indeterminate. Unchecked and Check are self-explanatory, but the Indeterminate member warrants explanation. The Indeterminate state can be used only when the CheckBox control's ThreeState property is set to true. When the CheckState is Indeterminate and the ThreeState property is true, the control is grayed out but still checked. This signifies that the check state cannot be determined. The control will not respond to user clicks as long as the AutoCheck property is false. When the AutoCheck property is set to true, a click with the stylus will enable the control.

The Apples.exe application is another simple survey that attempts to determine what type of apples the user likes. The topmost CheckBox control has the caption "I like apples." The other CheckBox controls are labeled with different types of apples and are in an indeterminate state until the CheckBox labeled "I like apples" is checked; only then can the user check which apples he or she actually likes. If the user does not like apples, then there is no reason for the other CheckBox controls to be available. Figure 3.8 shows the application running in the Pocket PC emulator. The full code for this application can be found in the source code of this book.

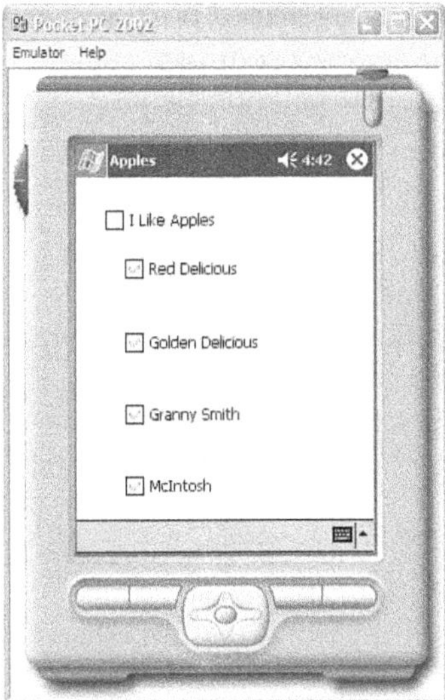

FIGURE 3.8 An application that showcases CheckBox controls running on the Pocket PC 2002 emulator.

Using the ComboBox Control

The ComboBox control is the ideal control to present a list of choices in a confined amount of screen space. The ComboBox appears as a TextBox control with an arrow on the right-hand side. A list of options drops down below the control when the user clicks the arrow. When the user selects an option or clicks the arrow again, the list of options rolls up again.

Adding items to the ComboBox control can be done both at design time and at runtime. To add items to the ComboBox at design time, simply select the ComboBox in the Form Designer. Then click the ellipsis next to the Items property in the Properties window. This will bring up the String Collection Editor (see Figure 3.9). In the String Collection Editor, enter the list of items to appear in the ComboBox. Each item must appear on a separate line.

Items can be added to the ComboBox control at runtime, as well. This can be accomplished in two different ways. First, call the Add method on the Items collection property of the ComboBox control. Items can be removed through the Remove method on the Items collection, or all items can be removed by calling the Clear method. The following code snippet adds three strings to a ComboBox control named comboBox1:

```
C#
comboBox1.Items.Add("Hi");
comboBox1.Items.Add("Howdy");
comboBox1.Items.Add("Wuz Up");

VB
comboBox1.Items.Add("Hi")
comboBox1.Items.Add("Howdy")
comboBox1.Items.Add("Wuz Up")
```

You can also add items to a ComboBox at runtime by binding the control to a collection object. This is done by setting the DataSource to the collection object. When the ComboBox attempts to add items to the drop-down list, it will call the ToString method on each item in the DataSource

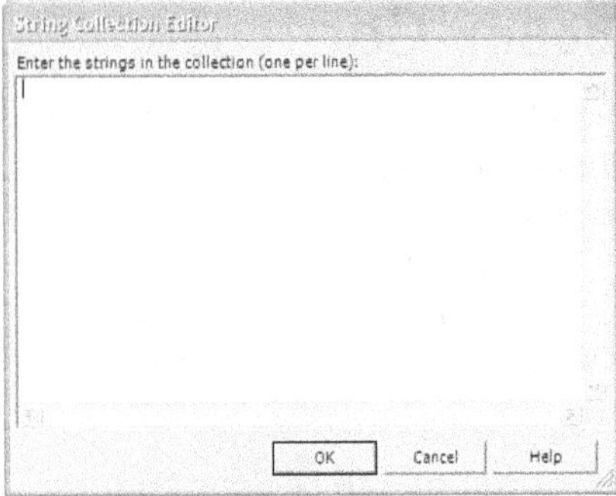

FIGURE 3.9 The String Collection Editor.

and add that string to the drop-down list. The string can be customized by setting the `ComboBox` control's `DisplayName` property. The `ComboBox` will call the property specified in the `DisplayName` property and add the returned string to the drop-down list.

Listing 3.1 demonstrates how to bind a `ComboBox` to a list of custom objects. The `Customer` class is a custom class that holds the name of a customer. The class has a property named `FullName` that properly formats the customer's full name. When the `ComboBox` is bound in the `LoadCustomer` method, the `FullName` property is set as the `DisplayName`.

LISTING 3.1

```
C#
class Customer {
  string m_First;
  string m_Middle;
  string m_Last;

  public Customer(string first, string middle, string last) {
    m_First = (first == null) ? string.Empty : first;
    m_Middle = (middle == null) ? string.Empty : middle;
    m_Last = (last == null) ? string.Empty : last;
  }

  public string FirstName {
    get { return m_First; }
  }

  public string MiddleName {
    get { return m_Middle; }
  }

  public string LastName {
    get { return m_Last; }
  }

  static string FullNameWithInitial = "{0} {1}. {2}";
  static string FullNameNoInitial = "{0} {1}";
  public string FullName {
    get {
```

```
      return (m_Middle.Length > 0) ?
        string.Format(FullNameWithInitial, m_First, m_Middle[0], m_Last) :
        string.Format(FullNameNoInitial, m_First, m_Last);
    }
  }
}

private void LoadCustomers() {
  if(customers != null)
    return;

    customers = new Customer[6];
    customers[0] = new Customer("Ronnie", "Donnell", "Yates");
    customers[1] = new Customer("Moya", "Alicia", "Hines");
    customers[2] = new Customer("Veronica", "Christine", "Yates");
    customers[3] = new Customer("Diane", "", "Taylor");
    customers[4] = new Customer("Kindell", "Elisha", "Yates");
    customers[5] = new Customer("Zion", "Donnell", "Yates");

    this.comboBox1.DataSource = customers;
    this.comboBox1.DisplayMember = "FullName";
}
```

VB

```
Public Class Customer
    Dim m_First As String
    Dim m_Middle As String
    Dim m_Last As String

    Public Sub New(ByVal first As String, ByVal middle As String,
            ByVal last As String)
      If first <> Nothing Then
        m_First = first
      Else
        m_First = String.Empty
      End If

      If middle <> Nothing Then
        m_Middle = middle
      Else
        m_Middle = String.Empty
      End If
```

```
      If last <> Nothing Then
        m_Last = last
      Else
        m_Last = String.Empty
      End If
    End Sub

    Public ReadOnly Property FirstName() As String
      Get
        Return m_First
      End Get
    End Property

    Public ReadOnly Property MiddleName() As String
      Get
        Return m_Middle
      End Get
    End Property

    Public ReadOnly Property LastName() As String
      Get
        Return m_Last
      End Get
    End Property

    Private Shared FullNameWithInitial = "{0} {1}. {2}"
    Private Shared FullNameNoInitial = "{0} {1}"

    Public ReadOnly Property FullName() As String
      Get
        If m_Middle.Length > 0 Then
          String.Format(FullNameWithInitial, m_First,
                m_Middle.Chars(0), m_Last)
        Else
          String.Format(FullNameNoInitial, m_First, m_Last)
        End If
    End Get
  End Property
End Class

Private Sub LoadCustomers()
  Dim customers(6) As Customer
  customers(0) = New Customer("Ronnie", "Donnell", "Yates")
```

```
    customers(1) = New Customer("Moya", "Alicia", "Hines")
    customers(2) = New Customer("Veronica", "Christine", "Yates")
    customers(3) = New Customer("Diane", "", "Taylor")
    customers(4) = New Customer("Kindell", "Elisha", "Yates")
    customers(5) = New Customer("Zion", "Donnell", "Yates")

    ComboBox1.DataSource = customers
    ComboBox1.DisplayMember = "FullName"
End Sub
```

There are two ways to obtain which item is currently selected in the `ComboBox`. First, the `SelectedIndex` item property returns the index of the currently selected item. This index can be used to access the selected item from the `ComboBox` control's `Items` property. The following code exemplifies the `SelectIndex` property:

```
C#
string selItem = comboBox1.Items[comboBox1.SelectedIndex].ToString();

VB
Dim selItem as string
selItem = comboBox1.Items(comboBox1.SelectedIndex).ToString()
```

The `ComboBox` control also provides the `SelectedItem` property that returns a reference to the currently selected item. Once you have a reference to the currently selected item, you do not need an index into the `Items` property. The following code demonstrates how to use the `SelectedItem` property:

```
C#
string selItem = comboBox1.SelectedItem.ToString();

VB
Dim selItem as string
selItem = comboBox1.SelectedItem.ToString()
```

Using the `ListBox` Control

The `ComboBox` is ideal for applications in space-constrained environments, but the `ListBox` should be used if you have enough screen space to display several options to the user at once.

The `ComboBox` and `ListBox` share almost the same exact set of properties and methods. This includes the `Items` collection property and the `Add`, `Remove`, and `Clear` methods on the `Items` property. For example, the following code adds strings to a `ListBox` control at runtime. This code is nearly identical to the code used to add string to a `ComboBox` control.

```
C#
listBox1.Items.Add("Hi");
listBox1.Items.Add("Howdy");
listBox1.Items.Add("Wuz Up");

VB
listBox1.Items.Add("Hi")
listBox1.Items.Add("Howdy")
listBox1.Items.Add("Wuz Up")
```

You can also add items to the `ListBox` control at runtime by binding the `ListBox` to a collection. The process of binding a `ListBox` control is identical to binding a `ComboBox` control. First, set the `DataSource` to the collection. Then, set the `DisplayMember` to the source item property that will be used as the display string. The following code is another version of the `LoadCustomers` method from the "Using the `ComboBox` Control" section. Instead of binding to a `ComboBox`, a list of `Customer` objects is bound to a `ListBox`.

```
C#
private void LoadCustomers() {
  if(customers != null)
    return;

    customers = new Customer[6];
    customers[0] = new Customer("Ronnie", "Donnell", "Yates");
    customers[1] = new Customer("Moya", "Alicia", "Hines");
    customers[2] = new Customer("Veronica", "Christine", "Yates");
    customers[3] = new Customer("Diane", "", "Taylor");
    customers[4] = new Customer("Kindell", "Elisha", "Yates");
    customers[5] = new Customer("Zion", "Donnell", "Yates");

    this.listBox1.DataSource = customers;
    this.listBox1.DisplayMember = "FullName";
}

VB
Dim customers(6) As Customer
customers(0) = New Customer("Ronnie", "Donnell", "Yates")
customers(1) = New Customer("Moya", "Alicia", "Hines")
customers(2) = New Customer("Veronica", "Christine", "Yates")
customers(3) = New Customer("Diane", "", "Taylor")
customers(4) = New Customer("Kindell", "Elisha", "Yates")
customers(5) = New Customer("Zion", "Donnell", "Yates")
```

```
ListBox1.DataSource = customers
ListBox1.DisplayMember = "FullName"
```

The ListBox also exposes the SelectedIndex and SelectedItem properties. These properties provide access to the currently selected item in the ListBox control. These properties are used exactly as they are with the ComboBox control.

Using the NumericUpDown Control

The NumericUpDown control is a simple way to give the user a way to select a number that falls between a minimum and a maximum value. The control can accept only integers, and decimal values will be truncated as opposed to rounded. On the Pocket PC, the maximum value cannot be greater than that of a 16-bit signed integer.

The NumericUpDown control is controlled by four integer properties: Minimum, Maximum, Value, and Increment. The Minimum and Maximum properties define the minimum and maximum values of the control. The Value property is the current value of the control. The Increment property defines the amount by which the current value is incremented or decremented when the user clicks the up or down arrow buttons. The current value is always incremented or decremented by the Increment value, unless the resulting value would be out of the range defined by the Minimum and Maximum values.

The user can also change the Value property by typing a new value into the control. If the value that the user types is between the Minimum and Maximum values, then both the Value and Text properties will be changed to reflect the newly entered value. If the new value is outside the set range, then the Text property takes the entered value, whereas the Value property becomes equal to the Maximum property. To stop users from typing data into the control altogether, set the ReadOnly property to true.

When a user changes the value of the NumericUpDown control, a ValueChanged event is fired. The ValueChanged event is fired only when the value is changed through code or via the up and down arrows. The event will not be fired when a user types input into the control. Listing 3.2 demonstrates how to use the NumericUpDown control and how to handle the ValueChanged event:

LISTING 3.2

```
C#
using System;
using System.Drawing;
using System.Collections;
using System.Windows.Forms;
using System.Data;
```

```
namespace NumericUpDown
{
  public class Form1 : System.Windows.Forms.Form
  {
    private System.Windows.Forms.NumericUpDown numericUpDown1;
    private System.Windows.Forms.Label label1;
    private System.Windows.Forms.Label label2;
    private System.Windows.Forms.MainMenu mainMenu1;

    public Form1()
    {
      InitializeComponent();
    }

    protected override void Dispose( bool disposing )
    {
      base.Dispose( disposing );
    }
    #region Windows Form Designer generated code
    /// <summary>
    /// Required method for Designer support - do not modify
    /// the contents of this method with the code editor.
    /// </summary>
    private void InitializeComponent()
    {
      this.mainMenu1 = new System.Windows.Forms.MainMenu();
      this.numericUpDown1 = new System.Windows.Forms.NumericUpDown();
      this.label1 = new System.Windows.Forms.Label();
      this.label2 = new System.Windows.Forms.Label();
      //
      // numericUpDown1
      //
      this.numericUpDown1.Location = new System.Drawing.Point(8, 56);
      this.numericUpDown1.Maximum =
              new System.Decimal(new int[] {2003, 0, 0, 0});
      this.numericUpDown1.Minimum =
              new System.Decimal(new int[] {1900, 0, 0, 0});
      this.numericUpDown1.Value =
              new System.Decimal(new int[] {190012, 0, 0, 131072});
      this.numericUpDown1.ValueChanged +=
        new System.EventHandler(this.numericUpDown1_ValueChanged);
```

```
            //
            // label1
            //
            this.label1.Location = new System.Drawing.Point(8, 24);
            this.label1.Size = new System.Drawing.Size(184, 16);
            this.label1.Text = "In what year were you born?";
            this. label1.ParentChanged +=
                    new System.EventHandler(this.label1_ParentChanged);
            //
            // label2
            //
            this.label2.Location = new System.Drawing.Point(8, 120);
            this.label2.Size = new System.Drawing.Size(224, 24);
            //
            // Form1
            //
            this.Controls.Add(this.label2);
            this.Controls.Add(this.label1);
            this.Controls.Add(this.numericUpDown1);
            this.Menu = this.mainMenu1;
            this.Text = "Form1";

        }
        #endregion

        static void Main()
        {
          Application.Run(new Form1());
        }

        static string msg = "You are ~{0} years young.";
        private void numericUpDown1_ValueChanged(object sender, System.EventArgs e)
        {
          int yearsOld = System.DateTime.Now.Year - (int)this.numericUpDown1.Value;
          this.label2.Text = String.Format(msg, yearsOld);
        }
    }
}

VB
Public Class Form1
  Inherits System.Windows.Forms.Form
```

```
  Friend WithEvents label2 As System.Windows.Forms.Label
  Friend WithEvents label1 As System.Windows.Forms.Label
  Friend WithEvents numericUpDown1 As System.Windows.Forms.NumericUpDown
  Friend WithEvents MainMenu1 As System.Windows.Forms.MainMenu

#Region " Windows Form Designer generated code "

  Public Sub New()
    MyBase.New()

    'This call is required by the Windows Form Designer.
    InitializeComponent()

    'Add any initialization after the InitializeComponent() call

  End Sub

  'Form overrides dispose to clean up the component list.
  Protected Overloads Overrides Sub Dispose(ByVal disposing As Boolean)
    MyBase.Dispose(disposing)
  End Sub

  'NOTE: The following procedure is required by the Windows Form Designer
  'It can be modified using the Windows Form Designer.
  'Do not modify it using the code editor.
  Private Sub InitializeComponent()
    Me.MainMenu1 = New System.Windows.Forms.MainMenu
    Me.label2 = New System.Windows.Forms.Label
    Me.label1 = New System.Windows.Forms.Label
    Me.numericUpDown1 = New System.Windows.Forms.NumericUpDown
    '
    'label2
    '
     Me.label2.Location = New System.Drawing.Point(8, 104)
     Me.label2.Size = New System.Drawing.Size(224, 24)
     '
     'label1
     '
     Me.label1.Location = New System.Drawing.Point(8, 8)
     Me.label1.Size = New System.Drawing.Size(184, 16)
     Me.label1.Text = "In what year were you born?"
     '
```

```
        'numericUpDown1
        '
        Me.numericUpDown1.Location = New System.Drawing.Point(8, 40)
        Me.numericUpDown1.Maximum = New Decimal(New Integer() {2003, 0, 0, 0})
        Me.numericUpDown1.Minimum = New Decimal(New Integer() {1900, 0, 0, 0})
        Me.numericUpDown1.Value =
                New Decimal(New Integer() {190012, 0, 0, 131072})
        '
        'Form1
        '
        Me.Controls.Add(Me.label2)
        Me.Controls.Add(Me.label1)
        Me.Controls.Add(Me.numericUpDown1)
        Me.Menu = Me.MainMenu1
        Me.Text = "Form1"
    End Sub

#End Region

    Private Sub _
    numericUpDown1_ValueChanged_
    (ByVal sender As System.Object, ByVal e As System.EventArgs)
    Handles numericUpDown1.ValueChanged
      Dim yearsOld As Int32
      yearsOld = System.DateTime.Now.Year - numericUpDown1.Value
      label2.Text = String.Format("You are ~{0} years young.",
              yearsOld.ToString())
    End Sub
End Class
```

Figure 3.10 displays the application running on the Pocket PC 2002 emulator.

Using the `DomainUpDown` Control

The `DomainUpDown` control is similar to the `NumericUpDown` control, the difference being that the `DomainUpDown` can display a list of strings, whereas the `NumericUpDown` control can display only a list of integers. The `DomainUpDown` control displays a list of options in a very space-efficient way because it can display only one option at a time to the user.

The control appears as a textbox with a pair of up and down arrows on the right side of the control. The user can move through the list in three different ways:

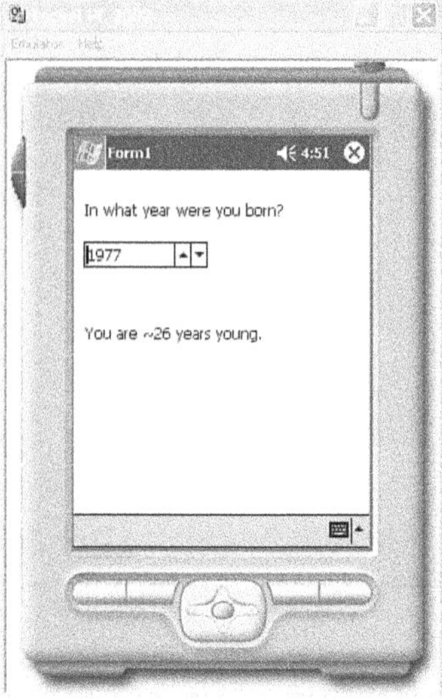

FIGURE 3.10 An application that showcases the NumericUpDown control running on the Pocket PC 2002 emulator.

- Clicking the up arrow moves you up the list of options; the down arrow moves you down the list.
- If the `ReadOnly` property is set to false, the user can type the name of the item in the list.
- On most Pocket PC devices, there is a button that can be used for navigating the `DomainUpDown` control. Pressing this button up and down moves you up and down the list items, respectively.

If `ReadOnly` is false, the user can enter text into the control that does not match an item in the list. The `Text` property will return the current text in the control. This means that you may need to do some input validation if `ReadOnly` is set to false. Because no input validation is performed by the `DomainUpDown` control, entering text into the control will not select a matching item. Setting `ReadOnly` to true restricts the user to selecting only items in the list, thus eliminating the need for input validation.

The `DomainUpDown` control exposes the `Items` property that represents the list of items in the control. You can add items to the list by using the `Add` method exposed on the `Items` property. The following code demonstrates how to add three items to a `DomainUpDown` control:

```
C#
DomainUpDown dud = new DomainUpDown();
dud.Items("TX");
dud.Items("LA");
dud.Items("WA");

VB
Dim dud as DomainUpDown
dud.Items("TX")
dud.Items("LA")
dud.Items("WA")
```

You can also add items to the `DomainUpDown` control at design time. First, select the control in the Form Designer. Then, in the Properties windows, click the ellipsis next to the `Items` property. This brings up the String Collection Editor dialog box (see Figure 3.9). You can enter the items there, one item per line.

The DomainUpDown exposes the two properties that allow you to determine the current text in the control. The Text property is the string that is currently being displayed in the control. The return value is not necessarily an item in the list. The SelectedIndex property is the index of the item in the list that is currently being display to the user. You can use this property in conjunction with the Items property to get the item's string value. When the Text property is changed, a TextChanged event is fired, and when the SelectedIndex property is changed, a SelectedIndexChanged event is fired. You can handle these events if your application needs to perform some operation when the item that is displayed in the control changes. The SelectedIndexChanged event is fired only when the index is changed through code or via the up and down arrows. The event will not be fired when a user types input into the control. The following code snippet demonstrates how to use handle both events:

```
C#
private void domainUpDown1_SelectedItemChanged(object sender,
        System.EventArgs e) {
  int selNdx = this.domainUpDown1.SelectedIndex;
  string selStr = this.domainUpDown1.Items[selNdx].ToString();
  MessageBox.Show("You selected " + selStr);
}

private void domainUpDown1_TextChanged(object sender, System.EventArgs e) {
  string selStr = this.domainUpDown1.Text;
  MessageBox.Show("You selected " + selStr);
}

VB
Private Sub _
DomainUpDown1_SelectedItemChanged(ByVal sender As Object,
        ByVal e As System.EventArgs)_ Handles DomainUpDown1.SelectedItemChanged
  Dim selNdx As Int32
  Dim selStr As String
  selNdx = DomainUpDown1.SelectedIndex
  selStr = domainUpDown1.Items[selNdx].ToString()
End Sub

Private Sub
DomainUpDown1_TextChanged(ByVal sender As Object,
        ByVal e As System.EventArgs) _Handles DomainUpDown1.TextChanged
  Dim selStr As String
  selStr = DomainUpDown1.Text
  MessageBox.Show("You selected " & selStr)
End Sub
```

Programming the ProgressBar Control

The ProgressBar control gives your users a graphical representation of how an operation is progressing. This is useful when your application needs to perform a very time-consuming operation and you do not want the user to think the application has crashed or entered an endless loop.

The ProgressBar control is controlled by three Int32 properties: Minimum, Maximum, and Value. The Minimum and Maximum values define the minimum and maximum values of the ProgressBar control. The Value property defines the ProgressBar control's current value. The ProgressBar will be filled in from left to right as the Value property moves increasingly away from the Minimum value toward the Maximum value. The ProgressBar control is empty when the Value property is equal to the Minimum property. And once the Value property is equal to the Maximum value, the ProgressBar will be full. Immediately after the Value property is changed, the ProgressBar control will be repainted to reflect the new value.

FIGURE 3.11 You win the BombSquad game by diffusing the bomb before the progress bar fills up.

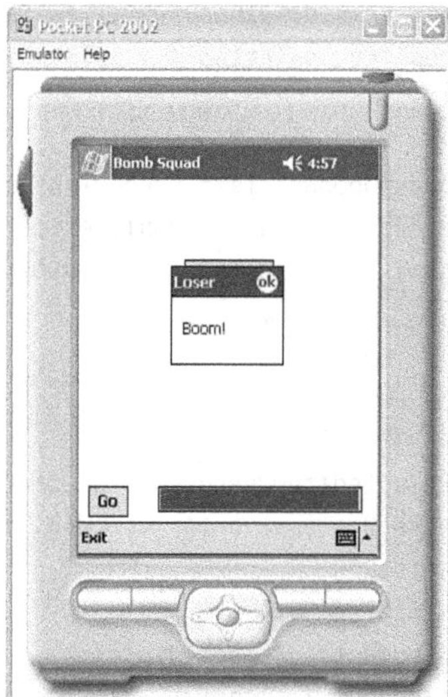

FIGURE 3.12 You lose the BombSquad game by failing to diffuse the bomb before the progress bar fills up.

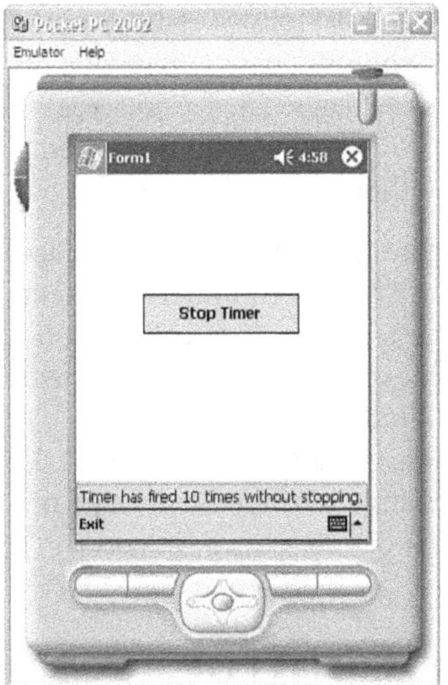

FIGURE 3.13 A sample application that showcases the StatusBar control running on the Pocket PC 2002 emulator.

Figures 3.11 and 3.12 show the BombSquad.exe game running in the Pocket PC 2002 emulator. The BombSquad game is a simple game that gives you a set amount of time to diffuse a dangerous bomb by clicking the Diffuse button before the ProgressBar control is filled. The Diffuse button's location continuously changes while the game is being played. Figure 3.11 shows the results of a successful diffusing, whereas Figure 3.12 shows what happens when you cannot diffuse the bomb in time. You will find the code for the BombSquad game in the source code for this book.

Using the StatusBar Control

The status bar is an area, at the button of a form, that displays status information to the user. This status bar can display only text information. A common use of a status bar is to provide users with a simple sentence that describes the state of the application or tell the user what operations can now be executed, given the current state of the application. It can also replace a ProgressBar control to display the status of time-consuming operations.

To update a StatusBar control, simply change the value of its Text property. When the text is changed, the control is automatically repainted to display the new text, just like a Label control. Figure 3.13 displays an application that starts and stops a Timer control. Every time the Timer control fires, the status bar is updated.

Using the TrackBar Control

The TrackBar control is a special slider control that allows the user to select a numeric value by changing the position of the slider. The TrackBar is controlled mainly by three properties: Minimum, Maximum, and Value. Given our earlier discussions, these values are self-explanatory.

The TrackBar exposes a few additional properties that control the appearance and behavior of the TrackBar. The Orientation property determines whether the TrackBar is oriented on the horizontal or vertical axis. The TickFrequency property defines the distance between the tick marks along the slider. By default the TickFrequency is set to one. The SmallChange and LargeChange

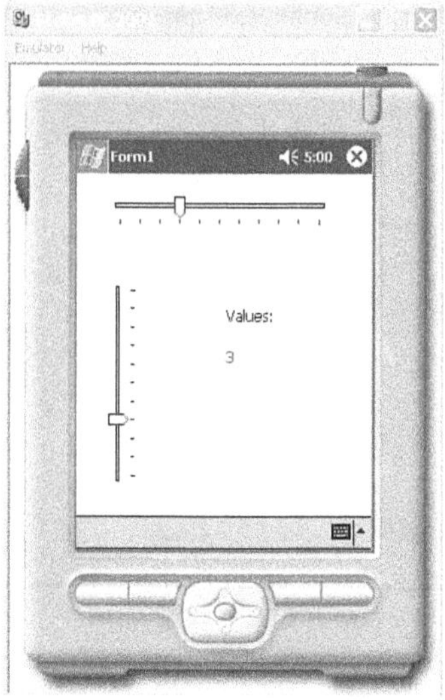

FIGURE 3.14 An application that showcases the TrackBar control running on the Pocket PC 2002 emulator.

properties define increment values. SmallChange defines how much the Value is incremented when the user presses one of the Pocket PC navigation pad arrow keys. The LargeChange property defines how much the Value is incremented when the user clicks the TrackBar on either side of the slider.

When the TrackBar controls value is changed, a ValueChanged event is fired. You can handle this event to update whatever properties that are linked to the TrackBar. Figure 3.14 shows two TrackBars that are identical except that one is oriented vertically and the other horizontally. The horizontal TrackBar controls the vertical, meaning that the value of the vertical TrackBar control will change to match the value of the horizontal TrackBar control.

Using the ToolBar Control

Navigating a menu with a stylus can soon become tedious, especially if the user is selecting the same menu item over and over again. The ToolBar control can reduce the number of clicks that a user must perform in order to execute menu options. Because the ToolBar control consists of images, adding a ToolBar can make your application more visually appealing.

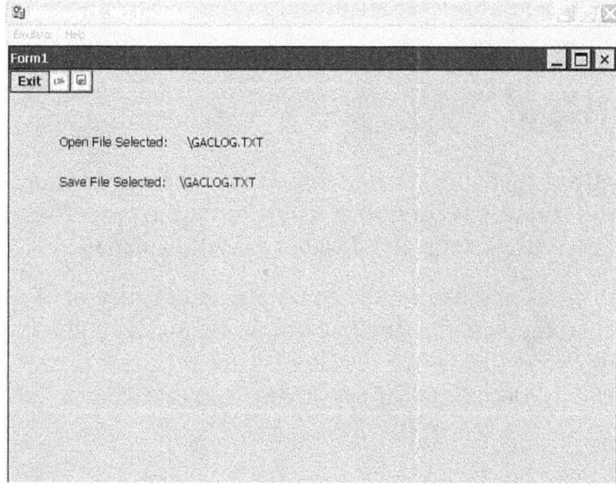

FIGURE 3.15 An application that showcases the ToolBar control running on the Windows CE .NET emulator.

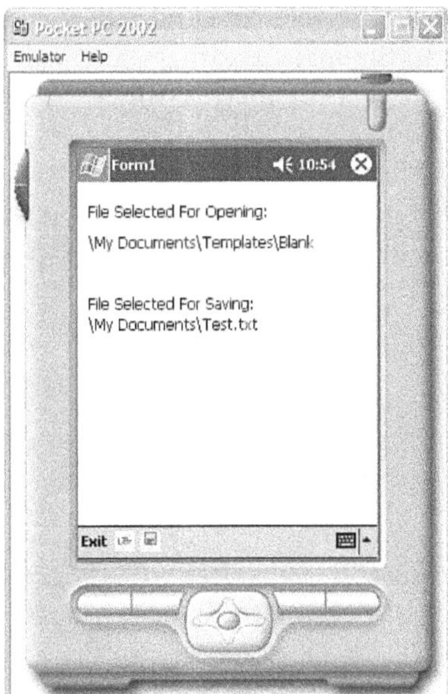

FIGURE 3.16 An application that showcases the ToolBar control running on the Pocket PC 2002 emulator.

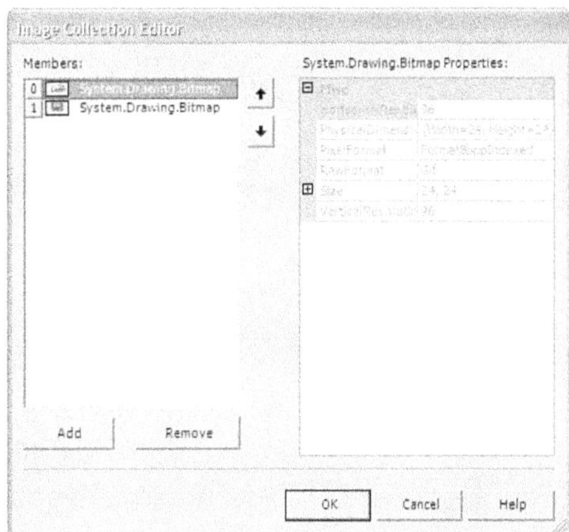

FIGURE 3.17 The Image Collection Editor.

The ToolBar is a graphical bar that consists of images that represent buttons. These buttons are usually shortcuts to commonly used application functionality. On the .NET Compact Framework, the ToolBar control cannot contain text. You must use an ImageList control to populate the ToolBar control.

The ToolBar control can appear in different locations, depending on whether the application is built for the Pocket PC or Windows CE. If it is a Windows CE application, then the ToolBar will appear along the top of the application form and to the right of any menu options (see Figure 3.15). For Pocket PC applications the ToolBar appears along the bottom of the form and to the right of any menu options (see Figure 3.16).

Adding a ToolBar Control to an Application

Adding a ToolBar control to an application can be done using the Visual Studio .NET Form Designer. Follow this list of steps to create a ToolBar control in your application:

1. Drag an ImageList onto the application form. This will create an ImageList icon at the bottom of the Form Designer.
2. In the Properties window for the ImageList, click the ellipsis button next to the Images property. This will bring up the Image Collection Editor (see Figure 3.17).

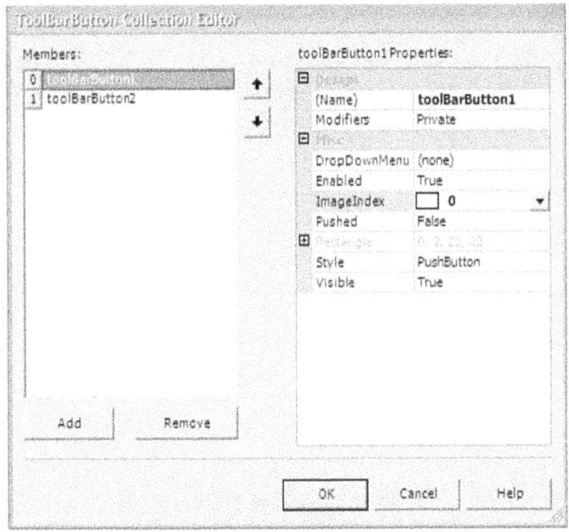

FIGURE 3.18 The `ToolBarButton` Collection Editor.

3. Use the `Image` Collection Editor to add the `ToolBar` images to the `ImageList`. It is important to note that the images will be resized to 16 × 16 pixels, which is the size of all `ToolBar` images. These images also get imported into the resource file for the application. You do not need to deploy the images along with your application.
4. Drag a `ToolBar` control onto the application form.
5. Set the `ToolBar` control's `ImageList` property to the name of the `ImageList` control that was created in step 1.
6. Bring up the `ToolBarButton` Collection Editor (see Figure 3.18) by clicking the ellipsis next to the `ToolBar` control's `Button` property in the Properties window.
7. Add all of the buttons that will appear on the `ToolBar` control, setting the `ImageIndex` property of each button. The `ImageIndex` is a zero-based index that corresponds to the index of an image in the `ImageList` control.
8. Change the button's `Style` property if you do not want the button to have the default `PushButton` style. Table 3.5 lists all of the possible `Style` values and a brief description.

TABLE 3.5
`ToolBarButtonStyle` Members and Their Descriptions

MEMBER NAME	DESCRIPTION
`DropDownButton`	When the button is clicked, a menu or other window is displayed.
`PushButton`	The regular, three-dimensional button (default).
`Separator`	A space or graphic between toolbar buttons.
`ToggleButton`	A button that appears pressed when clicked and remains pressed until it is clicked again.

Handling a `ToolBar`'s `ButtonClick` Event

When a user clicks a button on the `ToolBar` control, a `ButtonClick` event is fired. You can handle this event to execute the action assigned to the button that was clicked.

The event handler for the `ButtonClick` event receives a `ToolBarButtonClickEventArgs` object.

The ToolBarButtonClickEventArgs.Button property is a reference to the ToolBar button that was clicked. The following example demonstrates how bring up the OpenFileDialog control when the correct ToolBar button is clicked:

```
C#
private void
toolBar1_ButtonClick(object sender,
        System.Windows.Forms.ToolBarButtonClickEventArgs e) {
  if(e.Button == this.toolBarButton1) {
    OpenFileDialog dlg = new OpenFileDialog();
    if(dlg.ShowDialog() == DialogResult.OK) {
      this.lblOpenFile.Text = dlg.FileName;
    }
  }
  else if(e.Button == this.toolBarButton2) {
    SaveFileDialog dlg = new SaveFileDialog();
    if(dlg.ShowDialog() == DialogResult.OK ) {
      this.lblSaveFile.Text = dlg.FileName;
    }
  }
}
```

```
VB
Private Sub
toolBar1_ButtonClick(ByVal sender As System.Object,
        ByVal e As System.Windows.Forms.ToolBarButtonClickEventArgs)
        Handles toolBar1.ButtonClick
  If e.Button Is toolBarButton1 Then
    Dim dlg As OpenFileDialog
    dlg = New OpenFileDialog
    Dim res As DialogResult
    res = dlg.ShowDialog()

    If res = DialogResult.OK Then
      label3.Text = dlg.FileName
    End If
  ElseIf e.Button Is toolBarButton2 Then
    Dim dlg As SaveFileDialog
    dlg = New SaveFileDialog
    Dim res As DialogResult
    res = dlg.ShowDialog()
```

```
    If res = DialogResult.OK Then
      label4.Text = dlg.FileName
    End If
  End If
End Sub
```

Adding Menus with the MainMenu Control

Menus are an easy and efficient way to present options to your users. Menus are handled differently, depending on whether the application is running on the Pocket PC OS or the Windows CE OS. By default a Pocket PC application will contain a `MainMenu` control. A Pocket PC application displays menus at the bottom of the application. Windows CE applications have no default menu, and menus are displayed at the top of the application, like standard desktop applications.

The `MainMenu` control is a container control that holds all of the `MenuItem` controls in the application. Menus with sub-items can be created by adding multiple submenu items to a top-level `MenuItem`.

You can add menus to your application at design time through the menu designer or at runtime. The following code shows how to add a menu item with two submenu items to the `MainMenu` control named `mainMenu1`. The code also demonstrates how to add a separator to a menu. This is done by creating a new `MenuItem` object and setting its text to a string containing a single hyphen character, `"-"`.

```
C#
MenuItem fileMenu = new MenuItem();
MenuItem newItem = new MenuItem();
MenuItem sepItem = new MenuItem();
MenuItem exitItem = new MenuItem();

fileMenu.Text = "File";
newItem.Text = "New";
sepItem.Text = "-";
exitItem.Text = "Exit";

fileMenu.MenuItems.Add(newItem);
fileMenu.MenuItems.Add(sepItem);
fileMenu.MenuItems.Add(exitItem);
mainMenu1.MenuItems.Add(fileMenu);
```

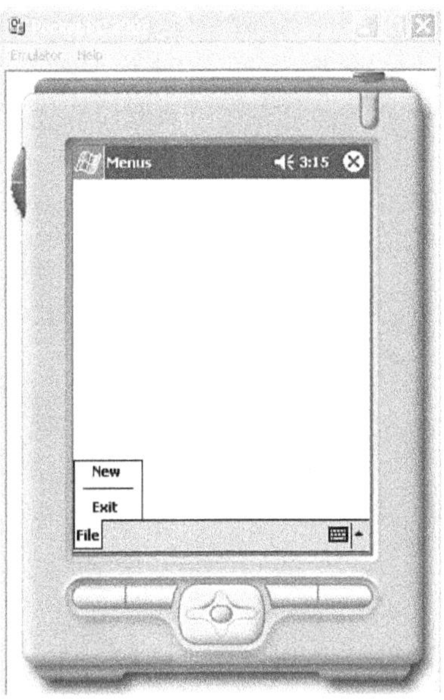

FIGURE 3.19 A sample application that showcases the MainMenu control running on the Pocket PC 2002 emulator.

```
VB
Dim fileMenu = new MenuItem()
Dim newItem = new MenuItem()
Dim sepItem = new MenuItem()
Dim exitItem = new MenuItem()

fileMenu.Text = "File"
newItem.Text = "New"
sepItem.Text = "-"
exitItem.Text = "Exit"

fileMenu.MenuItems.Add(newItem)
fileMenu.MenuItems.Add(sepItem)
fileMenu.MenuItems.Add(exitItem)
mainMenu1.MenuItems.Add(fileMenu)
```

Figure 3.19 shows the menu being displayed on the Pocket PC 2002 emulator.

When the user selects a menu item, a `Click` event is fired. You would handle this event just as you handled the `Button` control's `Click` event. See the "Handling a `ToolBar`'s `ButtonClick` Event" section for details.

Using a ContextMenu Control in an Application

The `ContextMenu` control is almost exactly like the `MainMenu` control except that the context menus are associated with other controls, whereas the `MainMenu` is associated with the application's form. `ContextMenus` are also known as pop-up menus that appear when you tap-and-hold a control on Pocket PC or right-click a control on Windows CE. Tap-and-hold can be simulated on the Pocket PC emulator by pressing and holding the left mouse button.

To add a `ContextMenu` to an application, drag the control from the `ToolBox` onto the application. The new `ContextMenu` control will appear in the panel at the bottom of the Form Designer. When the `ContextMenu` is selected, the Form Designer paints the control in the same place as the `MainMenu` control. This is done only for design purposes. At runtime the `ContextMenu` will appear above its respective control. To add `MenuItem` to the `ContextMenu`, click the `ContextMenu` on the Form or click the Edit Menu link in the Properties window. The Edit Menu link is visible only when the `ContextMenu` is selected.

You can also create `ContextMenu` controls at runtime. First, a `ContextMenu` control must be instantiated. Next add `MenuItem` objects to it, and then set the `ContextMenu` property of a control to the `ContextMenu` instance. The following code demonstrates how to add a `ContextMenu` at runtime:

```
C#
ContextMenu cMenu = new ContextMenu();
MenuItem menuItem1 = new MenuItem();
MenuItem menuItem2 = new MenuItem();

menuItem1.Text = "Default Item 1";
menuItem2.Text = "Default Item 2";

// Add menuItem2 as a child of menuItem1
menuItem1.MenuItems.Add(this.menuItem2);

// Add menuItem1 to the context menu
cMenu.MenuItems.Add(this.menuItem1);

// Add the context menu to a label control
label1.ContextMenu = cMenu;

VB
Dim cMenu = new ContextMenu()
Dim menuItem1 = new MenuItem()
Dim menuItem2 = new MenuItem()

menuItem1.Text = "Default Item 1"
menuItem2.Text = "Default Item 2"

' Add menuItem2 as a child of menuItem1
menuItem1.MenuItems.Add(menuItem2)

' Add menuItem1 to the context menu
cMenu.MenuItems.Add(menuItem1)

' Add the context menu to a label control
label1.ContextMenu = cMenu
```

When a `ContextMenu` is invoked, a `Popup` event is fired. This event can be handled to customize the `ContextMenu` before it is shown to the user. When a `MenuItem` is selected from the `ContextMenu`, a `Click` event is fired for the `MenuItem`. Listing 3.3 demonstrates how to handle both events:

LISTING 3.3

```
C#
private void contextMenu1_Popup(object sender, System.EventArgs e) {
  this.contextMenu1.MenuItems.Clear();
```

```
    if(this.checkBox1.Checked)
      this.contextMenu1.MenuItems.Add(this.menuItem1);
    if(this.checkBox2.Checked)
      this.contextMenu1.MenuItems.Add(this.menuItem2);
    if(this.checkBox3.Checked)
      this.contextMenu1.MenuItems.Add(this.menuItem3);

    // Always add the default menu
    this.contextMenu1.MenuItems.Add(this.menuItem4);
  }

  private void menuItem1_Click(object sender, System.EventArgs e) {
    MessageBox.Show("You selected MenuItem 1");
  }

  private void menuItem2_Click(object sender, System.EventArgs e) {
    MessageBox.Show("You selected MenuItem 2");
  }

  private void menuItem5_Click(object sender, System.EventArgs e) {
    MessageBox.Show("You selected MenuItem 3");
  }

  private void menuItem3_Click(object sender, System.EventArgs e) {
    MessageBox.Show("You selected MenuItem 3");
  }

  private void menuItem5_Click_1(object sender, System.EventArgs e) {
    MessageBox.Show("You selected Default Item 2");
  }
```

VB

```
Private Sub _
        contextMenu1_Popup(ByVal sender As Object,
        ByVal e As System.EventArgs) _Handles contextMenu1.Popup
  contextMenu1.MenuItems.Clear()

  If checkBox1.Checked Then
    contextMenu1.MenuItems.Add(menuItem1)

    If checkBox2.Checked Then
      contextMenu1.MenuItems.Add(menuItem2)
```

```
    End If

    If checkBox3.Checked Then
      contextMenu1.MenuItems.Add(menuItem3)
    End If

    ' Always add the default menu
    contextMenu1.MenuItems.Add(menuItem4)
  End If
End Sub

Private Sub menuItem1_Click(ByVal sender As Object,
        ByVal e As System.EventArgs) _Handles menuItem1.Click
  MessageBox.Show("You selected MenuItem 1")
End Sub

Private Sub menuItem2_Click(ByVal sender As Object,
        ByVal e As System.EventArgs) _Handles menuItem2.Click
  MessageBox.Show("You selected MenuItem 2")
End Sub

Private Sub menuItem3_Click(ByVal sender As Object,
        ByVal e As System.EventArgs) _Handles menuItem3.Click
  MessageBox.Show("You selected MenuItem 3")
End Sub

Private Sub menuItem4_Click(ByVal sender As Object,
        ByVal e As System.EventArgs) _Handles menuItem4.Click
  MessageBox.Show("You selected MenuItem 4")
End Sub

Private Sub menuItem5_Click(ByVal sender As Object,
        ByVal e As System.EventArgs) _Handles menuItem5.Click
  MessageBox.Show("You selected MenuItem 5")
End Sub
```

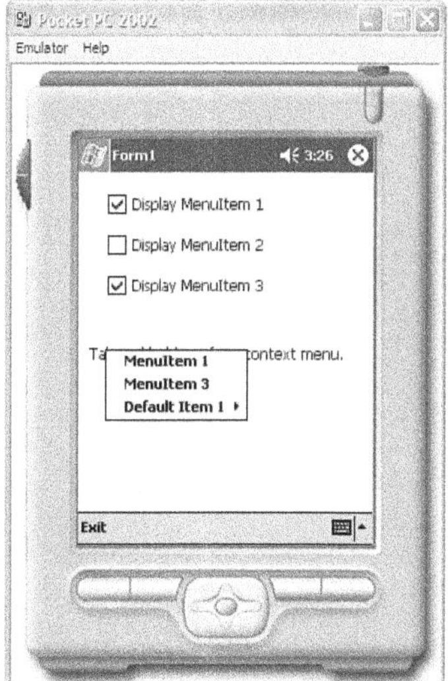

FIGURE 3.20 An application that showcases the `ContextMenu` control running on the Pocket PC 2002 emulator.

Figure 3.20 shows an application running that uses a `ContextMenu` control. The `MenuItems` on the control are configured when it is activated. The `CheckBox` controls determine which `MenuItems` will be displayed. The Default Item 1 menu is always displayed. You will find the code for this sample in the code listings for this book.

Using the `Timer` Control

The `Timer` control allows you to execute event-handling code repeatedly in defined intervals. The code will be executed on the same thread that the Windows code is running in. If you need code to run in a thread outside of the Windows thread, then you should use the `System.Threading` namespace.

The `Timer` control fires by raising a `Tick` event. A `Tick` event is raised only when the `Timer`'s `Enabled` property is set to true. You can stop a `Timer` control from firing by setting the `Enabled` property to false. The `Timer` control's `Interval` property defines the number of milliseconds between `Tick` events.

SHOP TALK

TRAPPING THE TAP-AND-HOLD EVENT

Pocket PC substitutes the tap-and-hold movement in place of the right mouse click. The tap-and-hold movement is simulated by placing the stylus on the screen and holding it in place. A control's `ContextMenu` control is activated this way.

You may want to respond to a tap-and-hold in other ways beyond bringing up a `ContextMenu`. To do this, you must use a little creativity. First create a `Timer` control whose interval is set to the amount of time a user would have to hold the stylus to the screen before triggering the tap-and-hold functionality. Then capture the `MouseDown` event on the control that will be handling the `tap-and-hold` event. In the `MouseDown` event, enable the Timer. Next trap the `MouseUp` event and disable the Timer in handle. Finally, in the Timer's `Tick` event handler, perform the action activated by the `tap-and-hold` event. The following code demonstrates how to handle the three events described above:

```
C#
private void Form1_MouseDown(object sender,
        System.Windows.Forms.MouseEventArgs e) {
```

SHOP TALK

```
  timer1.Enabled = true;
}

private void Form1_MouseUp(object sender,
        System.Windows.Forms.MouseEventArgs e) {
  timer1.Enabled = false;
}

private void timer1_Tick(object sender, System.EventArgs e) {
  label2.Text = string.Format("Message: {0}", "Hello, World!");
}

VB
private sub _Form1_MouseDown(sender as object,
        e As System.Windows.Forms.MouseEventArgs) _
        Handles form1.MouseDown
  timer1.Enabled = true
End Sub

private sub _Form1_MouseUp(sender as object,
        e As System.Windows.Forms.MouseEventArgs) _
        Handles form1.MouseMove
  timer1.Enabled = false
End Sub

private sub _timer1_Tick(sender as object,
        e As System.Windows.Forms.MouseEventArgs) _
        Handles timer1.Tick
  label2.Text = string.Format("Message: {0}", "Hello, World!")
End Sub
```

In the "Programming the `ProgressBar` Control" section, the `BombSquad.exe` game was presented. The object of the game was to click the Diffuse button before the `ProgressBar` control filled up. The trick was that the Diffuse button's location would continuously change, while the `ProgressBar`'s value increased. This application uses two timers. One `Timer` control determined how often the `ProgressBar`'s value was updated. And another `Timer` control handled moving the Diffuse button around the form. Listing 3.4 demonstrates how these two `Timers` are handled.

LISTING 3.4

```
C#
private void timerButtonMove_Tick(object sender, System.EventArgs e) {
```

```
    if(!gameOver) {
      this.button1.Location =
        new Point(rand.Next(0, 168), rand.Next(0, 216));
    }else {
      EndGame();
    }
  }
  private void timerProgressBar_Tick(object sender, System.EventArgs e) {
    if( counter < 60 ) {
      if( gameOver )
        return;

      if(this.progressBar1.Value < this.progressBar1.Maximum)
        this.progressBar1.Value += 1;

        ++counter;
    } else {
      if( !gameOver ) {
        EndGame();
        MessageBox.Show("Boom!", "Loser");
      }
    }
  }

  private void EndGame() {
    this.timerButtonMove.Enabled = false;
    this.timerProgressBar.Enabled = false;

    counter = 0;
    this.gameOver = true;
    this.button2.Enabled = true;
  }
```

VB

```
Private Sub timerButtonMove_Tick(ByVal sender As Object,
        ByVal e As System.EventArgs) _ Handles timerButtonMove.Tick
  If Not gameOver Then
    Me.button1.Location = _
      New Point(rand.Next(0, 168), rand.Next(0, 216))
  Else
    EndGame()
  End If
```

```
End Sub

Private Sub timerProgressBar_Tick(ByVal sender As Object,
        ByVal e As System.EventArgs) _ Handles timerProgressBar.Tick
  If counter < 60 Then
    If gameOver Then
      Return
    End If

    If Me.progressBar1.Value < Me.progressBar1.Maximum Then
      Me.progressBar1.Value = Me.progressBar1.Value + 1

      counter = counter + 1
    Else
      If Not gameOver Then
        EndGame()
        MessageBox.Show("Boom!", "Loser")
      End If
    End If
  End If
End Sub

Private Sub EndGame()
  Me.timerButtonMove.Enabled = False
  Me.timerProgressBar.Enabled = False

  counter = 0
  Me.gameOver = True
  Me.button2.Enabled = True
End Sub
```

Using the `OpenFileDialog` and `SaveFileDialog` Controls

A common task for an application is to open or save a file that represents its application data. In Windows applications the Open and Save file dialog boxes provide an easy way for the user to locate a file. The .NET Compact Framework provides the `OpenFileDialog` and `SaveFileDialog` controls to interact with these dialog boxes. On Windows CE and Pocket PC, these controls allow you to navigate only the `My Documents` folder and one folder underneath. Because of this, you may want to create a folder underneath `My Documents` that will be the default location for application data.

There are a few interesting properties that you may consider setting before displaying one of the file dialog boxes to the user. First, set the `Filter` property, which determines valid file types and their associated file extensions to be listed in either file dialog box. For example, the following string would allow only files with the `.dll` or `.exe` extensions to be listed in the dialog box:

```
"Dynamically Linked Library¦*.dll¦Executable¦*.exe"
```

Each file filter will be added to the Files Type drop-down menu in the Open or Save dialog box. You can change the currently selected file filter by setting the `FilterIndex` property. The default value of this property is `1` because the index is not zero-based.

Finally, the `InitalDirectory` property determines a folder under `My Documents` that will initially be enumerated by the dialog boxes. If this property is not set or the folder does not exist, then `My Documents` is enumerated.

To display an Open or Save file dialog box, call the `ShowDialog` method on either the `OpenFileDialog` or `SaveFileDialog` control. This method will block until the user clicks either the OK button or the Cancel button. `ShowDialog` returns a `DialogResult` enumeration value. The `DialogResult` enumeration has several members, but the `ShowDialog` method will only return either `DialogResult.OK` or `DialogResult.Cancel`. If `DialogResult.OK` is returned, the user has clicked the OK button, and you can check the `Filename` property to retrieve the complete path of the file the user selected. If `DialogResult.Cancel` is returned, the user clicked the Cancel button to discard the dialog.

These controls have different visual representations depending on the OS they are running on. On Pocket PC the dialog boxes will appear full screen. On Windows CE a dialog box will appear as a pop-up dialog window above the application, similar to its desktop counterpart.

The following code demonstrates how to display an `OpenFileDialog` control that lists files with the `.dll` or `.exe` extension. If a file is selected, then a message box appears with the complete filename.

```
C#
OpenFileDialog ofDlg = new OpenFileDialog();
ofDlg.Filter = "DLL¦*.dll¦Executable¦*.exe";
ofDlg.IntialDirectory = "\\My Documents";

if(DialogResult.OK == ofDlg.ShowDialog()) {
  MessageBox.Show("You Selected " + ofDlg.FileName);
} else {
  MessageBox.Show("Go ahead, select a file!");
}

VB
Dim ofDlg as OpenFileDialog
```

```
ofDlg = new OpenFileDialog()
ofDlg.Filter = "DLL¦*.dll¦Executable¦*.exe"
ofDlg.IntialDirectory = "\My Documents"

Dim res As DialogResult
If  DialogResult.OK = res Then
  MessageBox.Show("You Selected " & ofDlg.FileName)
Else
  MessageBox.Show("Go ahead, select a file!")
End If
```

Using the `Panel` Control

The `Panel` control is a container for other controls. You can add controls to the `Panel` at design time and runtime. Panels are customarily used to group related controls. For example, the .NET Compact Framework does not support the `GroupBox` control, which can be used to group together `RadioButton` controls. Instead, you can add your `RadioButton controls` to a `Panel` to display to the user that these `RadioButtons` are related.

Unsupported Functionality

None of the following properties is supported by the `Panel` control:

- `BorderStyle` property
- `BackGroundImage` property
- `AutoScroll` property

Using the `HScrollBar` and `VScrollBar` Controls

The `HScrollBar` and `VScrollBar` controls allow you to add scrolling capabilities to components that do not support scrolling by default. The controls are driven by several properties. The following list contains these properties and their meanings.

`Minimum` The value of the control when the scroll thumb is at the left end of the `HScrollBar` or top of the `VScrollBar`.

`Maximum` This is the largest possible value of the scroll bar. Note that the `Value` property of the scroll bar when the thumb is pulled all the way to the right for the `HScrollBar` or the bottom of the `VScrollBar` is not equal to the `Maximum` property. The `Value` property will be equal to this formula: `Maximum – LargeChange + 1`.

`SmallChange` Here's the increment value used when the user clicks one of the arrow buttons.

`LargeChange` This is the increment value used when the user clicks the scroll bar control on either side of the scroll thumb.

`Value` The current value of the scroll bar. This value represents the position of the top of the `HScrollBar`'s scroll thumb and the left side of the `VScrollBar`'s scroll thumb.

A `ValueChanged` event is fired when the `Value` property changes. Trap this event to perform some action, such as change the position of a control, when the scroll bar's value changes. The following code demonstrates how to handle the `ValueChanges` event:

```
C#
private void hScrollBar1_ValueChanged(object sender, System.EventArgs e) {
  this.label1.Text = string.Format("Scroll Bar Value: {0}",
          this.hScrollBar1.Value);
}

VB
Private Sub hScrollBar1_ValueChanged(object sender,
        System.EventArgs e) _Handles hScrollBar1.ValueChanged
  Me.label1.Text = string.Format("Scroll Bar Value: {0}",
          this.hScrollBar1.Value)
End Sub
```

Figure 3.21 shows an application with a `HScrollBar` control linked to a `Label` control. The `Label` control displays the `Value` of the `HScrollBar`. When you move the scroll bar, the label is updated with the new value. In this sample the `Minimum` is 0, the `Maximum` is 100, `SmallChange` is 10, and `LargeChange` is 20. The complete code for this sample can be found in the code list for this book.

Using the `ImageList` Control

The `ImageList` control is a non-graphical control that is meant to be a container for images. Other .NET Compact Framework controls use the `ImageList` control to retrieve images they need to display. This is true for the `ListView`, `TreeView`, and `ToolBar` controls. The usage of the `ImageList` with these controls is discussed in their respective sections.

Images can be added to the control at design time as well as at runtime. To add an `ImageControl` to a `Form`, drag an `ImageControl` from the `ToolBox` to the application form in the Form Designer. Now click the ellipsis next to the `Images` property in the Properties window. This will bring up the Image Collection Editor (see Figure 3.17). Use the Image Collection Editor to add the images to the `ImageList`. It is important to note that the

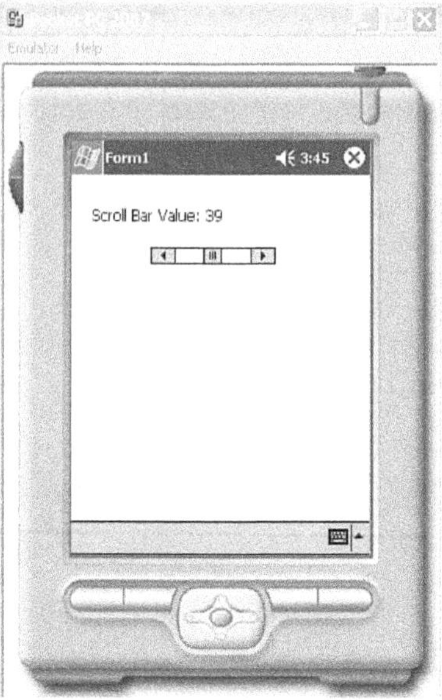

FIGURE 3.21 A Sample Application Showcasing the `HScrollBar`.

images will be resized to 16 × 16 pixels by default. You can set the `ImageSize` property if you want the images to be resized differently. These images also get imported into the resource file for the application. You need not deploy the images along with the application, because they will be embedded inside the application assembly.

Images can also be loaded at runtime. The `ImageList` control exposes an `Images` property, which represents the list of images stored in the control. The `Images` property provides the `Add` method, which allows you to add images to the list of images. Interestingly, this method will add the image to the head of the list (index 0)—not the tail of the list, like other methods that add objects to collections would. The `Add` method can accept either a `System.Drawing.Icon` or a `System.Drawing.Image`. Images loaded at runtime can be either loaded from the file system or loaded from the assembly's resource by using the `ResourceManager`. The following code demonstrates how to load an image from the assembly's resource file:

```
C#
BitMap image =
  new BitMap(Assembly.GetExecutingAssembly().GetManifestResourceStream("image1.jpg");
ImageList imgList = new ImageList();
imgList.Images.Add(image);

VB
Dim imgLst as new ImageList()
Dim image as BitMap
image = new _
BitMap(Assembly.GetExecutingAssembly().GetManifestResourceStream("image1.jpg")
imgList.Images.Add(image)
```

Using the `PictureBox` Control

The `PictureBox` control is a used to display an image. The `PictureBox` control has very limited functionality. The control always displays the image in the upper-left corner of the control. The `PictureBox` control does not provide any way to resize the image so that it can stretch to fill the control.

The PictureBox control exposes the Image property, which represents the current image the control will display. Let's discuss three ways to load an image into the PictureBox control.

An image can be loaded into the PictureBox by loading the image from a file. The following code demonstrates how this is done:

```
C#
pictureBox1.Image =
  new Bitmap(@"\Program Files\PictureBoxControl\tinyemulator_content.jpg");

VB
pictureBox1.Image = _
  new Bitmap("\Program Files\PictureBoxControl\tinyemulator_content.jpg")
```

When loading an image this way, you should add the image to your Visual Studio .NET project and set the Build Action property in the Properties window to Content. When your application is deployed or when a CAB file is built, the image will be included.

An image can also be loaded from your application's resource file. First add the image to your Visual Studio .NET project and set the Build Action to Embedded Resource. This will load the image into the application's resource file. You do not need to deploy the image file with the application. The following code demonstrates how to add a PictureBox control with an image located in the resource file:

```
C#
pictureBox1.Image =
  new Bitmap(Assembly.GetExecutingAssembly().
  GetManifestResourceStream("PictureBoxControl.tinyemulator_res.jpg"));

VB
pictureBox1.Image = _
  new Bitmap(Assembly.GetExecutingAssembly(). _
  GetManifestResourceStream("PictureBoxControl.tinyemulator_res.jpg"))
```

Finally, you can load a PictureBox control with an image located in ImageList control. See the "Using the ImageList Control" section for details on how to load an image into an ImageList. The advantage of using an ImageList control is that you can use the ImageList control to resize the image before loading it into a PictureBox control. By default all images loaded in to an ImageList are resized to 16 × 16. You can change this by setting the ImageList.ImageSize property to the desired size. Changing the ImageSize property will affect all images in the ImageList. When using the ImageList, you get the added benefit of the image's being loaded in the application's resource file. The following code demonstrates how to load a PictureBox control through the ImageBox control:

```
C#
// resize the image
imageList1.ImageSize = new System.Drawing.Size(92, 156);
// load the resized image
pictureBox1.Image = imageList1.Images[0];

VB
' resize the image
imageList1.ImageSize = new System.Drawing.Size(92, 156)
' load the resized image
pictureBox1.Image = imageList1.Images(0)
```

Figure 3.22 shows an application running in the Pocket PC 2002 emulator that allows the user to load a `PictureBox` from a file on the file system, the application's resource file, or an `ImageList`.

Using the `ListView` Control

The `ListView` control is similar to the `ListBox` control in that it can display a list of items to the user in a boxlike UI. The `ListView` control adds additional functionality over the `ListBox`. In addition to plain text items, the `ListView` control can display optional images and sub-items that add detail about the item.

FIGURE 3.22 This application showcases the `PictureBox` control running on the Pocket PC 2002 emulator.

The items in a `ListView` can be displayed in four different views: `Details` view, `LargeIcon` view, `List` view, and `SmallIcon` view. The `ListView` control exposes the `View` property that represents the current view. The `View` property is of type `System.Windows.Forms.View`, which is an enumeration. The following list describes the members of the `View` enumeration and how they affect the `ListView` control.

`Details` Items appear on separate lines, with information about the items arranged in columns.

`LargeIcon` Items appear as large icons with labels underneath them.

`List` Items appear as small icons with labels to their right. The items are arranged in columns, but no column headers are displayed.

`SmallIcon` Items appear as small icons with labels to their right.

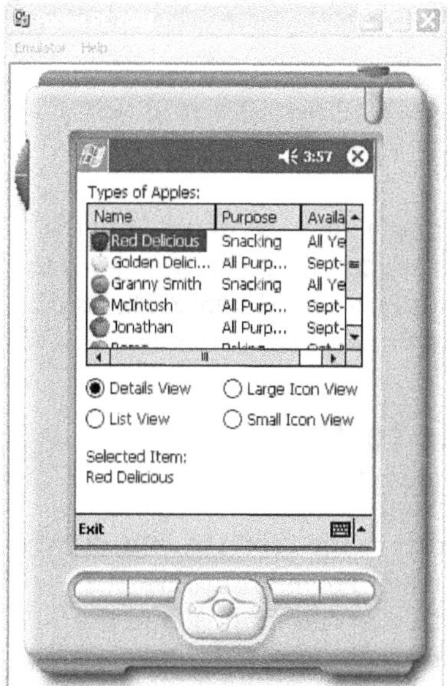

FIGURE 3.23 This application showcases the ListView control application running on the Pocket PC 2002 emulator.

When the View property is set to Details, the ListView can display sub-items, which act as additional information related to the item. When the ListView is in Details mode, the items appear to be listed in tabular format, with each row separated into columns. The first column contains text and an option icon. The subsequent columns contain the text of the sub-items. Figure 3.23 shows an example of the ListView running in details view on the Pocket PC emulator. The complete code listing for this sample can be found in the source code collection for this book.

Adding Columns to the ListView

Columns can be added to the ListView at design time as well as runtime. This is done by adding ColumnHeader objects to the ListView's Columns collection property. The columns will only be displayed only when the ListView is in Details mode. The following code demonstrates how to add columns to a ListView control:

```
C#
System.Windows.Forms.ColumnHeader columnHeader1 = new ColumnHeader();
System.Windows.Forms.ColumnHeader columnHeader2 = new ColumnHeader();
System.Windows.Forms.ColumnHeader columnHeader3 = new ColumnHeader();

columnHeader1.Text = "Name";
columnHeader2.Text = "Purpose";
columnHeader3.Text = "Availability";

listView1.Columns.Add(columnHeader1);
listView1.Columns.Add(columnHeader2);
listView1.Columns.Add(columnHeader3);

VB
Dim columnHeader1 = new ColumnHeader()
Dim columnHeader2 = new ColumnHeader()
Dim columnHeader3 = new ColumnHeader()

columnHeader1.Text = "Name"
columnHeader2.Text = "Purpose"
```

```
columnHeader3.Text = "Availability"

listView1.Columns.Add(columnHeader1)
listView1.Columns.Add(columnHeader2)
listView1.Columns.Add(columnHeader3)
```

Columns can also be added to the `ListView` at design time. First drag a `ListView` control from the `ToolBox` onto the applications form. Now click the ellipsis next to the `Columns` property name in the Properties windows. This will bring up the `ColumnHeader` Collection Editor (see Figure 3.24). Use the Add and Remove buttons to modify the list of `ColumnHeader` objects in the `ListView` control.

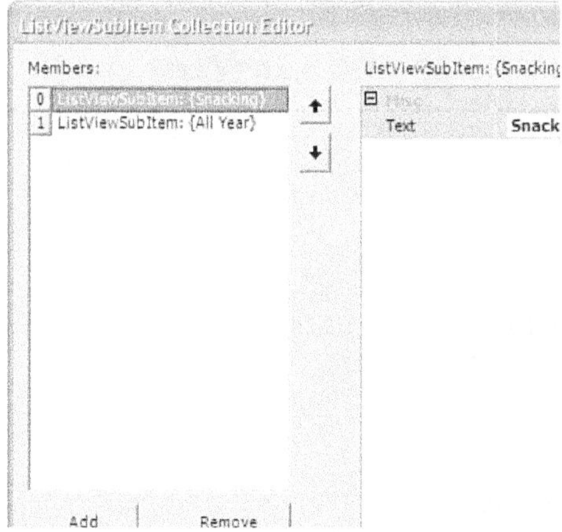

FIGURE 3.24 The `ColumnHeader` Collection Editor.

Adding Items to the `ListView` Control

After adding columns to the `ListView` control, you can begin to add items. The `ListViewItem` class represents items in a `ListView` control and the `ListViewSubItem` class represents a sub-item related to a `ListViewItem`. The `ListViewItem` class exposes the `SubItems` collection property that represents the list of sub-items related to an item. The number of sub-items should equal the number of columns minus one. The first column in a `ListView` control is populated with text and option image of the `ListViewItem` object.

A `ListViewItem` object can display an image along with text. An `ImageList` must be used to associate an image with a `ListViewItem` object. The `ListView` control exposes two `ImageList` properties: `SmallImageList` and `LargeImageList`. The `LargeImageList` `ImageList` contains the images to use when the `View` property is set to `View.LargeIcon`. The `SmallImageList` `ImageList` contains the images to use when the `View` property is set to anything other then `View.LargeIcon`. Once the `SmallImageList` and `LargeImageList` properties have been set, you can set the `ImageIndex` property of each `ListViewItem` in the `ListView` control to the index of the appropriate image in the `ImageList`. Because only one index can be specified for the `ListView.ImageIndex` property, both the `SmallImageList` and the `LargeImageList` should contain corresponding index positions. This means that if the `ImageIndex` is set to zero, then the image at index zero both in the `SmallImageList` and in the `LargeImageList` will be used for that `ListViewItem`. See the "Using the `ImageList` Control" section for detail on how to load an `ImageList`.

ListViewItem objects can be added to a ListView control at design time as well as runtime. The following code demonstrates how to add items to the ListView control configured in the "Adding Columns to the ListView" section.

```
C#
ListViewItem listViewItem1 = new ListViewItem();
ListViewSubItem listViewSubItem1 = new ListViewSubItem();
ListViewSubItem listViewSubItem2 = new ListViewSubItem();

listViewItem1.Text  = "Red Delicious";
listViewSubItem1.Text = "Snacking";
listViewSubItem2.Text = "All Year";

listViewItem1.SubItems.Add(listViewSubItem1);
listViewItem1.SubItems.Add(listViewSubItem2);
listView1.Items.Add(listViewItem1);
```

```
VB
Dim listViewItem1 = new ListViewItem()
Dim listViewSubItem1 = new ListViewSubItem()
Dim listViewSubItem2 = new ListViewSubItem()

listViewItem1.Text = "Red Delicious"
listViewSubItem1.Text = "Snacking"
listViewSubItem2.Text = "All Year"

listViewItem1.SubItems.Add(listViewSubItem1)
listViewItem1.SubItems.Add(listViewSubItem2)
listView1.Items.Add(listViewItem1)
```

ListViewItem objects can also be added at design time. First click the ListView control that will contain the items. If ImageList controls have been configured for the ListView control, then assign the ImageList controls to the LargeImageList and SmallImageList properties in the Properties window. Next click the ellipsis next to the Items property in the Properties window. The ListViewItem Collection Editor will appear (see Figure 3.25). Use the Add and Remove buttons to modify the items in the ListView. Set the ImageIndex property if an ImageList has been configured for the ListView. If you need to add sub-items to the ListViewItem objects, then click the ellipsis next to the SubItems property in the editor. The ListViewSubItem Collection Editor will appear (see Figure 3.26). Again, use the Add and Remove buttons to modify the collection of sub-items.

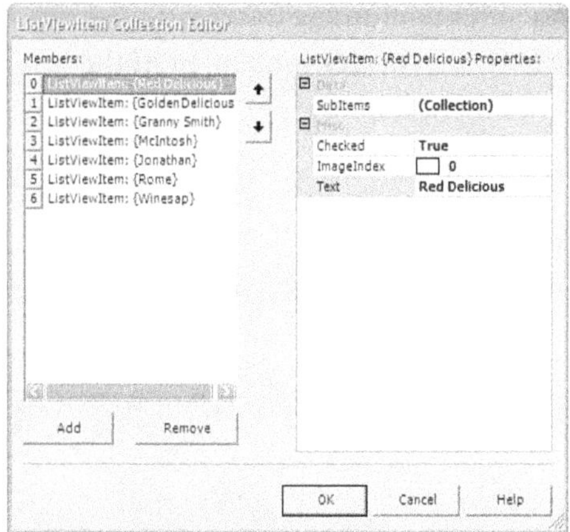

FIGURE 3.25 The `ListViewItem` Collection Editor.

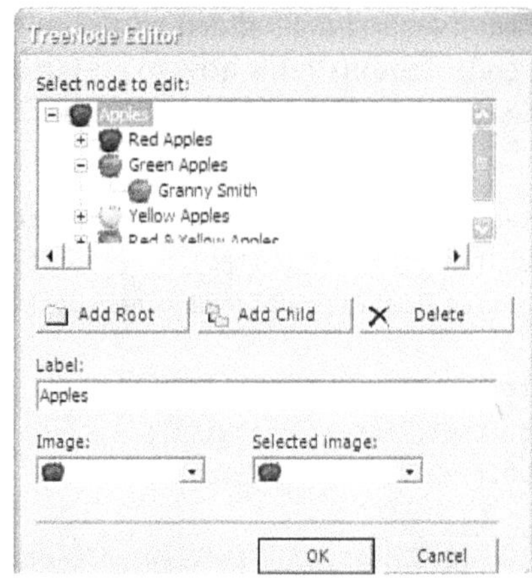

FIGURE 3.26 The `ListViewSubItem` Collection Editor.

Determining the Selected Item

On the full .NET Framework, the `ListView` control supports the `MultiSelect` property. This property determines whether multiple items can be simultaneously selected. The .NET Compact Framework does not support this property, and only one item can be selected at a time.

Despite the lack of support for the `MultiSelect` property, the `ListView` still exposes the `SelectedIndices` property. This property contains the list of indexes of all the select items. Because the `MultiSelect` property is not supported, this list will contain at most only one index.

A `SelectedIndexChanged` event is fired when the `SelectedIndices` property changes. You can trap this event to processing on the newly selected or deselected item. It is important to note that changing the selection from one item to another generates two `SelectedIndexChanged` events. The first event is fired to inform you that the previously selected index was removed from the `SelectedIndices` list. In this case the `SelectedIndices` collection will be empty. The second event informs you that the newly selected item index has been added to the `SelectedIndices` list. The following code demonstrates how to handle the `SelectedIndexChanged` event:

```
C#
private void listView1_SelectedIndexChanged(object sender,
        System.EventArgs e) {
  if(this.listView1.SelectedIndices.Count <= 0)
    return;
```

```
    int selNdx = this.listView1.SelectedIndices[0];
    label3.Text = listView1.Items[selNdx].Text;
}

VB
Private Sub _
        listView1_SelectedIndexChanged(ByVal sender As System.Object, _
        ByVal e As System.EventArgs) Handles listView1.SelectedIndexChanged
  If Me.listView1.SelectedIndices.Count <= 0 Then
    Return
  End If

  Dim selNdx = Me.listView1.SelectedIndices(0)
  If selNdx >= 0 Then
    label3.Text = listView1.Items(selNdx).Text
  End If
End Sub
```

You can also use `ListView.Selected` property to determine if the item is selected in the `ListView`. You would loop through the `ListView.Items` collection, checking the `Selected` property, to determine which item is selected.

Additional ListView Control Properties

Table 3.6 contains a list of addition `ListView` properties and a description of their effect on the `ListView` control.

TABLE 3.6

ListView Control Properties and Their Descriptions

PROPERTY	DESCRIPTION
`CheckBoxes`	This determines whether a check box appears next to each item in the control. When an item's checked state is about to change, an `ItemCheck` event is raised. The `View` property must be set to `View.Details`. Using check boxes is a great workaround for overcoming the lack of `MultiSelect` support.
`FullRowSelect`	This property determines whether clicking an item selects all its sub-items. The `View` property must be set to `View.Details`.
`HeaderStyle`	This determines the type of column headers to display. This property is a `ColumnHeaderStyle` enumeration. See Table 3.7 for a description of the `ColumnHeaderStyle` members.

TABLE 3.7

ColumnHeaderStyle Enumeration Properties and Their Descriptions

MEMBER	DESCRIPTION
Clickable	The column headers act as buttons. Clicking a header generates a ColumnClick event.
Nonclickable	Clicking the column headers do not generate ColumnClick events.
None	The column headers are not displayed.

Using the TabControl Control

The TabControl control is an ideal control for the small devices because it allows you to overlay multiple groups of controls. The TabControl stacks pages of control one on top of the other and then allows the user to switch between pages by clicking a pages tab.

Adding a TabControl control to your application can be done at design time using the Form Designer. First, drag a TabControl onto the application's form. Next, to add a TabPage to the TabControl, right-click the TabControl control and select the Add Tab menu option. This will add a TabPage to the TabControl. You can then use the Properties window to customize the TabPage.

Alternatively, you can use the TabPage Collection Editor to add TabPages. To bring up the TabPage, select the TabControl in the Form Designer. Then click the ellipsis next to the TabPages property name in the Properties window. Now the TabPage Collection Editor should be visible. Click the Add and Remove buttons to modify the TabPages in the TabControl control.

A third way to add and remove a TabPage is by clicking the Add Tab and Remove Tab links in the Properties window. This link appears only when the TabControl is selected.

On the Pocket PC the tabs will be displayed at the bottom of the TabControl. On Windows CE the tabs are displayed above the TabControl. Also, on the Pocket PC the TabControl will always be located in the upper-left corner of its container control. You can work around this by adding the TabControl to a Panel control and then placing the Panel control appropriately.

Once you have added TabPages to the TabControl, you can add controls to each TabPage. To make the TabPage visible in the designer, click the TabPage's tab, just as a user would to view the page.

The TabControl exposes the SelectIndex property to determine which TabPage is currently selected. When this property changes, a SelectedIndexChanged event is raised.

Figure 3.27 shows the four different tab pages of the MemberBrowser application. This application displays the fields, properties, and methods of a given type. Here is the code for the SelectedIndexChanged event handler for this application. The complete code can be found with the code samples for this book.

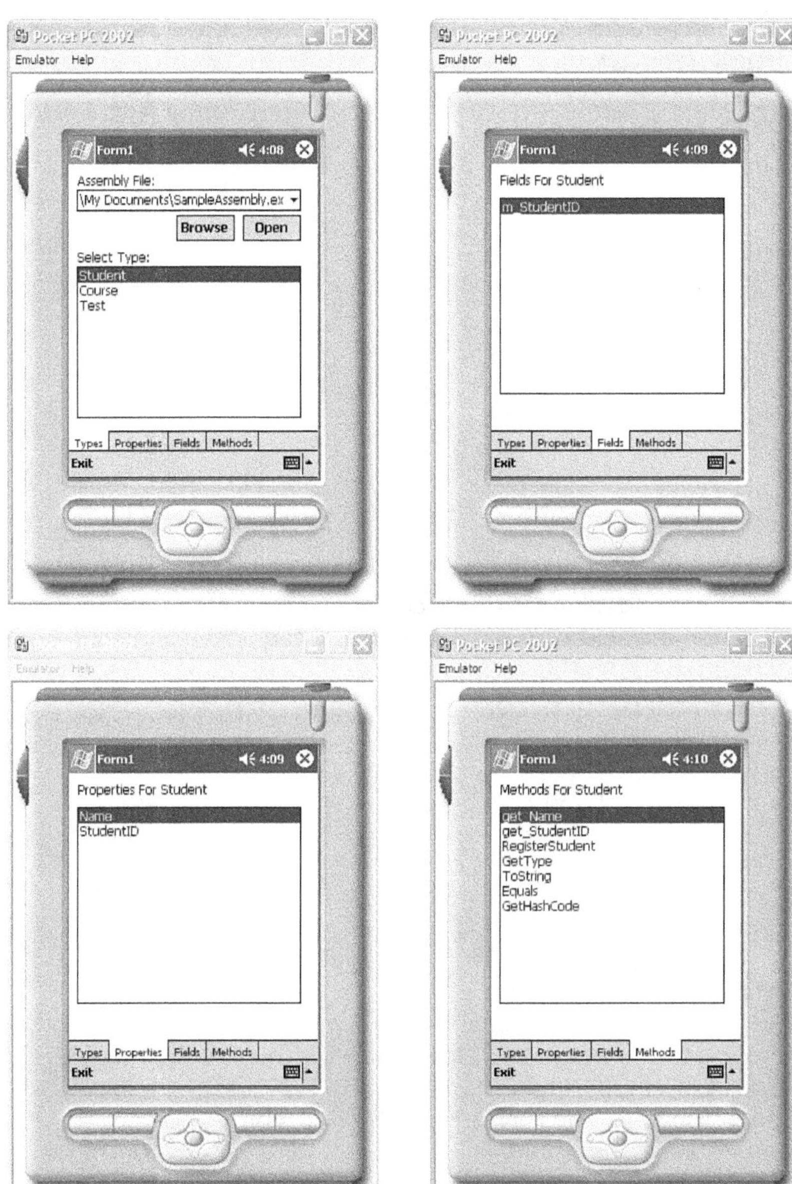

FIGURE 3.27 This application showcases the TabControl control running on the Pocket PC 2002 emulator.

```
C#
private void tpAssembly_SelectedIndexChanged(object sender,
        System.EventArgs e) {
switch(this.tpAssembly.SelectedIndex) {
  case 1: // Load Fields
    LoadFields();
    break;
  case 2: // Load Properties
    LoadProperties();
    break;
  case 3: // Load Methods
    LoadMethods();
    break;
  }
}

VB
Private Sub _
        tpAssembly_SelectedIndexChanged(sender As object,
        e As System.EventArgs) _ Handles tpAssembly.SelectedIndexChanged
  If Me.tpAssembly.SelectedIndex = 1 Then
    LoadFields()
  Else If Me.tpAssembly.SelectedIndex = 2 Then
    LoadProperties()
  Else If Me.tpAssembly.SelectedIndex = 3 Then
      LoadMethods()
  End If
End Sub
```

Using the `TreeView` Control

The `TreeView` control allows you to present data in a hierarchical view. The `TreeView` control is similar to the traditional tree data structure in that it is made up of a collection of nested node objects. The `TreeView` control exposes the `Nodes` property that is a collection `TreeNode` objects. These are the root nodes of the tree. Each `TreeNode` object exposes a `Nodes` property, as well. As you can see, the `TreeView` control will be composed of several levels of nested `TreeNode` objects.

Understanding TreeView Properties

The TreeView control allows quite a bit of control over how it is rendered. Table 3.8 contains a list of TreeView control properties that determine how the TreeView control will be rendered. A brief description that describes how the property affects the control is also provided.

TABLE 3.8
TreeView Control Properties and Descriptions

PROPERTY	DESCRIPTION
ImageList	This is the ImageList that contains the images that will be displayed next to the tree nodes.
ImageIndex	This is the index of the image that is displayed next to the tree nodes. Because only one index can be specified, every tree node will display the same image. You can override the image on a specific node by setting the TreeNode.ImageIndex property.
SelectedImageIndex	Here's the index of the image that that is displayed next to a tree node when it is selected. Because only one index can be specified, every tree node will display the same image. You can override the image on a specific node by setting the TreeNode.SelectedImageIndex property.
ShowLines	This determines whether lines are drawn between tree nodes.
ShowPlusMinus	This property determines whether plus-sign and minus-sign buttons are displayed next to the tree nodes that contain child tree nodes.
ShowRootLines	This determines whether lines are drawn between the tree nodes that are at the root of the tree.
CheckBoxes	This property determines whether check boxes are displayed next to the tree nodes. The Checked property of the TreeNode objects is set to true if the node is checked; false, otherwise. When the user checks a node, the AfterCheck event is raised.

TreeNode objects also expose properties that affect how the TreeNode will be rendered in the TreeView. Table 3.9 lists these properties and their descriptions.

TABLE 3.9
TreeNode Properties

PROPERTY	DESCRIPTION
Text	This is the text displayed in the label of the node.
ImageIndex	This is the index of the image that that is displayed next to the tree nodes. By default this value is equal to TreeView.ImageIndex. To specify another image index, you must set this property explicitly.
SelectedImageIndex	Here is the index of the image that that is displayed next to the node when the node is selected. By default this value is equal to TreeView.SelectedImageIndex. To specify another image index, you must set this property explicitly.

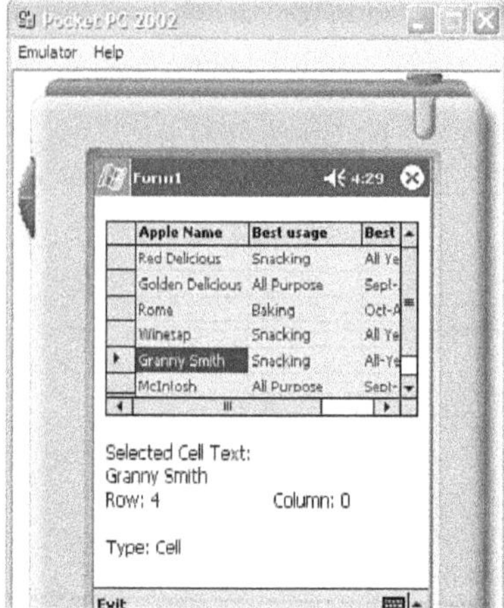

FIGURE 3.28 The TreeNode Editor.

Adding TreeNodes to a TreeView Control

Adding a TreeNode to the TreeView control can be done at both runtime and design time. Adding a TreeNode at runtime is done through the Form Designer. First, drag a TreeView control onto the application form. Next, select the TreeView control. If you have configured an ImageList control, set the ImageList and/or SelectedImageList properties in the Properties window. Now click the ellipsis next to Nodes property name in the Properties window. This displays the TreeNode Editor (see Figure 3.28).

The Add Root button in the TreeNode Editor can be used to add a TreeNode to the root of the TreeView control. The Add Child button will add a TreeNode to the currently selected node. The Delete button will remove the currently selected node. Changing the text in the Label textbox sets the TreeNode.Text property. If you have set the ImageList and/or SelectedImageList of the TreeView control and you want to override the default image of the TreeNode object, select the appropriate image in the Image and/or Selected Image drop-down controls.

TreeNode objects can also be added at runtime. The following code demonstrates how to add a TreeNode object through code:

```
C#
TreeNode treeNode1 = new TreeNode();
TreeNode treeNode2 = new TreeNode();

treeNode1.Text = "Red Apples";
treeNode2.Text = "Red Delicious";

treeNode1.Nodes.Add(treeNode2);
treeView1.Nodes.Add(treeNode1);

VB
Dim treeNode1 = new TreeNode()
Dim treeNode2 = new TreeNode()

treeNode1.Text = "Red Apples"
treeNode2.Text = "Red Delicious"

treeNode1.Nodes.Add(treeNode2)
treeView1.Nodes.Add(treeNode1)
```

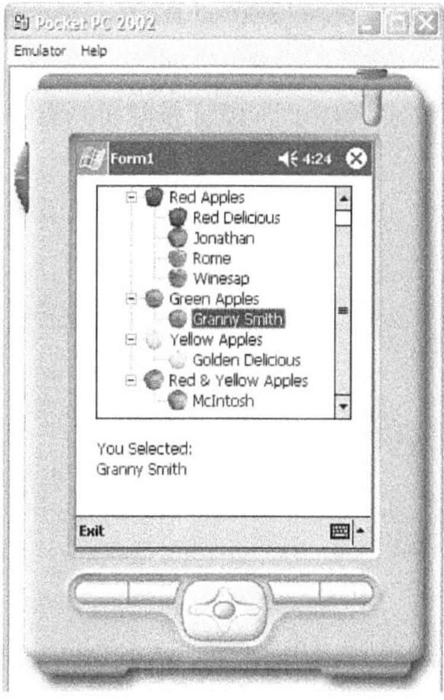

FIGURE 3.29 The `TreeView` control sample application.

Figure 3.29 shows an application with a `TreeView` control running on the Pocket PC 2002 emulator. The code for this application can be found in the code listing for this book.

Determining the Selected `TreeNode`

The `TreeView` control exposes the `SelectedNode` property. This property is a reference of the currently selected node in the control. The .NET Compact Framework's `TreeView` control allows only one control to be selected at once. An `AfterSelect` event is fired after a `TreeNode` is selected. The `AfterSelect` event handler receives a `TreeViewEventArgs` object, which carries information about the action.

The `TreeViewEventArgs` exposes the `Node` property. The `Node` property is a reference to the node that has been checked, expanded, collapsed, or selected. The `TreeViewEventArgs` also exposes the `Action` property. This property describes the type of action that raised the event. The `Action` property is a member of the `TreeViewAction` enumeration. Table 3.10 describes the members of the `TreeViewAction` enumeration.

TABLE 3.10
`TreeViewAction` Enumeration Members

MEMBER	DESCRIPTION
`ByKeyboard`	The event was raised by a keystroke.
`ByMouse`	The event was raised by a mouse click.
`Collapse`	The event was raised by the node's collapsing.
`Expand`	The event was raised by the node's expanding.
`Unknown`	The action that raised the event was unknown.

The following code demonstrates how to handle the `AfterSelect` event:

```
C#
private void
treeView1_AfterSelect(object sender,
        System.Windows.Forms.TreeViewEventArgs e) {
  TreeNode selNode = e.Node;
  label2.Text = selNode.Text;
}
```

```
VB
Private Sub _
treeView1_AfterSelect(sender As object,
        e As System.Windows.Forms.TreeViewEventArgs)
        Handles treeView1.AfterSelect
  Dim selNode As TreeNode
  selNode = e.Node
  label2.Text = selNode.Text
}
```

Working with the `DataGrid` Control

The `DataGrid` control displays data in a tabular form. The control can be bound to a data table or a collection. The .NET Compact Framework's `DataGrid` control provides a limited subset of the functionality that the full .NET Framework's `DataGrid` control provides. Despite this limitation, the `DataGrid` control is still quite useful. The control allows you to bind to different data sources; customize which columns will be displayed; customize the colors, row header text, and column header text; and handle events that are fired when the user selects rows and cells in the grid.

The `DataGrid` control does not provide editing support. You can work around this by providing your own editing user interface when users click inside the control. To learn how to handle user click events, see the "Determining the Selected Row or Selected Cell in the `DataGrid`" section. The `DataGrid` also provides limited support for customizing how columns are drawn. The "Customizing the `DataGrid` Control" section describes how to customize the columns of the `DataGrid`.

Populating the `DataGrid` Control

The `DataGrid` can be bound to the following data tables and collections:

- Any object that implements the `IList` or `IListSource` interface.
- An `ArrayList` object that contains the same type objects. If the `ArrayList` object contains objects of different types, then an exception will be thrown.
- A single `DataTable` object. The `DataGrid` control cannot handle displaying rows from multiple or related `DataTable` objects.
- A `DataView` object.

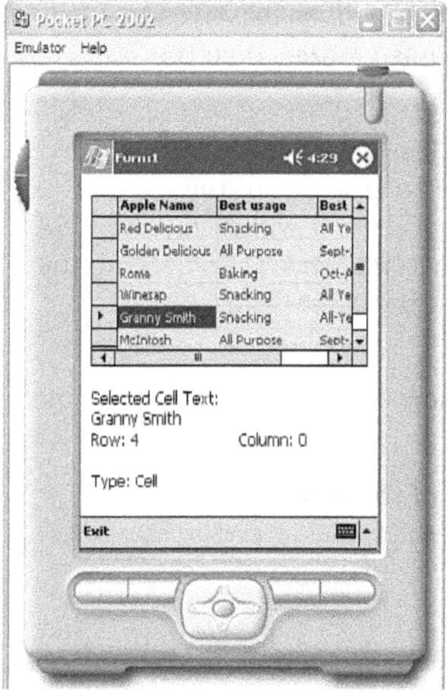

FIGURE 3.30 An application that showcases the DataGrid control running on the Pocket PC 2002 emulator.

Binding the DataGrid to its data source is simple. All you have to do is set the DataSource property on the DataGrid to a reference of the data source. The DataGrid will display all of the columns and rows in the data source by default. See the "Customizing the DataGrid Control" section to learn how to customize which columns will be displayed in the grid. The following code demonstrates how to bind a DataTable in a DataSet to a DataGrid control.

C#

```
dataGrid1.DataSource = dataset.Tables["Apples"];
```

VB

```
dataGrid1.DataSource = dataset.Tables("Apples");
```

Figure 3.30 demonstrates the DataGrid control running on the Pocket PC 2002 emulator. The complete listing for this application can be found in the code accompanying this book.

Customizing the DataGrid Control

The DataGrid control can be customized by changing the display colors and by customizing the style of each column. There are several properties that control the colors of the DataGrid control, as described in Table 3.11.

TABLE 3.11
DataGrid Properties That Format the Colors of the Grid

COLOR PROPERTY	DESCRIPTION
BackColor	The color of the non-row area of the grid
ForeColor	The color of the text in the grid
GridLineColor	The color of the lines in the grid
HeaderBackColor	The color of all the row and column headers
HeaderForeColor	The foreground color of the headers
SelectionBackColor	The back color of the selected rows
SelectionForeColor	The text color of the selected rows

The column headers and row-indicator headers are displayed by default. You can hide these headers by setting the ColumnHeadersVisible and RowHeaderVisible properties to false.

The columns of the DataGrid control can be further customized by using the DataGridTableStyle class, which represents how the DataGrid control is drawn. The

MappingName property represents the name of the data source that will be displayed in the DataGrid. For example, if the data source is a DataTable named Apples, then the MappingName should be set to "Apples".

The DataGridTableStyle class also exposes the GridColumnStyles property, which is a collection of DataGridColumnStyle objects. You should create an instance of the DataGridTableStyle class for each column that you want to display; it will dictate how the column will be drawn. Once you have added all of the DataGridColumnStyle objects to the DataGridTableStyle.GridColumnStyles property, you must add the DataGridTableStyle to the DataGrid's TableStyles property.

The DataGridColumnStyle is an abstract class that you can inherit from to customize how the column will be drawn.

The .NET Compact Framework provides an implementation of the abstract class: the DataGridTextBoxColumn class. Table 3.12 lists the properties of the DataGridTextBoxColumn and how they affect the DataGrid column.

TABLE 3.12
Properties of the DataGridTextBoxColumn Class

PROPERTY	DESCRIPTION
HeaderText	The text of the column header
MappingName	The name used to map the column style to a data member in the data source
NullText	The text that is displayed when the column contains a null reference
Width	The width of the column

Listing 3.5 demonstrates how to format columns through code using the DataGridTableStyle class:

LISTING 3.5

```
C#
DataGridTableStyle dgts = new DataGridTableStyle();
dgts.MappingName = "Apples";

DataGridTextBoxColumn nameColumn = new DataGridTextBoxColumn();
nameColumn.MappingName = "Name";
nameColumn.HeaderText = "Apple Name";
nameColumn.Width = 80;
dgts.GridColumnStyles.Add(nameColumn);

DataGridTextBoxColumn useColumn = new DataGridTextBoxColumn();
useColumn.MappingName = "Usage";
```

```
useColumn.HeaderText = "Best usage";
useColumn.Width = 80;
dgts.GridColumnStyles.Add(useColumn);

DataGridTextBoxColumn avalColumn = new DataGridTextBoxColumn();
avalColumn.MappingName = "Availability";
avalColumn.HeaderText = "Best Time to Buy";
avalColumn.Width = 80;
dgts.GridColumnStyles.Add(avalColumn);

this.dataGrid1.TableStyles.Add(dgts);

VB
Dim dgts = new DataGridTableStyle()
dgts.MappingName = "Apples"

Dim nameColumn = new DataGridTextBoxColumn()
nameColumn.MappingName = "Name"
nameColumn.HeaderText = "Apple Name"
nameColumn.Width = 80
dgts.GridColumnStyles.Add(nameColumn)

Dim useColumn = new DataGridTextBoxColumn ()
useColumn.MappingName = "Usage"
useColumn.HeaderText = "Best usage"
useColumn.Width = 80
dgts.GridColumnStyles.Add(useColumn)

Dim avalColumn = new DataGridTextBoxColumn()
avalColumn.MappingName = "Availability"
avalColumn.HeaderText = "Best Time to Buy"
avalColumn.Width = 80
dgts.GridColumnStyles.Add(avalColumn)

dataGrid1.TableStyles.Add(dgts)
```

Editing the TableStyles property can also be done through the Properties window. First, select the DataGrid in the Form Designer. Click the ellipsis next to the TableStyles property name in the Properties window. This will display the DataGridTableStyle Collection Editor (see Figure 3.31). You can click the Add and Remove buttons to add DataGridTableStyle objects. This dialog allows you to edit the MappingName property.

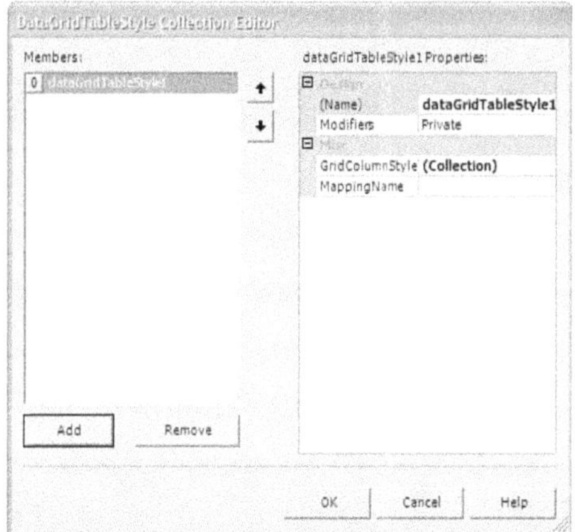

FIGURE 3.31 The `DataGridTableStyle` Collection Editor.

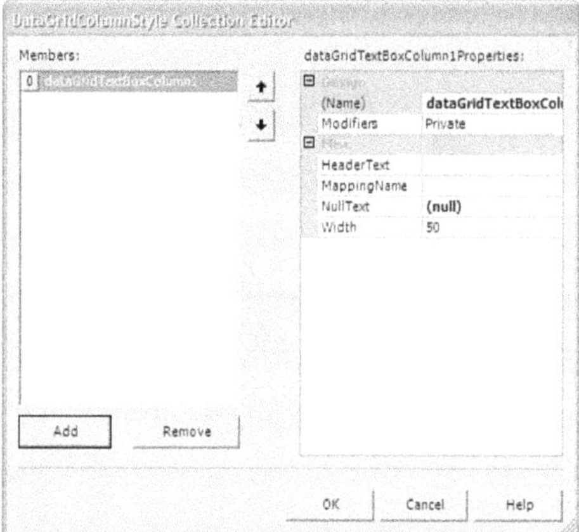

FIGURE 3.32 The `DataGridColumnStyle` Collection Editor.

You can also add `DataGridColumnStyle` objects to the `GridColumnStyle` property by clicking the ellipsis next to the property name. This will bring up the `DataGridColumnStyle` Collection Editor (see Figure 3.32). You can add and remove `DataGridColumnStyle` objects as well as edit their properties.

Determining the Selected Row or Selected Cell in the `DataGrid`

The `DataGrid` control can have either an entire row or a single cell selected. To determine which cell or row the user clicked, you should use the `HitTest` method in combination with the `DataGrid`'s `MouseUp` event. The `MouseUp` event will be fired when the user clicks anywhere inside the `DataGrid` control. You should retrieve the coordinates of the click event from the `MouseEventArgs` object (the `X` and `Y` properties).

Then pass the coordinates to the `HitTest` method. The method will return information about the `DataGrid` at the click coordinates. The method returns a `HitTestInfo` object. The object contains the `Column`, `Row`, and `Type` properties. The `Row` and `Column` properties are the indexes of the row and column the user clicks. The `Type` property specifies the part of the `DataGrid` control the user clicked. The `Type` property returns a member of the `DataGrid.HitTestType` enumeration. Table 3.13 defines the member of the `HitTestType` enumeration and the meanings.

TABLE 3.13
Members of the `HitTestType` Enumeration

MEMBER	DESCRIPTION
`Cell`	The user clicked a cell.
`ColumnHeader`	The user clicked a column header.
`ColumnResize`	The line between a column header was dragged to resize a column's width.
`None`	The user clicked the background area of the table.
`RowHeader`	The user clicked a row header.
`RowResize`	The line between a row header was dragged to resize a row's height.

The following code demonstrates how to handle the `MouseUp` event and use the `HitTest` method.

```
C#
private void dataGrid1_MouseUp(object sender,
        System.Windows.Forms.MouseEventArgs e) {
  HitTestInfo hitInfo = dataGrid1.HitTest(e.X, e.Y);

  if(hitInfo.Type == HitTestType.Cell) {
    string selCell = dataGrid1[hitInfo.Row, hitInfo.Column];
    if(cellData != null)
      this.label1.Text = selCell.ToString();
  }
}

VB
Private Sub DataGrid1_MouseUp _
        (ByVal sender As Object,
        ByVal e As System.Windows.Forms.MouseEventArgs)_
        Handles DataGrid1.MouseUp
   Dim hitInfo As System.Windows.Forms.DataGrid.HitTestInfo
   hitInfo = DataGrid1.HitTest(e.X, e.Y)

   Me.label2.Text = String.Empty
   Me.label3.Text = String.Format("Row: {0}", hitInfo.Row)
   Me.label4.Text = String.Format("Column: {0}", hitInfo.Column)
   Me.label5.Text = String.Format("Type: {0}", hitInfo.Type.ToString())

   If hitInfo.Type = System.Windows.Forms.DataGrid.HitTestType.Cell Then
     Dim selCell As Object
```

```
        selCell = DataGrid1.Item(hitInfo.Row, hitInfo.Column)
        If Not selCell Is Nothing Then
          Me.label2.Text = selCell.ToString()
        End If
      End If
    End Sub
```

In Brief

- The .NET Compact Framework supports a subset of functionality and Windows form controls.
- The Windows Form Designer provides a graphical way to design your applications user interface.
- .NET Compact Framework applications look and feel different when running on different target platforms.
- The .NET Compact Framework provides a rich set of controls, including the `DataGrid` control, the `TreeView`, and other advanced Windows controls.

Using Threads and Timers in the .NET Compact Framework

4

IN THIS CHAPTER

Threads, Timers, and Windows CE

Modern operating systems support multitasking and multithreading. *Multitasking* is the capability of the operating system to load many programs into memory simultaneously and share CPU time between the programs. In modern operating systems with memory protections, each task is set in its own virtual address space, which cannot be accessed by other tasks. This arrangement makes computers much more stable, but it also means that the cost of setting up a new, independent computing job as its own task is relatively expensive.

Multithreading is the capability of the operating system to support many independent units of execution in a single process. Each unit of execution is a thread, and each thread receives slices of CPU time. However, each thread occupies the same virtual address space. So, it is possible for one thread to interfere destructively with data used by another thread. However, setting up a new thread is far less expensive than setting up a whole new task, because no new address space is needed.

The .NET Compact Framework (CF) lets developers use the multithreading support available in the Windows CE operating system. However, there are some important limitations in Windows CE's support of multithreading that you should be aware of:

- The maximum number of processes is 32. Thus, developers should be very stingy about spawning a whole new process if spinning a thread can do the same work.
- Each process gets a maximum of 32MB of address space.
- Although threads are less expensive than processes, .NET CF programmers should consider everything expensive. Creating too many new threads will harm performance. One way to mitigate this problem is to use thread pools, discussed later in this chapter in the "Managing Multiple Threads with a Thread Pool" section.

The .NET Compact Framework supports threads in two ways. The first is through a `Thread` object. Through the `Thread` object, developers can spawn, terminate, and pass messages between threads. The second way to simulate threading behavior is through the `Timer`, an object that is familiar to Visual Basic developers. We discuss both threads and timers in great detail in this chapter.

THREADING LIMITATIONS ON THE .NET COMPACT FRAMEWORK

The threading support available in the .NET Compact Framework is much less deep than that available on the desktop. It is important to pay attention to which overloads and classes are available.

The Thread Class

The `System.Threading.Thread` class encapsulates all of the functionality threads offered on the Windows CE device that can be accessed with the .NET Compact Framework. Table 4.1 shows the methods and properties available in the `Thread` class on the .NET Compact Framework.

TABLE 4.1
The `Thread` Class

METHOD/PROPERTY NAME	DESCRIPTION
`Thread`	Constructor
`Priority`	Gets or sets the thread's priority
`void Start`	Starts a newly created or stopped thread; see "Creating and Starting a Thread" section
`void Sleep(int sleepTime)`	Suspends thread execution for at least `sleepTime` milliseconds, passing CPU control to other threads
`static Thread CurrentThread`	Returns the `Thread` object currently running

Understanding Thread Basics

To use threads effectively, there are several fundamental practices that developers must understand, such as how to start threads and put threads to sleep. Additionally, there are some obscurities involved with using threads in the .NET Compact Framework. For example, using threads with user interface controls requires special care. Also, shutting down a multi-threaded .NET Compact Framework application must be done in a specific fashion, or the application will not terminate.

In this section we discuss the basics that developers must master to use threads safely and effectively on the .NET Compact Framework. We'll wrap this section up with a sample application that demonstrates all of these concepts.

Creating and Starting a Thread

To create and start a thread with the .NET Compact Framework, follow these steps:

1. Create an instance of `System.Threading.ThreadStart`. Pass in the name of the method to execute in a thread. The method must be void with no input arguments.
2. Create an instance of `System.Threading.Thread`. Pass the `ThreadStart` into the `Thread` constructor. You now have a reference to the thread that will execute the method that was specified in step 1.
3. Call `Thread.Start()` on the reference to the thread created in step 2. This causes the code in the method you will run on the thread to be begin executing.

Writing Code to Start a Thread

> **STARTING A THREAD**
>
> Remember that the thread does not actually start running until `Thread.Start()` is called.

This sample code is from the SimpleThread sample application discussed later in the chapter. It demonstrates the launching of two threads, `m_SinglesThread` and `m_DecadeThread`. `m_SinglesThread` executes a method called `SinglesCounter`, which counts by ones, and `m_DecadeThread` executes a method called `DecadeCounter`, which counts by tens.

```
C#
System.Threading.Thread m_singlesThread;
System.Threading.ThreadStart m_singlesThreadStart;
System.Threading.Thread m_decadeThread;
System.Threading.ThreadStart m_decadeThreadStart;
```

```
m_singlesThreadStart = new
   System.Threading.ThreadStart(SinglesCounter);

m_singlesThread = new System.Threading.Thread(m_singlesThreadStart);

m_decadeThreadStart = new
   System.Threading.ThreadStart(DecadeCounter);

m_decadeThread = new System.Threading.Thread(m_decadeThreadStart);

m_singlesThread.Start();
m_decadeThread.Start();
```

```
VB
Dim m_singlesThread As System.Threading.Thread
Dim m_singlesThreadStart As System.Threading.ThreadStart

Dim m_decadeThread As System.Threading.Thread
Dim m_decadeThreadStart As System.Threading.ThreadStart

m_singlesThreadStart = New System.Threading.ThreadStart(AddressOf SinglesCounter)

m_singlesThread = New System.Threading.Thread(m_singlesThreadStart)

m_decadeThreadStart = New System.Threading.ThreadStart(AddressOf DecadeCounter)

m_decadeThread = New System.Threading.Thread(m_decadeThreadStart)
```

Suspending a Thread

To suspend a thread's execution, use the static `Thread.Sleep()` method, which suspends the thread running the block of code that called `Thread.Sleep`. Pass in the minimum number of milliseconds that the thread will sleep before it is placed in the ready queue to wait for more processor time.

If you pass 0 in as the argument for `Thread.Sleep`, then the current thread gives up the CPU but immediately goes back into the ready queue. It gets the processor again as soon as its turn comes up.

It is important to note that the number of milliseconds passed into `Thread.Sleep` specifies only the amount of time before the thread goes back into the ready queue. It could be considerably longer before the thread actually gets the CPU again, especially if the device is heavily loaded. Consider these snippets, for example:

```
C#
// The code running in myThread is suspended for at
// least 1000 milliseconds
Thread.Sleep(1000);

VB
' The code running in myThread is suspended for
' at least 1000 milliseconds
Thread.Sleep(1000)
```

Using Threads with User Interface Controls

On the .NET Compact Framework, controls cannot be updated by threads that do not own the control. Specifically, this means you cannot start a thread and then update elements in your user interface from that thread by directly interacting with the controls. If you try to do this, your application will lock up at seemingly random times. For example, running this code outside of the main application thread will cause problems:

```
C#
// This can lock up an app if called outside of
// the main thread
this.txtTextBox1.Text = "I set the text";

VB
' This can lock up an app if called outside of
' the main thread
Me.txtTextBox1.Text = "I set the text"
```

The desktop .NET Framework includes the `Form.BeginInvoke` and `Form.EndInvoke` methods to help cope with this situation, but these methods are not supported in the .NET Compact Framework.

To deal with this situation, first create a delegate for a method that does the user interface updates. Then use `Form.Invoke` to call the delegate from the thread. This is a confusing and cumbersome process, and it works only for calling methods that accept no parameters.

Rather than dealing with this problem directly, we present a custom class named `ControlInvoker`, which threads can use to call methods. The methods can safely update controls in the user interface of your application. `ControlInvoker` also allows developers to pass parameters to the methods they want to call. `ControlInvoker` is used with permission and is available from the Web site `http://samples.gotdotnet.com/quickstart/CompactFramework/`

`doc/controlinvoker.aspx` (special thanks to Microsoft's Seth Demsey, David Rasmussen, and Bruce Hamilton for providing `ControlInvoker`).

Note that the `ControlInvoker` and `MethodCallInvoker` classes are available in the `ControlInvokerSample` namespace. To use `ControlInvoker` to invoke a method, follow these steps:

1. Create an instance of the `ControlInvoker` class.
2. Structure the methods you want to call with the following signature: `void AnyMethod(object[] in_args)`.
3. Create an instance of a `MethodCallInvoker` class, passing the name of the method you want to call into the constructor.
4. Call the `ControlInvoker.Invoke()` method. Pass the `MethodCallInvoker` as the first argument. Pass additional parameters that you want passed to the method you will invoke as additional arguments to `ControlInvoker.Invoke()`.
5. The method will be called. The object array will hold the arguments passed into `ControlInvoker.Invoke()`. Extract the objects and cast them back to their original types before using them.

The following sample code is derived from the SimpleThread sample application. The `setTextBox` method is passed a string and a TextBox as an array of objects. It extracts the string and TextBox and sets the text of the TextBox to the string passed in. The `SinglesCounter` method is executed in a separate thread. It uses a `ControlInvoker` to call `setTextBox` and pass to it the `TextBox` whose text must be set and the string to set it to. By using the `ControlInvoker`, the code in `SinglesCounter` can update controls on the main form without risking locking up the application.

```
C#
private void setTextBox(object[] in_args)
{
   string in_text = (string)in_args[0];

   System.Windows.Forms.TextBox in_textBox =
      (System.Windows.Forms.TextBox)in_args[1];

   in_textBox.Text = in_text;
}


private void SinglesCounter()
{
   int l_currentSinglesCount = 0;
   while (!m_QuitRequested)
   {
```

```
        m_controlInvoker.Invoke (new MethodCallInvoker (setTextBox),
           Convert.ToString(l_currentSinglesCount), this.txtSingle);

        l_currentSinglesCount++;
        System.Threading.Thread.Sleep(200);
    }
}

VB
Private Sub setTextBox(ByVal in_args() As Object)
   Dim in_text As String
   in_text = in_args(0)
   Dim in_textBox As System.Windows.Forms.TextBox
   in_textBox = in_args(1)
   in_textBox.Text = in_text
End Sub

Private Sub SinglesCounter()
   Dim l_currentSinglesCount As Int32
   l_currentSinglesCount = 0
   While (m_QuitRequested = False)
      m_controlInvoker.Invoke(New MethodCallInvoker(AddressOf
              setTextBox), Convert.ToString(l_currentSinglesCount),
              Me.txtSingle)
      l_currentSinglesCount = l_currentSinglesCount + 1
      System.Threading.Thread.Sleep(200)
   End While

   m_SinglesThreadRunning = False

   ' Last thread out closes form
   If (m_DecadeThreadRunning = False) Then
      m_controlInvoker.Invoke(New MethodCallInvoker(AddressOf
         ShutDown))
   End If
End Sub
```

Quitting Threaded Applications

On the .NET Compact Framework, applications should not be considered stopped unless all threads have terminated. If a thread is sitting in a loop and is unaware that the application is closing, it can hold up the process of closing the application and can make the application appear as though it is not responding to close requests. Thus, it is important for developers to write their threads in such a way that they are aware whether the application is trying to close.

One way to accomplish this is to override the `Form.OnClosing()` method and set a flag so that threads can find out if the application is trying to close. If a thread determines that the application is trying to close, it can determine whether it is the last thread to exit, closing the application if needed. The following sample code, taken from the SimpleThread sample application, demonstrates this practice:

```
C#
private void ShutDown(object[] arguments)
{
   this.Close();
}

protected override void OnClosing(CancelEventArgs e)
{
   if (m_SinglesThreadRunning || m_DecadeThreadRunning)
   {
      e.Cancel = true;
      MessageBox.Show("Will wait for threads to stop, then quit");
      m_QuitRequested = true;
   }
   else
   {
      Close();
   }
}

// Runs on separate thread
private void SinglesCounter()
{
   int l_currentSinglesCount = 0;
   while (!m_QuitRequested)
   {
      m_controlInvoker.Invoke (new MethodCallInvoker (setTextBox),
         Convert.ToString(l_currentSinglesCount), this.txtSingle);

      l_currentSinglesCount++;
      System.Threading.Thread.Sleep(200);
   }
   m_SinglesThreadRunning = false;
```

```
    // Last thread out closes form
    if (m_DecadeThreadRunning == false)
        m_controlInvoker.Invoke(new MethodCallInvoker(ShutDown));
}

// Runs on separate thread
private void DecadeCounter()
{
    int l_currentDecadeCount = 0;
    while (!m_QuitRequested)
    {
        m_controlInvoker.Invoke (new MethodCallInvoker (setTextBox),
            Convert.ToString(l_currentDecadeCount), this.txtDecade);

        l_currentDecadeCount += 10;
        System.Threading.Thread.Sleep(200);
    }

    m_DecadeThreadRunning = false;

    // Last thread out closes form
    if (m_SinglesThreadRunning == false)
        m_controlInvoker.Invoke(new MethodCallInvoker(ShutDown));
}
```

VB

```
Private Sub ShutDown(ByVal arguments() As Object)
    Me.Close()
End Sub


Protected Overrides Sub OnClosing(ByVal e As
        System.ComponentModel.CancelEventArgs)
    If (m_SinglesThreadRunning Or m_DecadeThreadRunning) Then
        e.Cancel = True
        MessageBox.Show("Will wait for threads to stop, then quit")
        m_QuitRequested = True
    Else
        Close()
    End If
End Sub
```

```
Private Sub SinglesCounter()
   Dim l_currentSinglesCount As Int32
   l_currentSinglesCount = 0
   While (m_QuitRequested = False)
      m_controlInvoker.Invoke(New MethodCallInvoker(AddressOf
              setTextBox), Convert.ToString(l_currentSinglesCount),
              Me.txtSingle)
      l_currentSinglesCount = l_currentSinglesCount + 1
      System.Threading.Thread.Sleep(200)
   End While

   m_SinglesThreadRunning = False

   ' Last thread out closes form
   If (m_DecadeThreadRunning = False) Then
      m_controlInvoker.Invoke(New MethodCallInvoker(AddressOf
              ShutDown))
   End If
End Sub

Private Sub DecadeCounter()
   Dim l_currentDecadeCount As Int32
   l_currentDecadeCount = 0
   While (m_QuitRequested = False)
      m_controlInvoker.Invoke(New MethodCallInvoker(AddressOf
              setTextBox), Convert.ToString(l_currentDecadeCount),
              Me.txtDecade)
      l_currentDecadeCount += 10
      System.Threading.Thread.Sleep(200)
   End While

   m_DecadeThreadRunning = False

   ' Last thread out closes form
   If (m_SinglesThreadRunning = False) Then
      m_controlInvoker.Invoke(New MethodCallInvoker(AddressOf
              ShutDown))
   End If

End Sub
```

The `ShutDown` method wraps the `Form.Close` method so that it can be called using the `ControlInvoker`. The `OnClosing` method is overridden. It checks to see whether any child threads are running. If they are, a flag is set, indicating that the application wants to shut down. If none of the threads is running, then the application simply quits.

The `SinglesCounter` and `DecadeCounter` methods each run in a separate thread. Their loops break if the `m_QuitRequested` flag is set. When the threads fall out of the loop, each checks to see if another is running. The last thread still alive uses the `MethodInvoker` to call `ShutDown`, which closes the application.

Coding a Multithreaded Application: SimpleThread

The SimpleThread sample application is located in the folder `\SampleApplications\Chapter4`. There are C# and Visual Basic versions of SimpleThread.

This application demonstrates how to create and start threads by creating two new threads. One counts by ones, and the other by tens. Both threads sleep briefly after updating the count.

Each thread updates the user interface after each count update by using the `ControlInvoker`. SimpleThread also uses the `ControlInvoker` to shut down gracefully, as discussed in the previous section.

Coordinating Threads with the `Mutex` Class

The .NET Compact Framework includes the `Mutex` class, which is a useful entity for coordinating threads. The `Mutex` class's name is derived from the term *mutual exclusion*. The class creates a region of code that can be executed by only one thread at a time.

A `Mutex` works by tracking which thread "holds" it. Only one thread may hold a `Mutex` at a time. If another thread tries to acquire it, then it blocks until the first thread releases its hold.

To use a `Mutex` to force only one thread to execute code at a time, follow these steps:

1. Create an instance of the `Mutex` class. Pass `false` into the `Mutex` constructor so that the code that created the `Mutex` does not hold it.
2. At the beginning of the code, place a call to `Mutex.WaitOne()` that may only be executed by one thread at a time.
3. At the end of the code, place a call to `Mutex.ReleaseMutex()` that may only be executed by one thread at a time.

The following sample code demonstrates the steps:

```
C#
// Create instance of Mutex class.  The variable holding the
// reference to this Mutex must be visible to the block of code
// that will be protected.
m_Mutex = new System.Threading.Mutex(false);

// This is the code which we only want one thread at a time to execute

// This call blocks until we acquire the Mutex
m_Mutex.WaitOne();

// Put code here that only is executed by one thread at a time

// Done, let others enter this code block now
m_Mutex.ReleaseMutex();

VB
' Create instance of Mutex class.  The variable holding the
' reference to this Mutex must be visible to the block of code
' that will be protected.
m_Mutex = new System.Threading.Mutex(False)

' This is the code which we only want one thread at a time to execute

' This call blocks until we acquire the Mutex
m_Mutex.WaitOne()

' Put code here that only is executed by one thread at a time

' Done, let others enter this code block now
m_Mutex.ReleaseMutex()
```

There are some special considerations for using the `Mutex`:

- Using `Mutex` objects is a cooperative activity. As the developer, you must structure your code so that any block of code that should be run by only a single thread at a time is protected with a `Mutex`.
- You can use the same `Mutex` to protect two different regions of code. But this can cause unexpected behavior because of the following scenario. Thread 1 gets exclusive access to Code Block A through the `Mutex`. At the end of Code Block A, Thread 1 releases the `Mutex`. Soon after, Thread 1 gets to Code Block B and tries to acquire the same `Mutex`

again, but Thread 2 has already acquired it in the meantime. If Thread 1 expected no one else to touch any of the code blocks between Code Block A and Code Block B, there could be unexpected failures.

- The `Monitor` class is an alternative to the `Mutex` class. The `Monitor` class is very useful for making sure that only one thread has access to a specific data object at one time, while the `Mutex` class is well suited for protecting blocks of code. The `Monitor` class is described later in this chapter.

Suspending and Resuming Threads with Mutexes

The desktop .NET Framework includes support for the `Thread.Suspend` and `Thread.Resume` methods, which can force a thread to sleep indefinitely and then wake it back up. The .NET Compact Framework does not support these methods, so building applications with threads that can be suspended and resumed requires some thought and architecting. One way to simulate the `Thread.Suspend` and `Thread.Resume` methods is by using the `Mutex` class. The technique is especially useful on code running in a loop. To support suspending a thread that is running code in a loop, follow these steps:

1. Create an instance of a `Mutex` object. Pass `false` to the constructor.
2. At the top of the loop, call `Mutex.WaitOne()`.
3. At the bottom of the loop, call `Mutex.ReleaseMutex()`.
4. To suspend the thread, call `Mutex.WaitOne()`. This call blocks until the thread finishes the current iteration of its loop and releases the `Mutex`. The thread then blocks.
5. To resume the thread, call `Mutex.ReleaseMutex()`.

The following code, derived from the SuspendAndTerminateThreads sample application, demonstrates how to suspend and resume a thread running a loop:

```
C#
// This code runs in a loop.  The thread can be suspended by calling
// m_Mutex.WaitOne()
while (!l_quit)
{
   m_Mutex.WaitOne();

   // Insert useful code here

   m_Mutex.ReleaseMutex();
}
```

```
// This code runs on another thread, and will suspend the code above
// This call blocks until the other thread finishes its loop
// iteration, then that thread blocks
m_Mutex.WaitOne();

// Other thread is now blocked ...

// And now resume the other thread
m_Mutex.ReleaseMutex();

VB
' This code runs in a loop.  The thread can be suspended by calling
' m_Mutex.WaitOne()
while (l_quit = False)
   m_Mutex.WaitOne()

    ' Insert useful code here

   m_Mutex.ReleaseMutex()
End While



' This code runs on another thread, and will suspend the code above
' This call blocks until the other thread finishes its loop
' iteration, then that thread blocks
m_Mutex.WaitOne()

' Other thread is now blocked ...

' And now resume the other thread
m_Mutex.ReleaseMutex()
```

Examining a Sample Application: SuspendAndTerminateThreads

This sample application has C# and VB versions in the folder `\SampleApplications\Chapter4`. It demonstrates how to suspend and resume threads by using a `Mutex` and also how to set up a thread so that it can check to see if it should terminate.

The application has a button at the bottom labeled Start Threads. Clicking this button causes two threads to launch, one that counts by ones, and another by tens. Either of these threads can be independently suspended, resumed, or terminated by clicking the appropriately labeled button.

Blocking until a Thread Finishes

The desktop version of the .NET Framework features a method called `Thread.Join()`. Given a thread called `l_Thread`, if another thread calls `l_Thread.Join()`, then the other thread blocks until `l_Thread` terminates. The .NET Compact Framework does not support `Thread.Join()`.

One way to wait for another thread to die is to use a busy wait. For example, the thread method might update the Boolean class member `m_Alive` to `false` as it exits. Another thread can sit and wait for the first thread to finish by writing code that looks like this:

```
C#
// Wait for another thread to set m_Alive to false
while (m_Alive)
{
}


VB
' Wait for another thread to set m_Alive to false
while (m_Alive = True)
End While
```

This would be a very bad idea because the act of waiting consumes CPU cycles. The thread that is waiting continuously checks the `m_Alive` variable, using CPU resources that could be used for meaningful processing. Additionally, running such tight loops can slow down applications noticeably.

The right way to have one thread wait for another to finish is in such a way that the waiting thread does not consume CPU cycles. This feature can be realized using `Mutex` objects. To use this technique, developers must write support for a simulated join into their thread methods. Follow these steps to structure your thread so that other threads can wait for them to finish:

1. The thread that waits for another thread to finish is referred to as the *parent thread*. The thread that is being waited for is referred to as the *child thread*.
2. In the parent thread, create a new `Mutex` in a scope such that the parent and child threads can both see it. Pass `false` into the `Mutex` constructor so that the parent thread does not initially hold ownership of the `Mutex`.

3. Create a Boolean variable, such as `m_ThreadRunning`, and set it to `false` in the parent thread.
4. Start the child thread from the parent thread.
5. Structure the code run in the child thread so that the first thing it does is call `Mutex.WaitOne()` to acquire ownership of the `Mutex`. Immediately after, the child thread sets `m_ThreadRunning` to `true`.
6. The last thing the child thread should do before exiting is to release the `Mutex` by calling `Mutex.ReleaseMutex()`.
7. When the parent thread wants to block until the child is finished, effectively doing a `Join()`, have the parent loop and wait for `m_ThreadRunning` to turn `true`. Make sure to put a sleep statement in the loop to avoid wasting CPU cycles. Usually, this wait will not occur, but there occasionally will be a very short wait because as soon as the child thread starts, it will acquire the `Mutex` and set `m_ThreadRunning` to `true`.
8. When `m_ThreadRunning` is `true`, call `Mutex.WaitOne()` in the parent thread. The parent will block without consuming any CPU cycles until the child thread has finished.
9. Call `Mutex.ReleaseMutex()` from the parent.

The following sample code demonstrates these steps. The code is taken from the sample application JoinDemo.

```
C#
// This is the code that runs in the child thread
private void DoSomeWork()
{
   m_ThreadAliveMutex.WaitOne();
   m_isThreadRunning = true;

   // Do some work here.  We'll just sleep to simulate work that
   // took a while
   System.Threading.Thread.Sleep(10000);


   // Release the "alive" mutex on thread exit, indicating that
   // this thread is dying.
   m_ThreadAliveMutex.ReleaseMutex();
}

// This is the code that runs in the parent thread
// m_isThreadRunning is a bool class member
```

```
// m_ThreadAliveMutex is a Mutex class member
m_isThreadRunning = false;

m_ThreadAliveMutex = new System.Threading.Mutex(false);

m_threadStart = new System.Threading.ThreadStart(DoSomeWork);
m_workerThread = new System.Threading.Thread(m_threadStart);

MessageBox.Show("Starting thread and then waiting
   for it to finish");

m_workerThread.Start();

// We actually need to know that the thread has started before
// we wait for it.
while (m_isThreadRunning == false)
{
   System.Threading.Thread.Sleep(100);
}
m_ThreadAliveMutex.WaitOne();
m_ThreadAliveMutex.ReleaseMutex();

MessageBox.Show("Worker thread has stopped");

VB
Private Sub DoSomeWork()
   m_ThreadAliveMutex.WaitOne()
   m_isThreadRunning = True

   ' Do some work here.  We'll just sleep to simulate work that took a
   ' while
   System.Threading.Thread.Sleep(10000)


   ' Release the "alive" mutex on thread exit, indicating that
   ' this thread is dying.
   m_ThreadAliveMutex.ReleaseMutex()
End Sub


' This is the code that runs in the parent thread
' m_isThreadRunning is a bool class member
```

```
' m_ThreadAliveMutex is a Mutex class member

m_isThreadRunning = False
m_ThreadAliveMutex = New System.Threading.Mutex(False)

m_threadStart = New System.Threading.ThreadStart(AddressOf DoSomeWork)
m_workerThread = New System.Threading.Thread(m_threadStart)

MessageBox.Show("Starting thread and then waiting for it to finish")

m_workerThread.Start()

' We actually need to know that the thread has started before
' we wait for it.
While (m_isThreadRunning = False)
   System.Threading.Thread.Sleep(0)
End While

' This is like a join - no CPU cycles consumed while we wait
m_ThreadAliveMutex.WaitOne()
m_ThreadAliveMutex.ReleaseMutex()

MessageBox.Show("Worker thread has stopped")
```

Waiting for a Thread: The JoinDemo Application

The JoinDemo application demonstrates the process of simulating a call to `Thread.Join()` using the .NET Compact Framework. The main thread spins off a child thread, which simulates taking time to do some work by going to sleep for 10 seconds. Meanwhile, the main thread is blocked because it is waiting to acquire a `Mutex`, as discussed previously.

The JoinDemo sample is located in the folder `\SampleApplications\Chapter4`. There are C# and Visual Basic versions of the application.

Controlling Access to Data Objects with the `Monitor` Class

The `Monitor` class provides another synchronization mechanism available in the .NET Compact Framework. The `Monitor` class is useful for protecting access to specific objects. You can use a `Monitor` to make sure that only one thread accesses a specific object at a time. This intent is subtly different from using a Mutex, which you can use to prevent two threads from executing a specific block of code at one time.

To use a Monitor to prevent multiple threads from accessing the same object, follow these steps:

1. Acquire the object that you want to protect. When you want to access the protected object, first call System.Threading.Monitor.Enter(). Pass in the reference to the object to protect.
2. Access the object. Keep the lock for as little time as possible, since other threads may be blocking, waiting for access to the same protected object.
3. When you are finished with the protected object, call System.Threading.Monitor.Exit(). Pass in the reference to the object to stop protecting.

For example, assume that m_CustomerBankInfo is an object representing a customer's bank account information. You don't want two threads accessing it at the same time, since one thread might read the data while another is in the middle of writing to it. The code to access m_CustomerBankInfo safely looks like this:

```
C#
System.Threading.Monitor.Enter(m_CustomerBankInfo);
// Access m_CustomerBankInfo

// Now done with m_CustomerBankInfo
System.Threading.Monitor.Exit(m_CustomerBankInfo);

VB
System.Threading.Monitor.Enter(m_CustomerBankInfo)
' Access m_CustomerBankInfo

' Now done with m_CustomerBankInfo
System.Threading.Monitor.Exit(m_CustomerBankInfo)
```

Simulating the Producer-Consumer Problem

The ProducerConsumer sample application demonstrates the use of Monitor objects to simulate a derivative of the classic Producer-Consumer computer science problem. The sample application uses two threads. One thread produces objects by incrementing the class member called m_objectsAvailable each time it wakes up. The consumer thread consumes an object each time it wakes up by decrementing m_objectsAvailable. The two threads use a Monitor to prevent both threads from trying to access the m_objectsAvailable member at the same time.

To make things interesting, the application starts with four objects available, but the consumer thread consumes objects slightly faster than the producer thread produces them. This effect is simulated by making the producer thread sleep longer than the consumer thread. As the program runs, you will notice that the number of objects slowly decreases until it reaches zero. Then each time a producer produces an object, the count will briefly reach one, before a consumer consumes the object.

This sample application is available in the directory `\SampleApplications\Chapter4`. There are C# and Visual Basic versions.

Managing Multiple Threads with a Thread Pool

There are many real-world situations where the use of a thread makes sense but the thread's lifetime is short. For example, an architect might choose to spawn a thread to handle a short-lived socket connection to a remote party. Suppose the application has an average of 5 such connections occurring simultaneously, but each connection lasts an average of only 1 minute. In that case the application could easily spawn over 100 threads per hour, even though it really only needed 5 at a time.

This kind of behavior is wasteful and expensive and can be especially harmful on devices, with their limited resources. Particularly with devices, developers should always be stingy with resource allocation.

The .NET Compact Framework contains a very limited offering of the `ThreadPool` class, which is designed specifically to mitigate scenarios like the one just described. The `ThreadPool` class internally maintains a small number of threads and uses them to perform work that you pass in as specific methods to execute. When a method returns, the thread that executed returns to the pool until the `ThreadPool` class uses it again to execute another method.

Because the `ThreadPool` class minimizes thread creations, it causes less garbage to build up in the heap. Since there is a much smaller heap available on devices than on desktop computers, and since the slower processors take longer for garbage to build up, the performance gain can be significant.

To set up a thread pool, follow these steps:

1. Write a method that the `ThreadPool` will execute. The method must return void and take no parameters.

2. Create a new instance of the `WaitCallback` class, passing the name of the method to execute as an argument.

3. Call the static method `ThreadPool.QueueUserWorkItem`, and pass the `WaitCallback` instance into it. The `ThreadPool` will execute the method by using one of its threads.

The following sample code demonstrates setting up the ThreadPool to execute two methods, SinglesCounter and DecadeCounter. This sample code is derived from the ThreadPoolDemo sample application.

```
C#
m_singlesWaitCallBack = new WaitCallback(SinglesCounter);
m_decadeWaitCallBack = new WaitCallback(DecadeCounter);
ThreadPool.QueueUserWorkItem(m_singlesWaitCallBack);
ThreadPool.QueueUserWorkItem(m_decadeWaitCallBack);

VB
m_singlesWaitCallBack = New System.Threading.WaitCallback(AddressOf
        SinglesCounter)

m_decadeWaitCallBack = New System.Threading.WaitCallback(AddressOf
        DecadeCounter)
System.Threading.ThreadPool.QueueUserWorkItem(m_singlesWaitCallBack)
System.Threading.ThreadPool.QueueUserWorkItem(m_decadeWaitCallBack)
```

Controlling Threads with the ThreadPoolDemo Application

The ThreadPoolDemo sample application is located in the folder \SampleApplications\Chapter4. There are C# and Visual Basic versions.

Like the SimpleThread sample application, ThreadPoolDemo uses two threads to increment two counters. The first counter increments by one, and the other increments by tens. ThreadPoolDemo uses a ThreadPool to assign threads to the methods that increment the counters.

Timers

Timers are a familiar entity to Visual Basic programmers, as they have been part of that language for many years. The Timer class is available on the .NET Compact Framework and is thus available for both C# and Visual Basic .NET.

Timers are objects that have a Tick method into which developers insert code. The code in the Tick method is executed at regular intervals, which the developer gets to set.

Thus, the Timer is very useful for executing a body of code at specific intervals. The Timer is meant for performing light work at regular intervals, but it is not a good choice for executing code that will take a long time to execute. For such cases a Thread is a far better choice.

The advantages of using Timers on the .NET Compact Framework include these:

- Easy to set up
- Easy to control the wakeup interval
- Implicitly sleep on their own when not executing their code body
- Easy to enable and disable

In addition to these benefits, the code executing in a `Timer` can access control objects as long as the thread that owns the `Timer` also owns the control. Thus, timers can be used as a "poor man's" UI update thread in some circumstances.

Creating a `Timer`

The easiest way to create a `Timer` is to use the Smart Device Extensions in Visual Studio .NET to help. Follow these steps to create a `Timer`:

1. Select the design view of the form that will own the `Timer`.
2. Select the Timer control on the toolbox and drag it to the hidden object viewer. An icon representing the `Timer` appears. Figure 4.1 shows the design view for the form in the TimerDemo application. In this figure there are two timers that have already been created.

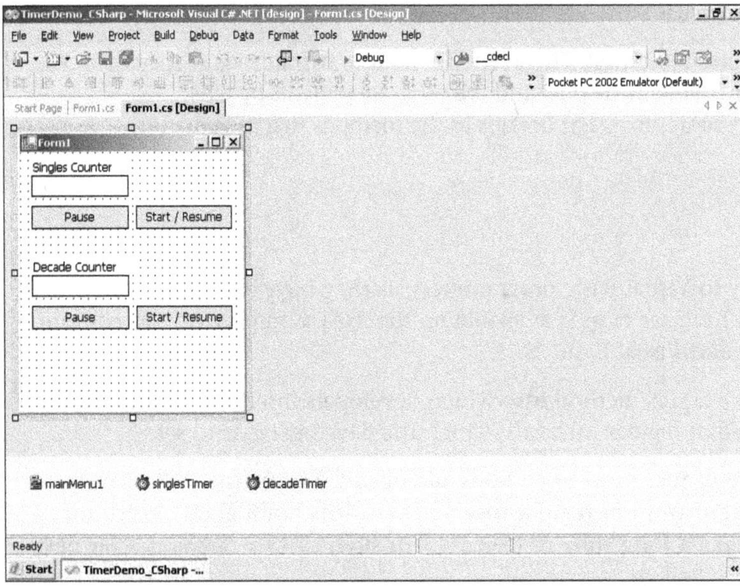

FIGURE 4.1 This `TimerDemo` application has singles and decade timers.

3. Double-click the icon for the `Timer`. Visual Studio responds by creating an empty `Tick` method for the `Timer`.
4. Insert code into the `Tick` method.

Writing `Tick` Method Code

The following sample code shows the contents of the `Tick` method for the `Timer` named `singlesCounter` in the TimerDemo sample application:

```
C#
private void singlesTimer_Tick(object sender, System.EventArgs e)
{
   this.txtSingles.Text = Convert.ToString(m_SinglesCount);
   m_SinglesCount = m_SinglesCount + 1;
}
```

```
VB
Private Sub singlesTimer_Tick(ByVal sender As System.Object, ByVal e As
        System.EventArgs) Handles singlesTimer.Tick
   Me.txtSingles.Text = Convert.ToString(m_SinglesCount)
   m_SinglesCount = m_SinglesCount + 1
End Sub
```

Setting the Tick Interval

The tick interval of a `Timer` controls how often the code in the Tick method is executed. The value of the tick interval is held in the `Timer.Interval` property, and it is measured in milliseconds. The property can be set in code simply by assigning a value to the `Interval` property, for example like so:

```
C#
// Code in the Tick method is executed every 1000 milliseconds
myTimer.Interval = 1000;
```

```
VB
' Code in the Tick method is executed every 1000 milliseconds
myTimer.Interval = 1000
```

You can also set the `Interval` property of the timer from within Visual Studio. To do so, select the icon representing the `Timer` in the designer and then press F4 to update its properties. Choose the desired value for the `Interval` property.

There is a very important potential pitfall when using `Timer` objects. If you use a `Timer` to handle a queue of data objects, it is possible for the queue to become populated faster than the `Timer` can process them. If this happens, your application's reaction times can appear slow, and you can even start to see exceptions if the queue overflows.

When the amount of time needed to execute the `Timer.Tick` method is longer than the interval, a variety of bad things can happen. If your code expects the timer to process a set of data at each interval but the timer doesn't get to it all in time, your application could appear to lose data objects. Alternately, your application could fail in a seemingly random manner, because the failure is timing sensitive. These kinds of bugs are among the worst to reproduce and diagnose, and it is a bad idea to work with a design that invites such troubles.

These two problems reinforce the notion that timers are not a good choice for executing complex code. They are wonderfully easy to set up and use for light "housekeeping"-type tasks, but if your `Tick` method code starts to get long, start thinking about using threads instead.

Suspending and Resuming Timers

The `Timer.Enabled` property controls whether the `Timer` wakes up and executes its `Tick` method. To suspend a `Timer`, set its Enabled Property to `false`. To resume a `Timer`, set the property to `true`. For example, these snippets accomplish this:

```
C#
// suspend the timer
myTimer.Enabled = false;

// Do some work with the timer suspended

// enable the timer again
myTimer.Enabled = true;

VB
' suspend the timer
myTimer.Enabled = False

' Do some work with the timer suspended

' enable the timer again
myTimer.Enabled = True
```

Manipulating Timers with a Sample Application

The TimerDemo sample application is located in the folder \SampleApplications\Chapter4. There are separate directories for the C# and Visual Basic versions.

This sample application demonstrates creating Timer objects, starting them, and toggling their Enabled properties so that individual Timer objects can be suspended and resumed. There are two Timer instances in the project, one to count off numbers individually and another to count off numbers by tens.

Updating Variables with the Interlocked Class

The System.Threading.Interlocked class provides means to increment, decrement, and set a variable in a thread-safe manner. When a variable is incremented or decremented with the Interlocked class, the process is atomic. The CPU cannot be taken away while the increment or decrement is in progress.

Why is this important? Consider an application with one thread that can update an integer and many threads that can read it. On the surface this seems like a safe arrangement. However, consider what would happen if a thread updated a variable in a process that took several CPU instructions, and the CPU was taken away before the update finished. At this instant the variable could have an unexpected, often bizarre value, which would be problematic if another thread read the variable at that time.

The Interlocked class protects against this scenario because it makes incrementing, decrementing, and updating a variable atomic. The Interlocked class supports integers and floating-point values.

Incrementing a Variable

Use System.Threading.Interlocked.Increment:

```
C#
System.Threading.Interlocked.Increment(ref my_integer);

VB
System.Threading.Interlocked.Increment(my_integer)
```

Decrementing a Variable

Use `System.Threading.Interlocked.Decrement`:

```
C#
System.Threading.Interlocked.Decrement(ref my_integer);

VB
System.Threading.Interlocked.Decrement(my_integer)
```

Exchanging a Value into a Variable

Use `System.Threading.Interlocked.Exchange`:

```
C#
System.Threading.Interlocked.Exchange(ref my_integer, new_value);

VB
System.Threading.Interlocked.Decrement(my_integer, new_value)
```

In Brief

- The `Thread` class is the means by which developers for the .NET Compact Framework implement multithreaded applications.
- Creating and starting a thread and putting a thread to sleep is done in much the same way on the .NET Compact Framework as it is on the desktop .NET Framework.
- There are special issues with suspending threads and quitting threaded applications on the .NET Compact Framework that require special attention from developers.
- Threads may not update user interface controls that they don't own, or they will likely lock up the application. The `ControlInvoker` class is an elegant work-around to this problem.
- The `Mutex` class provides a means for suspending and resuming threads, and it provides the infrastructure needed to simulate the `Thread.Join()` method, which is not available on the .NET Compact Framework.
- The `Monitor` class provides another model for controlling threads that is a good fit for controlling access to individual data objects by multiple threads.
- The `ThreadPool` class provides a convenient means for developers to use many short-lived threads efficiently, not wasting resources.

- Timers are supported by both C# and Visual Basic. They provide a handy way to regularly execute small amounts of code but are not suitable for regularly performing computationally intensive or heavily blocking algorithms.
- The `Interlocked` class is used to increment a variable, decrement a variable, or exchange values between variables as an atomic operation.

Network Connectivity with the .NET Compact Framework

5

IN THIS CHAPTER

Working with Sockets

Sockets are the de facto standard for communicating with other computers on both local area networks (LANs) and wide-area networks (WANs), such as the Internet. Two computers communicate to each other by using sockets that follow a general protocol where one computer is expecting to receive a connection, and the other makes the initial connection.

- The computer expecting to receive a connection, the *host* or *server*, listens for incoming connections on a specific *port*. The computer has a unique *IP address*, such as 172.68.112.34, and thousands of port numbers available, making it possible for many programs to listen for connections, each using a different port.
- The computer that makes the initial connection, the *client*, determines the IP address of the computer expecting the connection. The easiest way to find the IP address is to know it ahead of time. Alternately, if the computer making the initial connection knows the name of the computer, such as `www.mycomputer.org`, it can use *DNS lookup* to determine the IP address associated with the name.

- The client decides what port to connect to on the host. Usually, this is decided ahead of time. For example, Web servers always listen on port 80, so a computer that wants to connect to another computer's Web server process always knows that it needs to connect on port 80. A custom application usually uses a very large port number that is unlikely to be used by anyone else, such as 10998. The range of port numbers that custom applications can use varies by the operating system. Some operating systems reserve special port numbers, especially those below 1024. It is almost always safe to choose a port number between 2000 and 60000.
- The client connects to the IP address and port number. The host receives the connection. There is now a socket connection between the two computers.
- The client and host send data packets back and forth. Then one of the parties closes the connection.

The process of making socket connections is special because it is operating system agnostic—any two computers with socket support can communicate with each other. The pitfall is that the programming interfaces for socket programming are unique to each platform. The goal of this chapter is to learn how to manipulate socket connections in managed programs under the .NET Compact Framework.

Choosing a Protocol: TCP/IP versus UDP

In general, socket programming uses the *Internet Protocol* interface to send packets between two computers. There are two kinds of packets in common use for sending data through the Internet Protocol:

TCP packets This kind of packet is by far the most commonly used on the Internet—so commonly used, the protocol of using TCP packets on the Internet Protocol is called *TCP/IP networking*. The most important feature of a TCP packet is a guaranteed error-free delivery. If a computer sends a TCP packet over a socket connection, the data in the packet is guaranteed to arrive at the destination without errors. If the packet arrives but errors are detected, then the data is re-sent. If the packet doesn't arrive within a timeout period, then the function call used to send the packet reports an error. The way to check for errors varies by the platform, but we will examine this process in detail for the .NET Compact Framework.

UDP packets These packets differ from TCP packets because it is not guaranteed that a UDP packet will ever arrive at its destination or that its data will be error free. However, the lack of error checking means that using UDP packets incurs less overhead, so programs can transmit data more quickly. A good application for UDP packets is Internet telephony. If a packet gets lost or mangled, it means that there might be a crackle in the phone conversation. Conversations are in real time, so there is no use in correcting bad data, because it is too late by the time the errors are detected. You might as well just get the better throughput available with UDP packets.

Dealing with UDP packets, considered an advanced topic, is treated in the section "Using UDP Packets." Because it is the overwhelmingly more common scenario, the rest of the nuts-and-bolts discussion about socket programming in this chapter assumes the use of TCP packets.

Understanding IP Implementations: IPv4 versus IPv6

The process of a client's connecting to a host involves determining the IP address of the host and then making contact. The intricacies of dealing with addresses and forwarding Internet traffic to the correct address is one of the fundamental responsibilities of the Internet Protocol. This protocol has undergone a variety of revisions over the years. Version 4 of the Internet Protocol, IPv4, is the most commonly used on the Internet. An IPv4 address is composed of four 8-bit numbers. In human-readable form, the address looks like four numbers between 0 and 255 separated by dots, such as `172.68.112.34`.

In today's connected world, IPv4 does not provide enough addresses to give a unique address to each computer. There are tricks in widespread use to recycle IPv4 addresses so that they can be applied to more than one computer; the details of how this works is beyond the scope of this book, though.

The most recent version of the IP protocol is version 6, commonly referred to as IPv6. It is not yet in widespread use. IPv6 includes enhancements in security and addressing such that there will be enough unique IP addresses for the foreseeable future.

The .NET Compact Framework contains deep support for the older, far more common IPv4 protocol. It does not support the IPv6 protocol. Thus, this chapter relates all discussion to the IPv4 protocol.

Socket Programming with the .NET Compact Framework

The central class for socket programming with the .NET Compact Framework is the `System.Net.Sockets.Socket` class. The procedure for getting a `Socket` class connected to a remote computer depends on whether the socket connects as the host or as the client. Once the Socket is connected to a remote computer, however, the processes for reading and writing are exactly the same.

Making a Connection as a Client

To successfully make a connection as a client, we must first understand the `System.Net.EndPoint` class. The EndPoint holds the information about where to connect to the

IPv4 address of the host and the desired port. To set up a valid EndPoint and use it to connect a socket to a host, follow these steps:

1. Declare a variable for an EndPoint and a Socket.
2. Instantiate the EndPoint by passing the address and port number information into the constructor. There are two common ways to do this, depending on whether you know the address of the host, such as `172.68.25.34`, or only the DNS name of the host, such as `www.mycomputer.net`.

FINDING THE IP ADDRESS OF THE HOST

If you know the IP address of the host, use the `IPAddress` in the constructor. For example, this line of code instantiates a new EndPoint that describes the host at IP address `172.68.25.34`, port 9981:

```
C#
EndPoint l_EndPoint = new IPEndPoint( IPAddress.Parse(
        ) "172.68.25.34", Convert.ToInt16(9981));

VB
Dim l_EndPoint As System.Net.EndPoint

l_EndPoint = New
        System.Net.IPEndPoint(System.Net.IPAddress.
        Parse("172.68.25.34"), System.Int32.Parse(Me.txtPort.Text))
```

If you *don't* know the IP address of the host, then you must use DNS lookup to match the name of the host with its actual IP address. DNS lookup returns the IP address associated with a name. The code to do the lookup in a very simple case looks like this:

```
C#
IPHostEntry l_IPHostEntry = Dns.Resolve("www.mycomputer.net");
EndPoint l_EndPoint = new IPEndpoint(l_IPHostEntry.AddressList[0],
        9981);

VB
Dim l_IPHostEntry As System.Net.IPHostEntry
l_IPHostEntry = System.Net.Dns.Resolve("www.mycomputer.net")
l_EndPoint = New System.Net.IPEndPoint(l_IPHostEntry.
        AddressList(0), 9981)
```

3. Use the EndPoint to attempt to connect the socket to the host. Be sure to use a try/catch clause here because the attempt will throw an exception if there is a problem, such as the host's refusing to accept the connection.

The following code sample illustrates the three steps just described:

```
C#
try
{
    Socket l_Socket = new Socket(Socket(AddressFamily.InterNetwork,
            SocketType.Stream, ProtocolType.Tcp);
    l_Socket.Connect(l_EndPoint);
    if (l_Socket.Connected){
       // l_Socket is now ready to send and receive data
    }
}
catch (SocketException e)
{ /* do something about it */ }

VB
Dim l_Socket As System.Net.Sockets.Socket
Try
    l_Socket = New System.Net.Sockets.Socket(
            System.Net.Sockets.AddressFamily.InterNetwork,
            System.Net.Sockets.SocketType.Stream,
            System.Net.Sockets.ProtocolType.Tcp)
    l_Socket.Connect(l_EndPoint)

    If (l_Socket.Connected) Then
      ' l_Socket is now ready to send and receive data
    End If
Catch e as System.Net.Sockets.SocketException
    ' Do something about the problem
End Try
```

Receiving a Connection as a Host

You can acquire a socket connection to a remote computer by acting as the host. When a device acts as a host, it waits to receive a connection from another client. For your device to receive a connection as a host, you must set up a socket to listen on a specific port until someone sends a request to connect with your device. Follow these steps to listen on a socket for clients to connect to you:

1. Create a new socket that listens for new connections.
2. Bind the listening socket to a specific port so that it listens for connections on only that port.

3. Call Accept() on the listening socket to derive another socket when someone connects to you. Your code can read and write to the deriving socket, and the listening socket continues to wait for new connections.

The following code sample illustrates the three steps just described:

```
C#
m_listenSocket = new Socket(AddressFamily.InterNetwork,
        SocketType.Stream, ProtocolType.Tcp);

m_listenSocket.Bind(new IPEndPoint(IPAddress.Any, 8758));
m_listenSocket.Listen((int)SocketOptionName.MaxConnections);
m_connectedSocket = m_listenSocket.Accept();
if (m_connectedSocket != null)
{
    if (m_connectedSocket.Connected)
    {
       // Someone has connected to us.
    }
 }

VB
m_listenSocket = new Socket(AddressFamily.InterNetwork,
        SocketType.Stream, ProtocolType.Tcp)

m_listenSocket.Bind(new IPEndPoint(IPAddress.Any, 8758))
m_listenSocket.Listen((int)SocketOptionName.MaxConnections)
m_connectedSocket = m_listenSocket.Accept()
If (m_connectedSocket != null) Then
    If (m_connectedSocket.Connected) Then
       ' Someone has connected to us.
    End If
End If
```

Reading and Writing with a Connected Socket

Once a socket is connected to a remote computer, it can be used to send and receive data. The simplest way to do this is to call Socket.Send() to send data and Socket.Receive() to receive data.

Writing Data to a Socket with `Socket.Send`

`Socket.Send()` has four overloads, each offering a different level of control over what is sent:

`Send(Byte[] buffer)` This overload sends everything inside the buffer byte array.

`Send(Byte[] buffer, SocketFlags socketFlags)` This overload sends everything in buffer with special control over how the data is routed.

`Send(Byte[] buffer, Int32 size, SocketFlags socketFlags)` This overload sends everything in the buffer up to size bytes. If you want to send only part of a buffer, then you can specify `SocketFlags.None` to use the default routing behavior. For example, to send the first 16 bytes out of a byte array, you could use `l_Socket.Send(l_buffer, 16, SocketFlags.None)`.

`Send(Byte[] buffer, Int32 offset Int32 size, SocketFlags socketFlags)` This overload is like the previous one except you can specify which index within the byte array to start from. For example, to send bytes 3 through 7 of a byte array, you could use the following:

```
C#
l_Socket.Send(l_buffer, 2, 6, SocketFlags.None);

VB
l_Socket.Send(l_buffer, 3, 7, SocketFlags.None)
```

The `Send` method returns the number of bytes successfully sent. The problem with the `Send()` method is that it seems like a lot of work to convert everything you want to send into an array of bytes just to send it through a socket. Fortunately, the .NET Compact Framework supports two very useful classes, `System.Text.Encoding` and `System.Convert`, that make it easy to convert fundamental data types into byte arrays that can be sent through the socket. To send an entire object through a socket, you must serialize the object, as discussed in the section "Serializing Objects for Transmission through a Socket."

The easiest way to understand how to use the `Encoding` and `Convert` classes is to look at examples. The following examples assume that a socket named `l_Socket` exists and is connected:

C#

- Send a string by using ASCII byte encoding:

  ```
  l_Socket.Send(Encoding.ASCII.GetBytes("Send me")
  ```

- Send a string by using Unicode byte encoding:

  ```
  l_Socket.Send(Encoding.Unicode.GetBytes("Send me")
  ```

- Send the integer value 2003 as plain ASCII text:

  ```
  l_Socket.Send(Encoding.ASCII.GetBytes(Convert.ToString(2003))
  ```

- Send the floating-point value 2.71 as plain ASCII text:

```
l_Socket.Send(Encoding.ASCII.GetBytes(Convert.ToString(2.71))
```

VB

- Send a string by using ASCII byte encoding:

```
l_Socket.Send(System.Text.Encoding.ASCII.GetBytes("Send me")
```

- Send a string by using Unicode byte encoding:

```
l_Socket.Send(System.Text.Encoding.Unicode.GetBytes("Send me")
```

- Send the integer value 2003 as plain ASCII text:

```
l_Socket.Send(System.Text.Encoding.ASCII.GetBytes(
        Convert.ToString(2003))
```

- Send the floating-point value 2.71 as plain ASCII text:

```
l_Socket.Send(System.Text.Encoding.ASCII.GetBytes(
        Convert.ToString(2.71))
```

TRANSMITTING BINARY REPRESENTATION OF DATA

It is possible to write code that sends the byte-wise representation of numerical values instead of converting them to strings or converting them to network-ordered representations. This is a dangerous habit to get into, because different platforms have different internal representations for fundamental data types. If you are writing an application that transmits fundamental data types and the other party might not be a .NET application, then the bytes you send representing numerical values might be invalid to the remote party.

Reading Data from a Socket with `Socket.Receive`

Receive data from a socket through the `Socket.Receive` method. `Receive` has four overloads, similar to the overloads for `Socket.Send`. Each of the following overloads returns the number of bytes successfully read:

`Receive (Byte[] buffer)` This overload receives data into buffer.

`Receive (Byte[] buffer, SocketFlags socketFlags)` This overload receives data into buffer by using the flags to specify controlling how data is routed.

`Receive (Byte[] buffer, Int32 size, SocketFlags socketFlags)` This overload receives up to `size` bytes of data into the buffer. Even if more data is available, it is ignored. You can retrieve the remaining data by calling `Receive` again. If you want to receive only a specified number of bytes but you have no restrictions on the routing behavior, then you can specify `SocketFlags.None` to use the default routing behavior. For example, to receive only the first the 16 bytes of the available data, use `l_Socket.Receive(l_buffer, 16, SocketFlags.None)`

`Receive (Byte[] buffer, Int32 offset Int32 size, SocketFlags socketFlags)` This overload is like the previous one except you can specify which index within the byte array to use to start writing data into the array. For example, to receive up to 7 bytes of data into the buffer starting at position 3 in the buffer, use this code:

```
C#
l_Socket.Receive(l_buffer, 2, 6, SocketFlags.None);

VB
l_Socket.Receive(l_buffer, 3, 7, SocketFlags.None)
```

Just as there are techniques for converting the data to send out of a socket into byte arrays, there are simple techniques for converting byte arrays into fundamental data types. As before, the `Encoding` and `Convert` classes provide the means for conversion, and we will look at examples. These examples assume the data has been received into a Byte array named `l_Buffer`:

C#

- Convert received bytes into an ASCII string:

```
string l_ASCII = Encoding.ASCII.GetString(l_Buffer);
```

- Convert received bytes into a Unicode string:

```
string l_ASCII = Encoding.Unicode.GetString(l_Buffer);
```

- Convert the received bytes, which hold an encoded ASCII text integer, into a 32-bit-wide integer:

```
int l_Integer = Convert.ToInt32(Encoding.ASCII.GetString(l_Buffer));
```

- Convert the received bytes, which hold an encoded ASCII text integer, into a Double:

```
Double l_Double = Convert.ToInt32(Encoding.ASCII.GetString(l_Double));
```

VB

- Convert received bytes into an ASCII string:

```
Dim l_ASCII as string
l_ASCII = Encoding.ASCII.GetString(l_Buffer)
```

- Convert received bytes into a Unicode string:

```
Dim l_ASCII as string
string l_ASCII = Encoding.Unicode.GetString(l_Buffer)
```

- Convert the received bytes, which hold an encoded ASCII text integer, into a 32-bit-wide integer:

```
Dim l_Integer as Integer
l_Integer = Convert.ToInt32(Encoding.ASCII.GetString(l_Buffer))
```

- Convert the received bytes, which hold an encoded ASCII text integer, into a Double:

```
Dim l_Double as Double
l_Double = Convert.ToInt32(Encoding.ASCII.GetString(l_Double))
```

ILLEGAL CONVERSIONS

Although there are many conversions available, exceptions are thrown for conversions that are illegal. For example, trying to convert a large `long` to a `short` will cause an `OverFlowException`.

Table 5.1 lists all of the conversions supported by the `Convert` class on the .NET Compact Framework.

TABLE 5.1
The `Convert` Class on the .NET Compact Framework

METHOD	NAME ACCEPTS THESE INPUT TYPES
ToBoolean	object, bool, sbyte, char, byte, short, ushort, int, uint, long, String, float, double, decimal
ToChar	object, char, sbyte, byte, short, ushort, int, uint, long, ulong, String, float, double, decimal
ToSByte	object, bool, sbyte, char, byte, short, ushort, int, uint, long, ulong, float, double, decimal, String
ToByte	object, bool, byte, char, sbyte, short, ushort, int, uint, long, ulong, float, double, decimal, String
ToInt16	object, bool, char, sbyte, byte, ushort, int, uint, short, long, ulong, float, double, decimal, String
ToUInt16	object, bool, char, sbyte, byte, short, int, ushort, uint, long, ulong, float, double, decimal, String
ToInt32	object, bool, char, sbyte, byte, short, ushort, uint, int, long, ulong, float, double, decimal, String
ToUInt32	object, bool, char, sbyte, byte, short, ushort, int, uint, long, ulong, float, double, decimal, String
ToInt64	object, bool, char, sbyte, byte, short, ushort, int, uint, ulong, long, float, double, decimal, String
ToUInt64	object, bool, char, sbyte, byte, short, ushort, int, uint, long, UInt64, float, double, decimal, String
ToSingle	object, sbyte, byte, char, short, ushort, int, uint, long, ulong, float, double, decimal, String, bool

TABLE 5.1
Continued

METHOD	NAME ACCEPTS THESE INPUT TYPES
ToDouble	object, sbyte, byte, short, char, ushort, int, uint, long, ulong, float, double, decimal, String, bool
ToDecimal	object, sbyte, byte, char, short, ushort, int, uint, long, ulong, float, double, String, decimal, bool, DateTime
ToDateTime	object, String
ToString	Object, bool, char, sbyte, byte, short, ushort, int, uint, long, ulong, float, double, decimal, Decimal, DateTime
ToBase64String	byte[]
byte[] FromBase64String	String
ToBase64CharArray	byte[]
byte[] FromInt64CharArray	char[]

Sample Application: Remote Hello

This sample application pulls together the three most basic actions developers do with sockets: connect to a host, listen for a client, and exchange data. The application is split into two separate programs, a server and a client. The server always listens on port 8758 and uses DNS lookup to show what its IP address is. The client has a space for entering an IP address and a button to connect to the remote host.

When a connection is made, the server sends a string that says "Hello!" and the client displays it. Then the client closes the connection, and the host starts listening again.

The C# version of the Remote Hello client is in the directory `SampleApplications\Chapter5\RemoteHelloClient_CSharp`, and the VB version is in `SampleApplications\Chapter5\RemoteHelloClient_VB`. The server is in the folders `RemoteHelloServer_CSharp` and `RemoteHelloServer_VB`.

Observe that the code accesses the same classes from the .NET Compact Framework regardless of which language is used. Also, because sockets are platform agnostic, you can connect to the Visual Basic host with the C# client and vice versa.

Serializing Objects for Transmission through a Socket

The desktop version of the .NET Framework allows most types of objects to be serialized into an array of bytes, which can then be sent through a socket. For complex objects, developers implement the `ISerializable` interface, with code to serialize and deserialize their object data.

The .NET Compact Framework does not support this functionality. The only class that can serialize itself automatically is the `DataSet` class, which is discussed in detail in Chapter 6, "ADO.NET on the .NET Compact Framework." Normally, the `DataSet` class is used as an in-memory relational database. It is ideal for caching small amounts of data from a remote server while preserving the relational structure of the data. The `DataSet` can store every kind of fundamental data type available on the .NET Compact Framework, as outlined in Chapter 2, "Introducing the .NET Compact Framework." Additionally, the `DataSet` can serialize its contents to an XML string and repopulate itself from an XML string by using only a few lines of code. Chapter 8, "XML and the `DataSet`," is entirely devoted to explaining the `DataSet`'s support of XML.

To serialize and deserialize objects other than the `DataSet` with the .NET Compact Framework, developers must write their own serializing and deserializing code. We provide a simple class for serializing and deserializing primitive types in the next section. Also, Chapter 13, "Exploring .NET Reflection," provides a general-purpose class that can serialize and deserialize arbitrary object types.

Serializing Packets: Sample Application

We present here a sample application called ManagedChatSerialize. It is available in the folder `SampleApplications\Chapter5\ManagedChatSerialize_CSharp` and `SampleApplications\Chapter5\ManagedChatSerialize_VB`. The central class in ManagedChatSerialize is the `ChatData` packet. This packet is implemented as a separate class, and it has the following properties:

`Sender` A string holding the name of the person who sent the packet

`ChatText` The chat text message being sent

The program works by creating `ChatData` packets, populating them with data, and then extracting the state of the class by calling `ChatData.ToString()`. The program sends the string representation of the class through a socket. At the other side, the chat program populates a `ChatData` packet by calling `ChatData.FromString()`. Then the remote chat program accesses the chat data through the properties provided by the `ChatData` class.

The `ChatData` class uses the `SimpleSerializer` class to serialize its internal state to and from strings. The `SimpleSerializer` can serialize and deserialize objects to and from a `Stream`. Developers can use this code as a foundation for enabling their own objects to serialize and deserialize themselves on the .NET Compact Framework. A more advanced serializer that can serialize arbitrary data types by using reflection is presented in Chapter 13, "Exploring .NET Reflection."

You'll notice that there is only one instance of the `ChatData` packet ever created in the program. This is an optimization that is very important on the .NET Compact Framework. Creating and destroying objects is very expensive. Thus, we provide a method called `ChatData.Clear` to reset the packet to the same state as if it were just created, with no internal information. That makes it easy to retain the same instance of the `ChatData` class while creating the illusion of creating a new instance of the class whenever we need one.

Using UDP Packets

As mentioned earlier in the chapter, there are two kinds of packets commonly used to transmit information on IP networks. The most common type, the TCP packet, is the right choice for nearly all cases because it ensures that the data arrives uncorrupted or else signals an error if there is a problem that cannot be corrected.

The alternative, the UDP packet, is useful for real-time streaming applications. In such applications, if a packet arrives damaged, it does not matter whether the packet could be corrected or retransmitted, because there is no time to do so. The .NET Compact Framework supports the use of UDP packets.

There are other packet types in use today, but only the most advanced network experts are likely to care about them, and they are academic to everyone else. Thus, they are beyond the nuts-and-bolts programming approach of this book.

UDP packets differ from TCP packets in that they are *connectionless,* whereas the TCP protocol is a *connection-oriented protocol,* which means we need to connect a socket to a remote computer before we can start sending or receiving data using that socket. Connectionless protocols do not require any connection to be established before trying to send or receive data. If no one is listening to the IP address and port where a UDP packet is sent, then the packet is simply lost.

The easiest way to work with UDP packets is by using the `UdpClient` class, which is supported in the .NET Compact Framework. The `UdpClient` class allows programmers to send bytes to a remote party. Also, the `UdpClient` lets developers receive bytes either from a specific remote party or from any party who tried to send data to the port on which the `UdpClient` listens. The interesting constructors and methods for using the `UdpClient` are as follows:

> `void Connect(String hostname, Int32 port)` Sets up a connection to a computer whose IP address matches that specified by the hostname using the specified port number. Future calls using the `Send(Byte[] dgram, Int32 bytes)` method will send the data to the location specified by this connect call. This method returns `void` because there is no notion of a successful connection when using UDP packets. This method merely makes it easier for the programmer to set up the `UdpClient` for sending data to a specific IP address and port.

`void Connect(IPAddress addr, Int32 port)` Same as the previous method, except this one lets you specify the remote computer by passing an `IPAddress` and `port`, as with the following code:

C#

```
l_UdpClient.Connect(IPAddress.Parse("172.68.25.34"), 9981)
```

VB

```
l_UdpClient.Connect(IPAddress.Parse("172.68.25.34"), 9981)
```

`void Connect(IPEndpoint endPoint)` Same as the previous method, except this one lets you specify the remote computer by passing an `EndPoint`. You can set up the `EndPoint` in the same way `EndPoints` were set up to prepare to connect to a remote host using TCP packets (see the earlier section "Making a Connection as a Client").

`Int32 Send(Byte[] dgram, Int32 bytes, IPEndPoint endPoint)` Sends `bytes` total bytes of the `dgram` buffer to a computer whose IP address and port number are specified by `endPoint` parameter. You can set up the `endPoint` parameter in the same way `EndPoints` were set up to prepare to connect to a remote host by using TCP packets (see the earlier section "Making a Connection as a Client"). This method returns the number of bytes sent.

`Send(Byte[] dgram, Int32 bytes, String hostname, Int32 port)` Sends `bytes` total bytes of the `dgram` buffer to the computer whose IP address matches `hostname` and the specified `port`, as in the following, for example:

C#

```
Send(aBuffer, aBuffer.Length, "www.mycomputer.net", 9981)
```

VB

```
Send(aBuffer, aBuffer.Length, "www.mycomputer.net", 9981)
```

This method returns the number of bytes sent.

`Send(Byte[] dgram, Int32 bytes)` Sends `bytes` total bytes of the `dgram` buffer to the remote host that was specified in the `Connect` call. To use this overload, you must first call `Connect` so that the `UdpClient` knows where to send the UDP packet. This method returns the number of bytes sent.

`Receive(ref IPEndPoint remoteEP)` Waits to receive data from the `EndPoint` specified by EP. You can create an `EndPoint` that refers to a specific IP address and port number, as shown in the section "Making a Connection as a Client," or you can set up the `EndPoint` to receive the data from any IP address and port. The `EndPoint` is updated after data is received to indicate where the data came from, as shown in the following subsection, "Writing Code for `UdpClient`." This method returns the number of bytes received.

Writing Code for UdpClient

This sample code comes from the sample application from the next section, "Writing for the UdpClient: Sample Application." The code demonstrates how to set up a UdpClient that attempts to send UDP packets to a computer at IP address 192.168.0.200, port 8758. Note that although there is a call to UdpClient.Connect(), this is merely a convenience. The UdpClient now knows where to send the UDP packets when UdpClient.Send() is called, but there is no persistent connection established, as there is with TCP packets.

```
C#
IPEndPoint senderIP = new
        IPEndPoint(IPAddress.Parse("192.168.0.200"),
        Convert.ToInt32(8758));

UdpClient l_UdpClient = new UdpClient();
l_UdpClient.Connect(senderIP);

for (int i = 0; i < 20; i++)
{
    l_UdpClient.Send(Encoding.ASCII.GetBytes("Hello_UDP_1"),
            Encoding.ASCII.GetBytes("Hello_UDP_1").Length);

    System.Threading.Thread.Sleep(1000);
}

l_UdpClient.Close();

VB
Dim senderIP As System.Net.IPEndPoint

senderIP = New  System.Net.IPEndPoint(System.Net.IPAddress.
        Parse("192.168.0.200"), 8758))

Dim l_UdpClient As System.Net.Sockets.UdpClient
l_UdpClient = New System.Net.Sockets.UdpClient
l_UdpClient.Connect(senderIP)

Dim i As Integer
For i = 1 To 20
    l_UdpClient.Send(System.Text.Encoding.ASCII.GetBytes
        ("Hello_UDP_1"),
        System.Text.Encoding.ASCII.GetBytes("Hello_UDP_1").Length)
```

```
    System.Threading.Thread.Sleep(1000)
Next i

l_UdpClient.Close()
```

The following snippet of code creates a loop. Each iteration of the loop blocks and listens on port 8758 until it receives a UDP packet from any IP address. Note how this differs from standard TCP connections, where a persistent connection from a specific IP address must be established before data can be received.

```
C#
IPEndPoint listenerIP = new IPEndPoint(IPAddress.Any, 8758);
UdpClient listener = new UdpClient(listenerIP);

for (int i = 0; i < Convert.ToInt16(this.txtMaxPackets.Text); i++)
{
    // Now receive the three datagrams from the listener
    IPEndPoint receivedIPInfo = new IPEndPoint(IPAddress.Any, 0);

    byte[] data = listener.Receive(ref receivedIPInfo);

    this.textBox1.Text += ("GOT: " +
            Encoding.ASCII.GetString(data, 0,
            data.Length) + "  FROM: " + receivedIPInfo.ToString());
}

VB
Dim listenerIP As System.Net.IPEndPoint
listenerIP = New System.Net.IPEndPoint(System.Net.IPAddress.Any,
        8758)

Dim listener As System.Net.Sockets.UdpClient
listener = New System.Net.Sockets.UdpClient(listenerIP)

Dim i As Integer

For i = 1 To Convert.ToInt16(Me.txtMaxPackets.Text)

   ' Now receive the three datagrams from the listener
   Dim receivedIPInfo As System.Net.IPEndPoint
   receivedIPInfo = New
           System.Net.IPEndPoint(System.Net.IPAddress.Any, 0)
```

```
    Dim data As Byte()
    data = listener.Receive(receivedIPInfo)
    Me.textBox1.Text += ("GOT: " +
            System.Text.Encoding.ASCII.GetString(data, 0, data.Length)
            + " FROM: " + receivedIPInfo.ToString())
Next i
```

The same tricks for converting fundamental data types to and from byte arrays that were discussed in the section "Reading and Writing with a Connected Socket" are also useful when dealing with the `UdpClient`.

Writing for the `UdpClient`: Sample Application

The UDPHello sample application is composed of two separate programs. The first, `UDPHelloServer`, broadcasts simple UDP packets in a loop to a specified IP address and port. Because of the connectionless nature of UDP packets, there is no exception if no one is listening at the other end or if the packets are refused.

The second program, `UDPHelloClient`, listens on port 8758 for UDP packets broadcast from any source. After each UDP packet is received, the IPEndPoint named `receivedIPInfo` is populated with the IP address and port number of the source of the UDP packet. The contents of each packet received and the information in `receivedIPInfo` are painted into a TextBox.

The source code for UDPHelloClient is in `SampleApplications\Chapter5\UDPHelloClient_CSharp` and `SampleApplications\Chapter5\UDPHelloClient_VB`. The source code for UDPHelloServer is in `SampleApplications\Chapter5\UDPHellpServer_CSharp` and `SampleApplications\Chapter5\UDPHelloServer_VB`.

Multicasting with UDP Packets

The `UDPClient` can be easily configured to broadcast to a multicast IP address or to receive packets from a multicast IP address. A multicast IP address is operated by a server that maintains a list of multicast subscribers. When a packet is sent to a multicast IP address, the server sends a copy of the packet to the IP address of every client who has subscribed.

Sending Multicast Packets

To send UDP packets to a multicast IP address, no special action is required. Simply send your packets as shown in the sample code from the earlier section "Writing Code for `UdpClient`," but choose a multicast IP address as the destination for the packets.

Receiving Multicast Packets

To receive multicast packets, you must first subscribe with the server that is operating the multicast IP address. Once you have subscribed to the multicast IP address, you can listen for packets from the multicast IP address in the same way as for any other address, as shown in the section "Writing Code for `UdpClient`." When someone sends a packet to the multicast IP address, the server forwards it to everyone on the subscriber list, including you. To subscribe to a multicast IP address, follow these steps:

1. Create an `IPAddress` instance that points to the multicast IP address.
2. Call `UdpClient.JoinMultiCastGroup()`, passing the IPAddress as an argument.

Future attempts to receive from the multicast IP address will receive forwarded packets from the multicast server. Here are the `JoinMultiCastGroup` overloads supported on the .NET Compact Framework:

`JoinMultiCastGroup(IPAddress multicastAddr)` Joins a multicast group at `multicastAddr`

`JoinMultiCastGroup(IPAddress multicastAddr, int maxHops)` Joins a multicast group at `multicastAddr` but receives only packets that have made up to `maxHops` total travel hops

Here is an example of the code:

```
C#
IPAddress l_multicastAddress = new IPAddress("172.68.0.22");
// Only receive multicast packets that have traveled
// for 40 or less hops
l_UDPClient.JoinMulticastGroup(l_multicastAddress, 40);

VB
IPAddress l_multicastAddress = new IPAddress("172.68.0.22")
' Only receive multicast packets that have traveled
' for 40 or less hops
l_UDPClient.JoinMulticastGroup(l_multicastAddress, 40);
```

To unsubscribe from a multicast IP address, call `UDPClient.DropMulticastGroup()` as follows:

```
C#
l_UDPClient.DropMulticastGroup(l_multicastAddress);

VB
l_UDPClient.DropMulticastGroup(l_multicastAddress)
```

Communicating with Remote Servers through the HTTP Protocol

Up until now we have focused on how to work with socket programming that passes arbitrary data between a client and a host by using either TCP or UDP packets. In either case we have invented our own protocol for communication. For example, the managed chat application used a protocol where ChatPacket objects were serialized into bytes and sent through the network connection. Our simple Remote Hello and UDPHello examples simply sent strings back and forth. In each of these cases, the client and the host both knew what to expect to come down the wire at any given time. As developers, we are free to design our protocols and set those expectations as we see fit.

While this approach provides a lot of flexibility, there are many servers on the Internet that follow a much more specific communication protocol, HTTP, which is the protocol used on the World Wide Web. When using the HTTP protocol, there are very specific rules about how a client contacts a server and what the client can ask for at any specific time. The data that an HTTP server returns will arrive as a set of TCP packets, but wading through all of the protocol-related information is a very tedious task. A transaction with an HTTP server is structured as follows:

1. The client connects to the HTTP server.
2. The HTTP server responds.
3. The client requests data by using GET or asks to place data by using a POST command.
4. The server responds to the request and hands down an error code if the client request cannot be fulfilled. For example, the famous error code 404 is returned if the client tries to GET a file that does not exist.
5. Step 4 is repeated an arbitrary number of times.
6. The client closes the connection.

Each time the client makes a request or the server responds, a brand-new socket connection is made with the server, and the data that comes down the wire must be parsed and processed by the client. Setting up all of the socket code by hand would be quite tedious. The `HttpWebRequest` class streamlines all of the processes involved with interacting with an HTTP server. Specifically, `HttpWebRequest` can perform these tasks for you:

- Initialize a connection with an HTTP server
- Receive the response from an HTTP server
- Return a stream that holds the data that the HTTP server sent back as a result of your request

To summarize, the value of the HttpWebRequest class is that it frees the developer from having to do any TCP-style socket programming when interacting with a Web server. Developers can instantiate an instance of an HttpWebRequest and get the response from the server with only a few lines of code.

Nuts-and-Bolts Usage of HttpWebRequest

To use an HttpWebRequest class to download information from an HTTP server, follow these steps:

1. Create an instance of the Uri class to hold the location of the HTTP server.
2. Instantiate an HttpWebRequest by using the Uri of step 1.
3. Ask the HttpWebRequest to return the response from the Web server in the form of a Stream class.
4. Consume the contents of the Stream.

Coding Sample for HttpWebRequest

The HttpWebRequest class reduces the complex task of contacting an HTTP server into just four steps, as performed by this code snippet:

```
C#
Uri l_Uri = new Uri("http://www.myserver.com");

HttpWebRequest l_WebReq = (HttpWebRequest)WebRequest.Create(l_Uri);

HttpWebResponse l_WebResponse =
        (HttpWebResponse)l_WebReq.GetResponse();

Stream l_responseStream = l_WebResponse.GetResponseStream();

StreamReader l_SReader = new StreamReader(l_responseStream);

// Do something with l_SReader. For example, if you downloaded a
// Web page, you could
// extract the HTML code that came in the response and paint it on
// the screen.

VB
Dim l_Uri As Uri
l_Uri = New Uri(Me.txtRemoteIP.Text)
```

```
Dim l_WebReq As System.Net.HttpWebRequest
l_WebReq = System.Net.WebRequest.Create(l_Uri)

Dim l_WebResponse As System.Net.WebResponse
l_WebResponse = l_WebReq.GetResponse()

Dim l_SReader As System.IO.StreamReader
l_SReader = New
        System.IO.StreamReader(l_WebResponse.GetResponseStream())

' Do something with l_SReader. For example, if you downloaded a Web
' page, you could
' extract the HTML code that came in the response and paint it on
' the screen.
```

Writing Applications with the `HttpWebRequest`: WebHello Sample

Our sample application uses an `HttpWebRequest` to contact the server and an `HttpWebResponse` to create a stream holding the response. The application acquires a stream from the WebResponse and paints the text into a TextBox. The source code for Web Hello is in `SampleApplications\Chapter5\WebHello_CSharp` and `SampleApplications\Chapter5\WebHello_VB`.

Communicating with Remote Servers through the HTTPS Protocol

The HTTPS protocol allows secure transactions to occur at Web sites. The .NET Compact Framework implementation of `HttpWebRequest` is capable of accessing servers by using the HTTPS protocol with no special effort from the developer (see the earlier section "Communicating with Remote Servers through the HTTP Protocol").

Communicating through the Device IrDA Port

Many Pocket PC and other Windows CE devices include a built-in infrared data transmission port. The .NET Compact Framework includes classes to program against the IrDA port.

IrDA communication occurs between two computers, a client and a server. The server computer offers a connection to any client that comes within range of its infrared port. The connection that a server offers is identified by a device name and device ID.

Client computers may be in range of many computers offering IrDA connections. It is assumed that each computer in range that offers a connection has a unique device ID and device name. The client enumerates through the available connections, chooses one, and communicates with the desired remote computer. Thus, infrared communication occurs with the passing of these events:

1. A device offers one or more services to all other devices within range of its IrDA port. The device is identified by a device name and device ID. The offered services are identified by name.
2. A client device that wishes to open a communication enumerates through all possible devices in range of the client device.
3. The client chooses one of the available devices and connects to one of the services offered by the chosen device.

Using the `IrDAClient` to Access the Device IrDA Port

The central object for IrDA connectivity on the .NET Compact Framework is the `IrDAClient`. With the help of several support classes discussed in the sidebar, the `IrDAClient` can act as a server or a client. That is, the `IrDAClient` can be used to look for available connections or to offer connections to other devices.

The `IrDAClient` and related IrDA classes reside in the library named `System.Net.IrDA.dll`. You must add a reference to this library in your project before you can use these classes. To add the library, use the pull-down menu to view the Solution Explorer: View, Solution Explorer. Find the icon labeled References in the Solution Explorer and right-click it. From the context menu, select Add Reference. In the new dialog that opens, double-click the entry labeled `System.Net.IrDA` and then click OK.

Once a connection is made with a remote party, the `IrDAClient` offers the `GetStream()` method, which exposes a `Stream` instance with which programs can read and write data.

Connecting to an IrDA Port as a Client

When connecting as an IrDA client, it is presumed that as the developer, you know the name of the device with which you want to connect. The program must iterate through all of the available devices and choose one with the desired service. Specifically, follow these steps:

1. Create a new instance of an `IrDAClient`.
2. Retrieve the list of available devices offering a connection by calling `IrDAClient.DiscoverDevices`. Pass in the maximum number of devices to look for. The `DiscoverDevices` method returns an array of `IrDADeviceInfo` objects.

3. Scan each `IrDADeviceInfo` in the array to find out whether one of the available devices is the one the application should connect with.
4. If a desired device is found, then connect to it by calling `IrDAClient.Connect()`. Pass in the name of the service to connect to.
5. Use the `IrDAClient` to communicate with the remote party.

Connecting as a Client: Sample Code

The following sample code is derived from the IrDAChat sample application. The code enumerates all of the devices available and tries to connect to one that offers a service named IRDA_CHAT_SERVER. It is this connection that has a chat server at the remote end waiting for someone to connect and start chatting. The following code shows the user each connection it finds using a MessageBox:

```
C#
m_IrDAClient = new IrDAClient();

bool l_foundAnyDevice = false;
int MAX_DEVICES = 5;

// Find out who's out there to connect with...
IrDADeviceInfo[] l_DevsAvailable =
        m_IrDAClient.DiscoverDevices(MAX_DEVICES);

// Show a MessageBox telling user every device we see out there
foreach (IrDADeviceInfo l_devInfo in l_DevsAvailable)
{
    l_foundAnyDevice = true;
    MessageBox.Show(l_devInfo.DeviceName, "Discovered IrDA device");

    // Now try to connect to the devices, hoping it offers a service
    // named "IRDA_CHAT_SERVER"

   try
   {
       // Assume that first device is offering a service that we
       // want
       IrDAEndPoint chatEndPoint = new IrDAEndPoint(
               l_DevsAvailable[0].DeviceID, "IRDA_CHAT_SERVER");
       m_IrDAClient.Connect(chatEndPoint);
```

```
            MessageBox.Show("Connected to chat server!", "Ready to chat");
            m_Connected = true;
            break;
        }
        catch (SocketException exc)   {    }
    }

    // m_IrdaClient can now be read from or written to.

    VB
    m_IrDAClient = New System.Net.Sockets.IrDAClient

    Dim l_foundAnyDevice As Boolean
    Dim MAX_DEVICES as Integer

    l_foundAnyDevice = False

    ' Find out who's out there to connect with...
    Dim l_DevsAvailable() As System.Net.Sockets.IrDADeviceInfo
    l_DevsAvailable = m_IrDAClient.DiscoverDevices(MAX_DEVICES)

    ' Show a MessageBox telling user every device we see out there
    Dim i As Integer
    i = 0
    While ((i < l_DevsAvailable.Length) And (m_Connected = False))
        Dim l_devInfo As System.Net.Sockets.IrDADeviceInfo
        l_devInfo = l_DevsAvailable(i)

        l_foundAnyDevice = True
        MessageBox.Show(l_devInfo.DeviceName, "Discovered IrDA device")

        ' Now try to connect to the devices, hoping it offers a service
        ' named "IRDA_CHAT_SERVER"
        Try
            ' Assume that first device is offering a service that we want
            Dim chatEndPoint As System.Net.IrDAEndPoint
            chatEndPoint = New
                    System.Net.IrDAEndPoint(l_DevsAvailable(0).DeviceID,
                    "IRDA_CHAT_SERVER")
             m_IrDAClient.Connect(chatEndPoint)
```

```
        MessageBox.Show("Connected to chat server!", "Ready to chat")
        m_Connected = True

    Catch exc As System.Net.Sockets.SocketException
End Try

' m_IrDAClient can now be read from and written to
```

Establishing an IrDA Connection as a Server

To establish an IrDA connection as a server device, follow these steps:

1. Create an instance of an `IrDAListener`, passing the name of the device into the constructor.
2. Call `Start()` on the `IrDAListener`.
3. Call `IrDAListener.AcceptIrDAClient()` to receive an instance of an `IrDAClient` when someone connects.
4. Use the `IrDAClient` to communicate with the remote party.

Establishing an IrDA Connection as a Server: Sample Code

This sample code is derived from the IrDAChat sample application. It demonstrates how to use an `IrDAListener` to offer a connection called IRDA_CHAT_SERVER to other devices and then wait for someone to connect.

```
C#
IrDAListener l_IrDAListener = new IrDAListener("IRDA_CHAT_SERVER");

// Listen for anyone who wants to connect
l_IrDAListener.Start();

// And now pull the first queued connection request out as an
// IrDAClient
m_IrDAClient = l_IrDAListener.AcceptIrDAClient();

MessageBox.Show("Accepted a connection", "Ready to chat");

VB
Dim l_IrDAListener As System.Net.Sockets.IrDAListener
l_IrDAListener = New
        System.Net.Sockets.IrDAListener("IRDA_CHAT_SERVER")
```

```
' Listen for anyone who wants to connect
l_IrDAListener.Start()

' And now pull the first queued connection request out as
' an IrDAClient
m_IrDAClient = l_IrDAListener.AcceptIrDAClient()

MessageBox.Show("Accepted a connection", "Ready to chat")
```

Reading Data from an `IrDAClient`

Once an `IrDAClient` is connected to a remote party, reading data is achieved the same way whether connected as a server or as a client, as follows:

1. Create a `StreamReader` by passing in the `Stream` associated with the `IrDAClient` into the `StreamReader` constructor.
2. Read data from the `StreamReader`.

Reading Data from an `IrDAClient`: Sample Code

The following example code is derived from the IrDAChat sample application. This code expects a single line at a time to come from the remote party.

```
C#
l_StreamReader = new StreamReader(this.m_IrDAClient.GetStream(),
        System.Text.Encoding.ASCII);
// Read a line of text and paint it into a GUI
this.lbInText.Items.Add(l_StreamReader.ReadLine());
l_StreamReader.Close();

VB
l_StreamReader = New
        System.IO.StreamReader(Me.m_IrDAClient.GetStream(),
        System.Text.Encoding.ASCII)
Me.lbInText.Items.Add(l_StreamReader.ReadLine())

l_StreamReader.Close()
```

Writing Data to an IrDAClient

Once an IrDAClient is connected to a remote party, writing data is achieved the same way whether connected as a server or as a client, as follows:

1. Create a StreamWriter by passing in the Stream associated with the IrDAClient into the StreamWriter constructor.
2. Write data to the StreamWriter.

Writing Data to an IrDAClient: Sample Code

The following example code is derived from the IrDAChat sample application. This code writes a single line of text, which it acquires from the user interface, to the stream acquired from the IrDAClient.

```
C#
// Grab a reference to the stream in the m_IrDAClient and send the
// text to it.
StreamWriter l_StreamWriter = new StreamWriter(this.m_IrDAClient.GetStream(),
        System.Text.Encoding.ASCII);
l_StreamWriter.WriteLine(this.txtSendText.Text);
l_StreamWriter.Close();
```

```
VB
' Grab a reference to the stream in the m_IrDAClient
' and send the text to it.
Dim l_StreamWriter As System.IO.StreamWriter
l_StreamWriter = New System.IO.StreamWriter(Me.m_IrDAClient.GetStream(),
        System.Text.Encoding.ASCII)
l_StreamWriter.WriteLine(Me.txtSendText.Text)
l_StreamWriter.Close()
```

Sample Application: IrDAChat

The IrDAChat sample application is available in the directories SampleApplications\Chapter5\IrDAChat_CSharp and SampleApplications\Chapter5\IrDAChat_VB. This sample application pulls together all of the ideas presented in this "Communicating through the Device IrDA Port" section by implementing a simple chat program. The application uses an IrDAClient object to connect with a remote party as either the client or the server. It transmits and receives data by accessing the stream associated with the IrDAClient by calling IrDAClient.GetStream().

In Brief

- Socket programming is the de fato standard by which computers running different operating systems communicate with one another.
- The two most important protocols for socket programming are TCP/IP and UDP. TCP/IP guarantees packet integrity but is slower than UDP. UDP packets are useful for such applications as streaming audio, where a bad packet doesn't harm the application much.
- Most of the chapter focuses on TCP/IP because it is the protocol used in the vast majority of applications.
- There are two popularly supported versions of the IP protocol on which TCP/IP and UDP rest: IPv4 and IPv6. Presently, the .NET Compact Framework supports only IPv4.
- The `Socket` class is the central socket connectivity object on the .NET Compact Framework. You must set up a connection with a `Socket` either as a host or a client before you can transmit data through a `Socket`.
- Once a `Socket` is connected, programs can read and write bytes using the `Socket`. With custom code you can serialize your .NET Compact Framework objects into byte arrays and transmit them through a `Socket`.
- You can use the `UdpClient` object to communicate using the UDP protocol. The `UdpClient` is also capable of multicasting, the act of sending packets to more than one destination.
- The `HttpWebRequest` encapsulates all of the logic needed to connect to a remote HTTP server, make a request, and receive a reply. It makes contacting HTTP servers with the .NET Compact Framework very easy.
- The `HttpWebRequest` is also capable of interacting with secure HTTPS servers without any special effort from the developer.
- The `IrDAClient` class encapsulates the logic needed to send and receive data through the device IrDA port.
- The `IrDAClient` can connect to a remote device's IrDA port either as a host or as a client. Establishing a connection requires special steps depending on whether the device acts as the host or client.
- Once an `IrDAClient` is connected, developers can access the underlying `Stream` object and use it to read and write data.

ADO.NET on the .NET Compact Framework

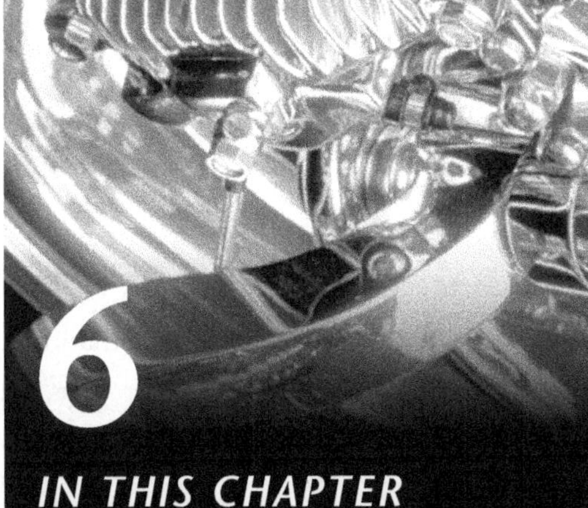

6

IN THIS CHAPTER

Introducing ADO.NET on the .NET Compact Framework

ADO.NET is the name given to the set of classes used for database access in the .NET programming world. The set of classes available in the ADO.NET world is rich and can seem overwhelming. However, simple data manipulation and even connections to remote database servers can be performed with just a few lines of code. This chapter is devoted to showing how to manipulate data locally on the device. It lays the foundation for later ADO.NET-related chapters, which cover how to work with data stored on servers.

Caching Data with the `DataSet`

The `DataSet` is the fundamental framework class for manipulating data with the .NET Compact Framework. The `DataSet` can be thought of as a tiny relational database engine in itself. It holds its tables in memory organized as tables, rows, and columns and allows developers to perform standard database operations, such as adding and removing data, sorting, and checking constraints.

Developers who fully understand how to work effectively with the `DataSet` on the .NET Compact Framework will write effective ADO.NET applications on the framework.

Working with the .NET Compact Framework is more difficult than working with the desktop .NET Framework because many overloads are missing on the .NET Compact Framework and it is easy to do things that bring performance to a crawl. There are more details on performance strategies in Chapter 15, "Measuring the Performance of a .NET Compact Framework Application."

The overall strategy for handling data in ADO.NET programming is to fill up a `DataSet` from a large database, work with the data held in the `DataSet`, and then write back any changes to the database. This chapter discusses how to fill the `DataSet` by inserting the data programmatically and to perform simple manipulations on the data.

Looking Inside the `DataSet`: `DataTables`, `DataRows`, and `DataColumns`

The `DataSet` is a container for one or more `DataTables`. Each `DataTable` corresponds to a table in a relational database. It has a set of `DataRows`, and each `DataRow` has a set of `DataColumns` that actually hold the data. To make `DataSets`, `DataTables`, and `DataColumns` easier to work with, they can have names. Figure 6.1 diagrams the overall structure of how a `DataSet` stores the data for a simple phonebook.

You can use a `DataTable` alone to store the data associated with a single table, but the `DataSet` provides methods and properties that add utility and truly make it a miniature in-memory relational database. For example, with the `DataSet` you can do all of the following:

- Manipulate the information inside a `DataSet` as a small relational database. For example, you can set up parent-child relationships, cascading updates and deletes, and create data columns that are computed from other data fields.
- Save or load the contents of all of the `DataTables` to an XML file with just one line of code (see Chapter 8, "XML and the `DataSet`").
- Pass it to the SQL CE engine, which will fill it with tables from a relational database stored on the device or with replicated data from a remote server (see Chapter 7, "Programming with Microsoft SQL Server CE").
- Pass it to the SQL provider to be filled with tables from a remote server (see Chapter 7).
- Receive populated `DataSets`, which are the return values of Web services, or pass `DataSets` back to a Web service (see Chapter 9, "Using XML Web Services").

This chapter focuses mainly on understanding how the `DataSet` fits into the big picture and learning the basics of how to manipulate the data it holds.

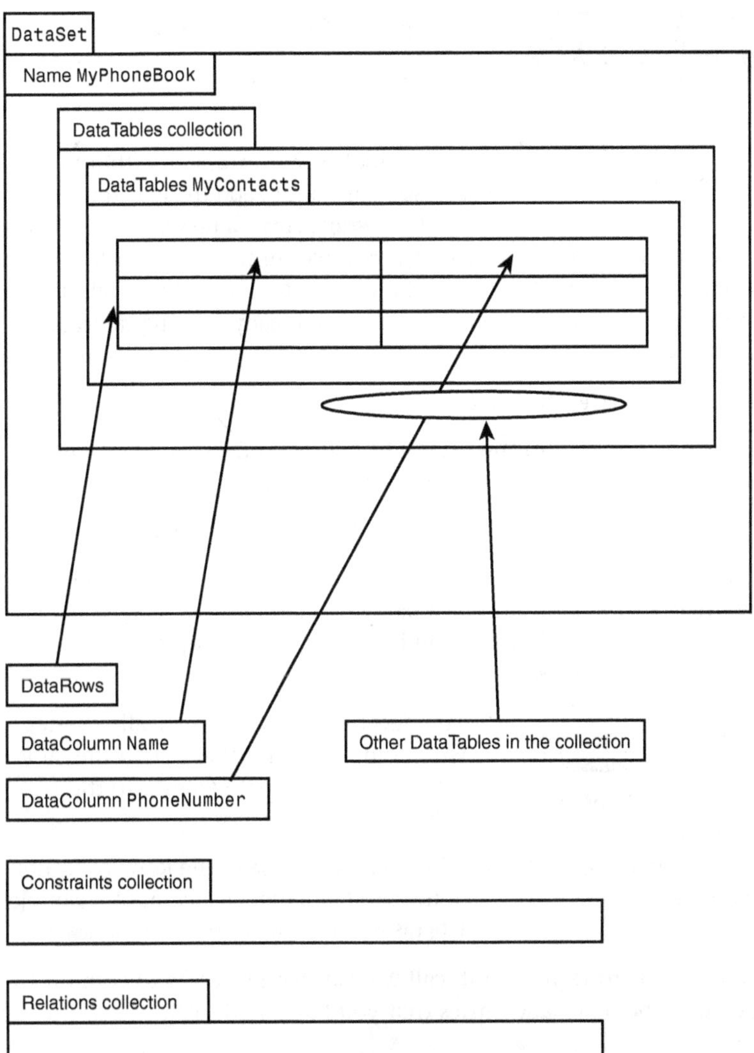

FIGURE 6.1 This DataSet represents the structure of a phone book.

Inserting Data into a DataSet

To insert new data into a DataSet, we follow these steps:

1. Get a handle to the DataTable within the DataSet you want to add the new row of data in. If necessary, create a new DataTable. The collection of DataTables that the DataSet manages is accessible through the DataSet.Tables property. If you must create a new DataTable and add it to the DataSet.Tables collection, follow these substeps (usually, you will be adding data to an already existing table, so you can skip these substeps):

   Create a DataTable via the DataTable constructor.

   Create new DataColumns and add them to the DataTable.Columns collection. For each new DataColumn, you must specify the name of the column and its data type.

   Add the new DataTable to the DataSet.Tables collection by calling its .Add method.

2. Create a new DataRow by asking the DataTable to create it for you. For example, to create a new row for the first DataTable in a DataSet, use `l_newRow = l_DataSet.Tables[0].NewRow`. You can also specify the table by name, for example `l_newRow = l_DataSet.Tables["Customers"].NewRow`.

3. The new DataRow is automatically created with column entries that match the table that created it. Insert values into the columns of the DataRow.

4. Add the new DataRow to the Rows collection of the DataTable that created it, for example `l_DataSet.Tables[0].Rows.Add(l_newRow);`.

5. After you have added all of the rows you want, call DataSet.AcceptChanges to commit to the changes. To undo all of the new data entries that you have made, call DataSet.RejectChanges.

> **SOME DEVELOPERS CAN SKIP THIS SECTION**
>
> If you are an experienced ADO.NET developer, you can skip the following subsections, which cover inserting data into a DataSet and extracting it.

> **SPECIFYING TABLES BY NAME**
>
> Specifying the table by name is a more expensive operation and can hurt performance in large applications running on the .NET Compact Framework. Use numeric indexing whenever possible.

Building a DataSet to Hold a Phone Book

To demonstrate how to create a DataSet capable of holding relational data, we show some sample code taken directly out of the PhoneBook sample application. This sample application, which is explained in greater detail in the next section, sets up a DataSet capable of holding a simple phone book. The DataSet holds a single DataTable, which is set up with two DataColumns, one to hold a name and the other for the phone number. Listing 6.1 demonstrates each of the five steps needed to add data to a DataSet, including creating a brand-new table.

LISTING 6.1 Creating and populating a DataSet

```
C#
DataSet     l_DataSet = new DataSet();

// Create a DataTable that holds a "Name" and a "PhoneNumber"
DataTable  l_newTable = new DataTable("Phone Contacts");
        l_newTable.Columns.Add(new DataColumn("Name",
        typeof(System.String)));

l_newTable.Columns.Add(new DataColumn("PhoneNumber",
        typeof(System.String)));

// Add the DataTable to the DataSet's table collection
l_DataSet.Tables.Add(l_newTable);

// Now put a few names in...
// GEORGE WASHINGTON
DataRow l_newRow = l_DataSet.Tables[0].NewRow();
l_newRow[0] = "George Washington";
l_newRow[1] = "555 340-1776";
l_DataSet.Tables[0].Rows.Add(l_newRow);

// BEN FRANKLIN
l_newRow = l_DataSet.Tables[0].NewRow();
l_newRow["Name"] = "Ben Franklin";
l_newRow["PhoneNumber"] = "555 336-3211";

l_DataSet.Tables[0].Rows.Add(l_newRow);

// Commit the changes
l_DataSet.AcceptChanges();

VB
' Create a DataTable that holds a "Name" and a "PhoneNumber"
Dim l_newTable As New DataTable("Phone Contacts")
l_newTable.Columns.Add(New DataColumn("Name",
        System.Type.GetType("System.String")))
        l_newTable.Columns.Add(New DataColumn("PhoneNumber",
        System.Type.GetType("System.String")))

' Add the DataTable to the DataSet's table collection
l_DataSet.Tables.Add(l_newTable)
```

```
' Now put a few names in...
' GEORGE WASHINGTON
Dim l_newRow As DataRow
l_newRow = l_DataSet.Tables(0).NewRow()
l_newRow(0) = "George Washington"
l_newRow(1) = "555 340-1776"
l_DataSet.Tables(0).Rows.Add(l_newRow)

' BEN FRANKLIN
l_newRow = l_DataSet.Tables(0).NewRow()
l_newRow("Name") = "Ben Franklin"
l_newRow("PhoneNumber") = "555 336-3211"
l_DataSet.Tables(0).Rows.Add(l_newRow)

' ALEXANDER HAMILTON
l_newRow = l_DataSet.Tables(0).NewRow()
l_newRow(0) = "Alexander Hamilton"
l_newRow(1) = "555 756-3211"
l_DataSet.Tables(0).Rows.Add(l_newRow)

' Commit the changes
l_DataSet.AcceptChanges()
```

Extracting Data from a DataSet

Extracting data from a DataSet is as simple as accessing the desired DataTable in the DataSet.Tables collection and looking at the desired rows in the table. Each row has an indexer to make it easy to access the desired columns. Remember that the indexers are zero-based, as in these examples:

l_DataSet.Tables[0].Rows[0][0] Access the first column in the first row of the first DataTable.

l_DataSet.Tables[0].Rows[0][9] Access the tenth column in the first row of the first DataTable.

l_DataSet.Tables[0].Rows[29][9] Access the tenth column in the thirtieth row of the first DataTable.

Extracting PhoneBook Data from a DataSet

The following code sample is also taken directly out of the PhoneBook sample application. It loops through all of the rows in the first DataTable in a DataSet and paints the 0th and 1st DataColumn values into a ListBox:

```
C#
for (int i = 0; i < phonebookEntriesDataSet.Tables[0].Rows.Count; i++)
{
   this.listBox1.Items.Add(
        phonebookEntriesDataSet.Tables[0].Rows[i][0] + "  " +
        phonebookEntriesDataSet.Tables[0].Rows[i][1]);
}

VB
Dim i As Integer
For i = 0 To phonebookEntriesDataSet.Tables(0).Rows.Count - 1
   Me.listBox1.Items.Add(phonebookEntriesDataSet.Tables(0)
        .Rows(i)(0) + "  " +
        phonebookEntriesDataSet.Tables(0).Rows(i)(1))
Next i
```

Altering Data in a DataSet

To alter data held inside the DataSet, access the DataColumn that you wish to alter and set it to a new value. When all of the alterations are finished, call AcceptChanges to commit to the changes.

For example, the following code snippet sets the second column in the first row of the first table in the DataSet collection to a random number provided by l_randomGenerator, which is an instance of the Random class:

```
C#
// Column 1 is the phone number.
//                                 ¦
//                                 V
m_phonebookDS.Tables[0].Rows[0][1] = randomGenerator.Next().ToString();

VB
' Column 1 is the phone number.
'                                  ¦
'                                  V
m_phonebookDS.Tables(0).Rows(i)(1) = randomGenerator.Next().ToString()
```

An alternative would be to index using names as follows, but this way is much slower on the .NET Compact Framework when large amounts of data are involved:

C#
```
m_phonebookDS.Tables["Phone Contacts"].Rows[0]["PhoneNumber"] =
        l_randomGenerator.Next().ToString();
```

VB
```
m_phonebookDS.Tables("Phone Contacts").Rows(i)("PhoneNumber") =
        l_randomGenerator.Next().ToString()
```

Designing a PhoneBook Application with a DataSet

PhoneBook is a very simple application that fills a DataSet with a series of entries from a phone book and paints the entries into a ListBox. This application uses some of the WinForms controls introduced in Chapter 3, "Designing GUI Applications with Windows Forms," while showing how to programmatically populate a DataSet and extract and alter the data inside. The source code is on the CD-ROM under the directory SampleApplications\Chapter6\PhoneBook_CSharp and PhoneBook_VB.

In this simple incarnation, the PhoneBook is not very useful, because all of the information held in the DataSet is placed there programmatically and there is no way to update the information. The section "Altering Data in a DataSet" describes how to update existing data.

Troubleshooting Common DataSet-Related Errors

The most common error users encounter occurs when they attempt to access or alter an object in a DataSet, such as a DataColumn or DataRow, that does not exist. For example, if a DataSet has only two tables, then this piece of code would be faulty because the DataSet.Tables collection uses 0-based indexing:

C#
```
m_phonebookDS.Tables[2].Rows[0][1] = l_randomGenerator.Next().ToString();
```

VB
```
m_phonebookDS.Tables(2).Rows(0)(1) =
        l_randomGenerator.Next().ToString();
```

In these situations the most common exception that developers will see is the IndexOutOfRangeException. A special pitfall with the .NET Compact Framework is that the strings describing exceptions are held in a stand-alone DLL assembly called System.SR.dll. If this DLL is not installed on the device, strange exceptions that beginner developers don't expect can show up when the runtime wants to throw a descriptive exception. If this .DLL assembly is installed on the device, it is stored in the device Global Assembly Cache (GAC) in the \Windows directory. For example, the file is typically named \GAC_System.SR_v1_0_5000_0_cneutral_1.dll.

Understanding Constraints

The `DataSet` allows you to specify special rules that the data held inside the `DataSet.Tables` collection must follow. The `Constraint` base class specifies the rules that the data within a `DataTable` must follow in order to maintain database integrity. Two descendents from the `Constraint` class represent specific restrictions that the data must follow. The `UniqueConstraint` specifies that a particular value for a `DataColumn` must be unique in the table.

The `ForeignKeyConstraint` is used to enforce specifc behavior when altering or deleting the primary key column for a table. Because the `ForeignKeyConstraint` is intended for use in modeling parent-child relationships across tables, it is discussed in more detail in the "Enforcing Foreign Key Relationships with the `ForeignKeyConstraint`" section later in this chapter.

Adding Constraints to a `DataSet`

Each `DataTable` held in the `DataSet.Tables` collection holds a `ConstraintCollection` in the `Constraints` property. For example, to access the `ConstraintCollection` in the first table of a `DataSet`, use

```
C#
m_phonebookDS.Tables[0].Constraints
```

```
VB
m_phonebookDS.Tables(0).Constraints
```

The steps involved in creating and initializing constraints differ according to what kind of constraint you are trying to create. Once the constraint is created and initialized, you must perform these steps to make it active:

1. Add the constraint to the `Constraints` collection of the appropriate table(s).
2. Set the `DataSet.EnforceConstraints` flag to true to turn on enforcement. When you set the flag back to true, each constraint in each `DataTable.Constraints` collection is checked, and an exception is thrown if a check fails.

BE CAREFUL WITH DataSet.EnforceConstraints

When DataSet.EnforceConstraints is true, you may see an exception at any time that you insert, delete, or update data in the DataSet. Whether an exception is thrown depends on the specifics of the constraints that are in the DataSet. If an exception is thrown, remember that you can revert the changes you made by catching the exception and calling DataSet.RejectChanges(). The contents of the DataSet roll back to their previous valid state.

TEMPORARILY DISABLING CONSTRAINT ENFORCEMENT

If you want to alter a DataSet significantly and you know that you will violate some constraints while you are making changes, you can temporarily set DataSet.EnforceConstraints to false and then set it back to true when you are ready. It is a good practice to turn off constraint checking while manipulating large DataSets on the .NET Compact Framework, because you will avoid the performance penalty of checking constraints until the very end of the updates.

Adding a UniqueConstraint

To add a UniqueConstraint to a DataSet, follow these steps:

1. Create a UniqueConstraint by using one of the four meaningful constructors available on the .NET Compact Framework. Each constructor is now discussed in detail:

   UniqueConstraint(String name, DataColumn col) Creates a UniqueConstraint with specified name that enforces uniqueness on a single DataColumn.

   UniqueConstraint(DataColumn col) Creates a UniqueConstraint that enforces uniqueness on a single DataColumn.

   UniqueConstraint(String name, DataColumn[] cols) Creates a UniqueConstraint that enforces uniqueness for multiple columns in a row. The columns are specified by passing them as an array.

   UniqueConstraint(DataColumn[] cols) Same as above except the UniqueConstraint is nameless.

   UniqueConstraint(String name, string[] colNames, bool isPrimaryKey) This fifth public constructor is useful only to the Smart Device Extensions environment.

2. Add the UniqueConstraint to the Constraints collection of the desired DataTable.
3. Set DataSet.EnforceConstraints to true to turn on enforcement.

Working with a UniqueConstraint by Example

This code sample comes out of an updated PhoneBook sample application that is described in the section "Updating the PhoneBook Application with Constraints and Autoincremented Fields." This sample application is available in the folder SampleApplications\Chapter6\PhoneBook_Constraint_AutoIncrement_CSharp and PhoneBook_Constraint_AutoIncrement_VB. It sets up a UniqueConstraint that prevents any two contacts from sharing the same phone number.

```
C#
// Add a UniqueConstraint to the phone number column
// Note: Using indexing by the string "PhoneNumber" is slower
UniqueConstraint l_UniqueConstraint = new UniqueConstraint(l_DataSet.Tables[0].
        Columns["PhoneNumber"]);
l_DataSet.Tables[0].Constraints.Add(l_UniqueConstraint);

VB
' Add a UniqueConstraint to the phone number column
Dim l_UniqueConstraint As UniqueConstraint
l_UniqueConstraint = New UniqueConstraint(l_DataSet.Tables(0).
        Columns("PhoneNumber"))

l_DataSet.Tables(0).Constraints.Add(l_UniqueConstraint)
```

Preventing `NULL` Values in a `DataColumn`

The `DataColumn.AllowDBNull` property is very useful for disallowing a `DataColumn` from having a `DBNull` value. If you create a new `DataRow` and don't assign a value to one of the columns, it receives the default value of `DBNull`.

The following code sample is taken out of the updated PhoneBook sample application. This sample disallows the `Name` field of the PhoneBook from being `DBNull`.

```
C#
l_newTable.Columns["Name"].AllowDBNull = false;

VB
l_newTable.Columns("Name").AllowDBNull = False
```

If a DataColumn that has `AllowDBNull false` is set to `DBNull`, a `System.Data.NoNullAllowed` exception is thrown when the new row is added to a DataTable held inside the `DataSet`. For example, the following code attempts to add a row where the `Name` column was never set, and so executing this code would result in a `System.Data.NoNullAllowed` exception:

```
C#
DataRow l_newRow = m_phonebookDS.Tables[0].NewRow();
l_newRow[0] = "Violator"
l_newRow[1] = "5555587";
```

```
// This is going to throw an exception because the "Name"
// DataColumn was never set, so it is DBNull, and that is
// not allowed for the DataColumn
m_phonebookDS.Tables[0].Rows.Add(l_newRow);

VB
l_newRow = m_phonebookDS.Tables(0).NewRow()
l_newRow(0) = "Violator"
l_newRow(1) = "5551212"

' This is going to throw an exception because the "Name"
' DataColumn was never set, so it is DBNull, and that is not
' allowed for the DataColumn
m_phonebookDS.Tables(0).Rows.Add(l_newRow)
```

Setting Up Autoincremented Fields

When a new row is added to a `DataTable`, the empty row is created by calling `DataTable.NewRow`. The `DataTable` knows the schema for the row that it must create and instantiates the new row to match the schema. That means the new row holds the right `DataColumns` with the correct data types ready for you to set the data.

The `DataColumn.AutoIncrement` property can be set to tell the `DataTable` to set the value for a `DataColumn` automatically when a new row is created. This is an especially useful feature for `DataColumns` that represent the primary key of a table, since the key can be automatically created for you.

There are three important properties in the `DataColumn` that relate to autoincremented fields:

- `DataColumn.AutoIncrement` Set to true to make this DataColumn autoincrement.
- `DataColumn.AutoIncrementSeed` The starting value for autoincrementing.
- `DataColumn.AutoIncrementStep` The step amount for each new value.

If the `DataColumn` is a computed column, then trying to set it as an autoincremented column will cause an `ArgumentException`. Computed columns are discussed in great detail in the section "Deriving `DataColumn` Values with Expressions and Computed Fields."

If the data type of the `DataColumn` is not an `Int16`, `Int32`, or `Int64`, then it is coerced back into an `Int32`. This can cause a loss of precision if the `DataColumn` is a floating-point data type. If the `DataColumn` is a string DataType, setting it to autoincrement will coerce the type of the `DataColumn` back to integer.

Creating Autoincremented Field Code by Example

This code sample comes out of an updated PhoneBook sample application which is described in the section "Updating the PhoneBook Application with Constraints and Autoincremented Fields." This code snippet sets the AutoIncrement property for the ContactID DataColumn in the table named l_newTable. The starting value is 10, and the step value is 5.

```
C#
l_newTable.Columns["ContactID"].AutoIncrement = true;
l_newTable.Columns["ContactID"].AutoIncrementSeed = 10;
l_newTable.Columns["ContactID"].AutoIncrementStep = 5;

VB
l_newTable.Columns("ContactID").AutoIncrement = True
l_newTable.Columns("ContactID").AutoIncrementSeed = 10
l_newTable.Columns("ContactID").AutoIncrementStep = 5
```

Updating the PhoneBook Application with Constraints and Autoincremented Fields

This sample application is based on the PhoneBook application presented in the earlier section "Designing a Phone Book Application with a DataSet." It is located in the folder SampleApplications\Chapter6\PhoneBook_Constraint_AutoIncrement_CSharp and PhoneBook_Constraint_AutoIncrement_VB. This sample application demonstrates the use of UniqueConstraints, autoincremented fields, and checking DataColumns that may not have DBNull values.

Modeling Relational Data with the DataSet

You now understand enough about the DataSet to populate it with a DataTable, access the data, and enforce some forms of constraints on the data. In this section we build on that knowledge and learn to perform common relational database operations with data inside a populated DataSet.

Deriving DataColumn Values with Expressions and Computed Fields

The value for a DataColumn can be computed based on the value of another DataColumn in the same DataRow. To do this, use the DataColumn.Expression property to describe the computed value for the DataColumn. The Expression property is a string value that describes the computation that derives the value for the DataColumn.

The syntax for the expression is rich and supports a wide variety of arithmetic and string operators. Table 6.1 shows all of the expression operators that are supported by the .NET Compact Framework.

TABLE 6.1
Framework-Supported Expression Operators

OPERATOR	FUNCTION
Sum	Computes the sum of the arguments
Avg	Computes the average of the arguments
Min	Selects the minimum of the arguments
Max	Selects the maximum of the arguments
+, -, *, /	Addition, subtraction, multiplication, division
%	Modulus (remainder)
+	String concatenation operator

Creating Expressions by Example

The easiest way to understand how to create an Expression is to look at an example. For our first example, consider a DataTable called l_newTable that has three columns named FirstName, LastName, and FullName. The goal is to create an Expression that sets the FullName column to be the string concatenation of the FirstName and LastName columns. The code to do that appears below:

```
C#
l_newTable.Columns["FullName"].Expression = "FirstName + ' ' + LastName";

VB
l_newTable.Columns("FullName").Expression = "FirstName + ' ' + LastName"
```

For our second example, consider a DataTable named l_newTable. We want the TotalPrice column to be the value of the MSRP column minus the value of the Discount column:

```
C#
l_newTable.Columns["TotalPrice"].Expression = "MSRP - Discount";

VB
l_newTable.Columns("TotalPrice").Expression = "MSRP - Discount"
```

Finally, imagine that l_newTable is a DataTable with four numeric columns: FinalGrade, Exam1, Exam2, and Exam3. We want to set the value for FinalGrade to the average of Exam1, Exam2, and Exam3, as follows:

```
C#
l_newTable.Columns["FinalGrade"].Expression = "Avg(Exam1, Exam2, Exam3)";

VB
l_newTable.Columns("FinalGrade").Expression = "Avg(Exam1, Exam2, Exam3)"
```

Expressing Parent-Child Relationships in a DataSet

The essential ingredients of a relational database are tables with rows and the ability to create a parent-child relationship, or a relation, between two tables. A relation between two tables is made by linking the two tables by one or more data columns called the `primary key`. In the parent table, the `primary key` is unique to all of the rows in the table. The rows in the child table have a column called the `foreign key`, which does not have to be unique in the child table. It points back to the corresponding row in the parent table.

For example, consider Table 6.2, a parent table that describes the main contacts for a physician's office.

TABLE 6.2
MainContactTable

COLUMN NAME	DATA TYPE
CustID	Integer, PRIMARY KEY
FirstName	String
LastName	String

A child table that holds cholesterol readings might look like Table 6.3.

TABLE 6.3
CholesterolTable

COLUMN NAME	DATA TYPE
CustID	Integer, FOREIGN KEY
Reading1	Decimal
Reading2	Decimal
Reading3	Decimal
Average	Decimal

In the `CholesterolTable`, the `CustID` references a specific, unique entry in the `MainContactTable`. Tables 6.4 and 6.5 show what these parent-child tables might look like when populated.

TABLE 6.4

MainContactTable

CustID	FirstName	LastName
001	George	Washington
002	Ben	Franklin
003	Alexander	Hamilton

TABLE 6.5

CholesterolTable

CustID	Reading1	Reading2	Reading3	Average
001	87	78	66	77.0
001	99	54	89	80.667
002	90	88	55	77.667

In this parent-child table example, there were two entries in the `CholesterolTable` for George Washington and one entry for Ben Franklin. What would happen if the entry for George Washington were deleted from the `MainContactTable`? The system should delete all of the corresponding entries in the `CholesterolTable`, or the database would be in an invalid state.

Good database systems understand the notion of a relation between two tables and can delete child rows automatically if the parent row is deleted. Alternately, they should at least report an error if the user does something that puts the data into an invalid state, such as deleting George Washington's record from `MainContactTable` without also removing his entries from the `CholesterolTable`.

The .NET Compact Framework offers two classes that can do this bookkeeping automatically: `DataRelation` and `ForeignKeyConstraint`.

Creating a DataRelation to Express Parent-Child Relationships

When you set up a `DataRelation` between two tables, you specify which `DataColumn` serves as the primary key and which serves as the foreign key. After the `DataRelation` has been created, it will ensure that the `DataSet`'s relational data, as described by the `DataRelation`, remains valid. For example, if you delete the first row from the `MainContactTable`, the `DataRelation` automatically deletes all of the child rows in the `CholesterolTable`.

To set up a `DataRelation` between two tables in a `DataSet`, first create an instance of the `DataRelation` using its constructor to pass in the `DataColumns` that comprise the primary and foreign keys. The `DataRelation` constructor overloads that are available on the .NET Compact Framework are as follows:

- `DataRelation(String relName, DataColumn parent, DataColumn child)` creates a named `DataRelation` between the parent and child `DataColumns`.
- `DataRelation(String relName, DataColumn[] parent, DataColumn[] child)` creates a named `DataRelation` between two tables by using multiple columns per table for the relation.
- `DataRelation(String relName, DataColumn parent, DataColumn child, bool createConstraints)` creates a named `DataRelation` between the parent and child `DataColumns`, with the option to create associated constraints to enforce the relation.
- `DataRelation(string relName, DataColumn[] parent, DataColumn[] child, bool createConstraints)` creates a named `DataRelation` between two tables by using multiple columns per table for the relation, with the explicit option to create associated constraints to enforce the relation.
- `DataRelation(string relName, string parentTableName, string childTableName, string[] parentColNames, string[] childColNames, bool isNested)` is the constructor used by the Smart Device Extensions environment.

Then simply add the `DataRelation` to the `DataSet.Relations` collection.

Writing Code to Create a `DataRelation`

This sample code is from an updated PhoneBook example application that demonstrates the use of the `Expression` and `DataRelation` classes. The code creates a new `DataRelation` that binds the `ContactID` column from the `PhoneContactsMainTable` and the `Cholesterol` table inside the `DataSet`. The order in which the columns are passed to the constructor dictates which is the parent table and which is the child table. It is important to be careful about the ordering to avoid strange behavior and exceptions from occurring in applications.

```
C#
DataRelation l_newRelation = new DataRelation(
        "MainContactToCholesterolRelation",
        l_DataSet.Tables["PhoneContactsMainTable"].Columns["ContactID"],
        l_DataSet.Tables["Cholesterol"].Columns["ContactID"]);
l_DataSet.Relations.Add(l_newRelation);

VB
Dim l_newRelation As DataRelation
l_newRelation = New
        DataRelation("MainContactToCholesterolRelation",
        l_DataSet.Tables("PhoneContactsMainTable"). Columns("ContactID"),
        l_DataSet.Tables("Cholesterol").Columns("ContactID"))
l_DataSet.Relations.Add(l_newRelation)
```

Enhancing the PhoneBook Application with DataRelations and Expressions

The full sample application that demonstrates using a DataRelation and an Expression is located in the folder SampleApplications\Chapter6\PhoneBook_Relations_Expressions_CSharp and PhoneBook_Relations_Expressions_VB. This sample application sets up two tables in the DataSet named PhoneContactsMainTable and Cholesterol. A DataRelation links the tables, with PhoneContactsMainTable as the parent and Cholesterol as the child. The tables are populated such that the entry for George Washington in PhoneContactsMainTable has a child row in the Cholesterol table.

The Cholesterol table has a DataColumn named AverageReading, which is computed by averaging the values of the three other columns. This demonstrates how to use a DataExpression to create a computed column.

When the button labeled "Delete row, trigger DataRelation" is clicked, the application deletes the row for George Washington from the PhoneContactMainTable. Because there is a child entry for this row in the Cholesterol table, the DataRelation forces the child rows also to be deleted. On the other hand, pressing the button labeled "Delete row—don't trigger" deletes the row for Alexander Hamilton from the PhoneContactMainTable. This row has no children, so deleting it does not require deleting any child rows.

Enforcing Foreign Key Relationships with the ForeignKeyConstraint

The ForeignKeyConstraint is very much like the DataRelation, but it provides extra flexibility. As with a UniqueConstraint, the ForeignKeyConstraint is added to a DataTable.Constraints collection. Specifically, the ForeignKeyConstraint gets added to the Constraints collection of the child table.

When a row with children is deleted from a parent table, the ForeignKeyConstraint can cause the following behaviors:

- It can cause all child rows to be deleted. This behavior is thus identical to using a DataRelation.
- It can set the child column values, that is, the foreign keys, to NULL. Thus, they no longer point to a parent row that does not exist.
- It can set the child column value to a default value. This is useful because it makes it easy to see all of the "orphaned" child rows by pointing to a default "orphan parent" in the parent table.
- It can throw an exception.

To set up a ForeignKeyConstraint, first create a new ForeignKeyConstraint through one of the constructors available on the .NET Compact Framework. The available constructors on the .NET Compact Framework are as follows:

- `ForeignKeyConstraint(DataColumn parentCol, DataColumn childCol)` creates a `ForeignKeyConstraint` between parent and child `DataColumns`.
- `ForeignKeyConstraint(String name, DataColumn parentCol, DataColumn ChildCol)` creates a `ForeignKeyConstraint` between a parent and child, but the constraint gets a name.
- `ForeignKeyConstraint(DataColumn[] parentCols, DataColumn[] childCols)` creates a `ForeignKeyConstraint` between two tables by using multiple `DataColumns` for the constraint.
- `ForeignKeyConstraint(String name, DataColumn[] parentCols, DataColumn[] childCols)` creates a `ForeignKeyConstraint` between two tables by using multiple `DataColumns` for the constraints, but every constraint gets a name.
- `ForeignKeyConstraint(string cName, string pName, string[] pColNames, string[] cColNames, AcceptRejectRule arRule, Rule dRule, Rule uRule)` is used internally by the Smart Device Extensions environment.

Next set the `ForeignKeyConstraint's DeleteRule`, `UpdateRule`, and `AcceptRejectRule`. The `DeleteRule` controls what happens when a parent row is deleted. The `UpdateRule` controls what happens when a parent row is modified. The `AcceptRejectRule` controls what happens when a parent row is modified and `DataSet.AcceptChanges()` is called. The `UpdateRule` and `DeleteRule` are `Rule` types, while the `AcceptRejectRule` is a `AcceptRejectRule` type.

For example, consider a `ForeignKeyConstraint` that is used to express a parent-child relationship between two tables. Imagine that a row from the parent table is deleted. In this case the value for the `Delete` rule is examined to determine what happens to the child tables:

`Rule.Cascade` The delete is cascaded, so the child rows are also deleted.

`Rule.SetDefault` The child rows' values are set to a default value.

`Rule.SetNull` The child rows' values are set to `DBNull`.

`Rule.None` An exception is thrown.

The `AcceptRejectRule` is examined only when `DataSet.AcceptChanges` is called. The `AcceptRejectRule` type has two values: `Cascade` and `None`. If the `AcceptRejectRule` is set to the `Cascade` value, then the `DataSet` attempts to cascade changes made in a parent row to its children when `DataSet.AcceptChanges` is called.

The `ForeignKeyConstraint` is now set up. To use it, add it to the `ForeignKeyConstraint` to the `Constraints` collection of the child table.

Creating a `ForeignKeyConstraint` with Sample Code

The following sample code creates a `ForeignKeyConstraint` that cascades when a parent row is deleted. This means that when a parent row is deleted, the children rows are also deleted. The code is taken from the modified PhoneBook sample application using `ForeignKeyConstraints`.

```
C#
ForeignKeyConstraint l_ForeignKC = new
ForeignKeyConstraint("MainToCholesterolFKConstraint",
        l_DataSet.Tables["PhoneContactsMainTable"].Columns
        ["ContactID"], l_DataSet.Tables["BloodPressure"].
        Columns["ContactID"]);

l_ForeignKC.DeleteRule = Rule.Cascade;
l_ForeignKC.UpdateRule = Rule.Cascade;
l_ForeignKC.AcceptRejectRule = AcceptRejectRule.Cascade;

l_DataSet.Tables["BloodPressure"].Constraints.Add(l_ForeignKC);
l_DataSet.EnforceConstraints = true;

VB
Dim l_ForeignKC As ForeignKeyConstraint
l_ForeignKC = New ForeignKeyConstraint
        ("MainToCholesterolFKConstraint",
        l_DataSet.Tables("PhoneContactsMainTable").Columns
         ("ContactID"), l_DataSet.Tables("BloodPressure").
        Columns("ContactID"))

l_ForeignKC.DeleteRule = Rule.Cascade
l_ForeignKC.UpdateRule = Rule.Cascade
l_ForeignKC.AcceptRejectRule = AcceptRejectRule.Cascade

l_DataSet.Tables("BloodPressure").Constraints.Add(l_ForeignKC)
l_DataSet.EnforceConstraints = True
```

Enhancing the PhoneBook Application with a `ForeignKeyConstraint`

The full sample application that demonstrates `ForeignKeyConstraints` is located in the folders `SampleApplications\Chapter6\PhoneBook_ForeignKeyConstraint_CSharp` and `PhoneBook_ForeignKeyConstraint_VB`. This sample application sets up two tables in the `DataSet`, named `PhoneContactsMainTable` and `BloodPressure`. A `ForeignKeyContraint` links the tables, with `PhoneContactsMainTable` as the parent and `BloodPressure` as the child. The tables are populated such that the entry for George Washington in `PhoneContactsMainTable` has a child row in the `BloodPressureTable`.

The `BloodPressure` table has a DataColumn named `AverageReading`, which is computed by averaging the values of the three other columns. This demonstrates how to use a `DataExpression` to create a computed column.

When the button labeled Trigger ForeignKeyConstraint is clicked, the application deletes the row for George Washington from the PhoneContactMainTable. Because there is a child entry for this row in the BloodPressure table, the ForeignKeyContraint triggers, throwing a ConstraintException.

On the other hand, pressing the button labeled "Delete row—no trigger" deletes the row for Alexander Hamilton from the PhoneContactMainTable. This row has no children, so deleting it does not put the database into an invalid state. Thus, no exception is thrown.

Creating Bindable Views of Data with a DataView

Up until now, the means by which data is extracted from a DataTable is to access the DataTable.Rows collection. Each DataRow in the DataTable.Rows collection has DataColumn objects that can be indexed by name or number. The first sample application in this chapter demonstrates the process of iterating through the DataTable.Rows collection to examine the data.

While this is an effective way of accessing the data inside a DataTable, it does not provide an easy means of sorting the data against a specific column or filtering the rows so that only specific DataRows in the DataTable.Rows collection are visible.

The DataView class is the correct way to view the contents of a DataTable in a sorted and/or filtered format. From the outside the DataView looks very much like a DataTable. There is an indexer through which the rows in a DataView are accessible, and the DataColumns inside each row can be accessed and manipulated in the usual way.

The DataView is also a *data-bindable* object. This means that it implements interfaces that allow the DataView to become the data source for other objects that can manipulate data-bindable objects. Specifically, you can use the DataView to populate a DataGrid with the contents of a DataView. After you have set up a relationship between a DataView and a DataGrid, the contents of the DataGrid are bound to the DataView. The DataGrid shows the contents of the DataGrid, and the display is automatically kept up to date when the DataView changes.

For example, you can create a DataView that sorts by the Age column of a DataTable and bind it with a DataGrid. The DataGrid displays the contents of the DataView. If the application inserts more data into the underlying DataTable, then the DataGrid updates its view automatically.

LIMITATIONS OF THE DataGrid

The DataGrid is present in the .NET Compact Framework, but it is a strongly cut-down version compared to the desktop DataGrid. The DataGrid is treated in great detail in Chapter 3, "Designing GUI Applications with Windows Forms." Instructions on how to perform data binding with a DataGrid and a DataView are given in the "Binding to a DataGrid" section later in this chapter.

Sorting with the DataView

To sort the data in a DataTable using a DataView, follow these steps:

1. Create a new DataView.
2. Set the Sort property of the newly instantiated DataView. The Sort property is a string that states which columns to sort by. Add the term **DESC** to sort a column in descending order and separate the columns with a comma.

The following sample code is pulled from the DataView_Sort_And_Filter sample application. It creates a new DataView from the first table in a DataSet. It sorts by age in descending order, then by name in descending order as a secondary sort key. Finally, it inserts the DataRows, which are now sorted, into a ListBox.

```
C#
DataView l_sortAgeView = new DataView(in_DataSet.Tables[0]);
l_sortAgeView.Sort = "Age DESC, Name DESC";
for (int i = 0; i < l_sortAgeView.Count; i++)
{
   this.listBox2.Items.Add(l_sortAgeView[i]["Name"] + "    " +
            l_sortAgeView[i]["Age"]);
}
```

```
VB
Dim l_sortAgeView As DataView
l_sortAgeView = New DataView(in_DataSet.Tables(0))
l_sortAgeView.Sort = "Age DESC, Name DESC"
Dim i As Integer
For i = 0 To l_sortAgeView.Count - 1
   Me.listBox2.Items.Add(l_sortAgeView(i)("Name") + "    " +
            Convert.ToString(l_sortAgeView(i)("Age")))
Next i
```

Sorting and Filtering Data in a Sample Application

The sample application called DataView_Sort_And_Filter demonstrates how to use a DataView to sort and filter the contents of a DataTable. The source code is located in SampleApplications\Chapter6\DataView_Sort_And_Filter_CSharp and DataView_Sort_And_Filter_VB.

This simple application creates a DataSet with a single table. The table rows hold an entry for a name and an age. The table is populated with six entries.

The main window has three ListBoxes. The topmost ListBox shows the contents of the DataTable as they are stored in the DataTable. The middle ListBox shows the contents of a DataView that is configured to sort the entries in the DataTable by age in descending order, and then by name in descending order as a secondary sort key. That is, if two entries have the same age, then they are alphabetically sorted in descending order by their names. The entries in the DataSet are doctored so that it is obvious that sorting occurs.

The bottom ListBox shows the contents of a DataView configured with a filter so that it shows only entries whose age is 21 or greater.

Tracking Changes in a DataRow

Every DataRow pulled from a DataTable or a DataView has a property called RowState. The RowState is of type DataRowState, which has one of nine possible values:

Unchanged This DataRow has not been changed since the last time AcceptChanges was called on the DataSet that owns the DataRow.

Original Refers to all original rows that are either unchanged or were deleted since the last time AcceptChanges was called on the DataSet that owns the DataRow.

CurrentRows All of the rows available in the DataRow or DataView, regardless of whether they are newly added, deleted, or modified since the last time AcceptChanges was called on the DataSet that owns the DataRow. If the rows have been modified, this value refers to the modified value.

Added This DataRow has been added to the DataTable.Rows collection since the last time AcceptChanges was called on the DataSet that owns the DataRow.

Deleted This DataRow has been deleted from the DataTable.Rows collection since the last time AcceptChanges was called on the DataSet that owns the DataRow.

ModifiedCurrent Refers to the current data in rows that have been modified since the last time AcceptChanges was called on the DataSet that owns the DataRow.

ModifiedOriginal Refers to the original data of DataRows that have been modified since the last time AcceptChanges was called on the DataSet that owns the DataRow.

None Refers to none of the DataRows.

Detached This DataRow is not part of the DataTable.Rows collection. This happens, for example, if the DataRow is created but never added to the collection.

DataSet.AcceptChanges COMMITS ALL CHANGES

Remember that calling `DataSet.AcceptChanges` commits to changes made to the data, such as adding, removing, or modifying `DataRows`.

Filtering with the `DataView`

There are situations in which it would be useful to filter the rows in a `DataTable` such that, for example, it shows only those rows that have been modified in some way. This is easy to do using the `DataView`. To set up a `DataView` to filter the data according to its `RowState` status, set the `DataView.RowStateFilter` property to one of the `DataRowState` values discussed in the preceding list.

The following code demonstrates how to use the `DataRowState` to show only those rows that have been added since the last call to `AcceptChanges()`. The code is taken from the DataView_SortByRowState_AddTables sample application:

```
C#
// This DataView shows only rows that have been added since
// the last call to AcceptChanges
m_addedRowsView = new DataView(l_DataSet.Tables[0]);
m_addedRowsView.RowStateFilter = DataViewRowState.Added;

VB
' This DataView shows only rows that have been added since
' the last call to AcceptChanges
m_addedRowsView = New DataView(l_DataSet.Tables(0))
m_addedRowsView.RowStateFilter = DataViewRowState.Added
```

MATCHING MULTIPLE RowState VALUES IN A DataView

To view `DataTables` that match different `DataRowState` values, or those values together—for example, to view deleted and newly added rows only—use the following code:

```
C#
l_DataView.RowStateFilter = DataViewRowState.Added | DataViewRowState.Deleted;

VB
l_DataView.RowStateFilter = DataViewRowState.Added Or DataViewRowState.Deleted
```

Adding Data into a `DataView`

New `DataRows` can be added to a `DataView` and then percolated back to the source `DataTable`. This can be a convenient feature in business applications that manipulate their data almost exclusively through DataView classes. To add a new DataRow to an existing DataView and percolate it back to the original DataTable, follow these steps:

1. Create an instance of DataRowView by calling DataView.AddNew() on the DataView that is being used to add a new row.
2. Insert values for the DataColumns of the DataRowView. The DataRowView can index its DataColumns only by the column name.
3. Call DataRowView.EndEdit(). This signals that you have finished setting up the new DataRowView but does *not* call AcceptChanges on the parent DataSet.

The following sample code demonstrates how to add data using a DataView. The code is borrowed from the sample application DataView_SortByRowState_AddTables, which is described in detail in the next section. It uses an instance of the Random class to provide pseudo-random integers to append to the name of the person and for the age of the person.

```
C#
int l_maxRandom = 100;

// Step 1 - Create instance of the DataRowView
DataRowView l_firstNewDataRowView = m_addedRowsView.AddNew();

// Step 2 - Set column values
l_firstNewDataRowView["Name"] = "NewPerson" +
        m_Random.Next(l_maxRandom).ToString();
l_firstNewDataRowView["Age"] = m_Random.Next(l_maxRandom).ToString();

// Step 3 - call EndEdit()
l_firstNewDataRowView.EndEdit();
```

```
VB
Dim l_maxRandom As Integer
l_maxRandom = 100

' Step 1 - Create instance of the DataRowView
Dim l_firstNewDataRowView As DataRowView
l_firstNewDataRowView = m_addedRowsView.AddNew()

' Step 2 - Set column values
l_firstNewDataRowView("Name") = "NewPerson" +
        m_Random.Next(l_maxRandom).ToString()
l_firstNewDataRowView("Age") = m_Random.Next(l_maxRandom)
```

```
' Step 3 - call EndEdit()
l_firstNewDataRowView.EndEdit()
```

Using a DataView in a Sample Application: DataView_SortByRowState_AddTables

The DataView_SortByRowState_AddTables sample application is located in the folder `\SampleApplications\Chapter6\`. There is a C# and a Visual Basic version. It demonstrates filtering `DataRows` based on their RowState values and adding tables to a `DataView`.

The application starts by creating and populating a `DataSet` with a single `DataTable`. The `DataTable` schema has `DataColumns` called `Name` and `Age`. The application fills the `DataTable` with six records and creates two `DataView` classes. One `DataView` reveals only those records that have been deleted since the last call to `AcceptChanges`, while the other reveals only those records that have been added since the last call to `AcceptChanges`.

There are three ListBoxes. The first is painted with data directly from the `DataTable`. Only those `DataRows` with `RowState` value set to `Unchanged` are painted into the ListBox. The second ListBox is painted with data from the `DataView` that reveals the added rows. The third ListBox is painted with data from the `DataView` that reveals the deleted rows.

When the application first launches, there are no added or deleted rows, so only the topmost ListBox shows any entries. Clicking the button labeled Add and Delete some rows causes two rows to be deleted from the `DataTable` and two rows to be added to the `m_AddedRowsView` `DataView` by using the code in Listing 6.2:

LISTING 6.2 Adding and deleting `DataRows` through a `DataView`

```
C#
// Delete two rows
m_DataSet.Tables[0].Rows[0].Delete();
m_DataSet.Tables[0].Rows[1].Delete();

// Add two new rows
int l_maxRandom = 100;
DataRowView l_firstNewDataRowView = m_addedRowsView.AddNew();

l_firstNewDataRowView["Name"] = "NewPerson" +
        m_Random.Next(l_maxRandom).ToString();
l_firstNewDataRowView["Age"] = m_Random.Next(l_maxRandom);

l_firstNewDataRowView.EndEdit();
```

```
DataRowView l_secondNewDataRowView = m_addedRowsView.AddNew();

l_secondNewDataRowView["Name"] = "NewPerson" +
        m_Random.Next(l_maxRandom).ToString();

l_secondNewDataRowView["Age"] = m_Random.Next(l_maxRandom);

l_secondNewDataRowView.EndEdit();
```

VB

```
' Delete two rows
m_DataSet.Tables(0).Rows(0).Delete()
m_DataSet.Tables(0).Rows(1).Delete()

' Add two new rows
Dim l_maxRandom As Integer
l_maxRandom = 100
Dim l_firstNewDataRowView As DataRowView
l_firstNewDataRowView = m_addedRowsView.AddNew()

l_firstNewDataRowView("Name") = "NewPerson" +
        m_Random.Next(l_maxRandom).ToString()

l_firstNewDataRowView("Age") = m_Random.Next(l_maxRandom)

l_firstNewDataRowView.EndEdit()

Dim l_secondNewDataRowView As DataRowView
l_secondNewDataRowView = m_addedRowsView.AddNew()
l_secondNewDataRowView("Name") = "NewPerson" +
        m_Random.Next(l_maxRandom).ToString()

l_secondNewDataRowView("Age") = m_Random.Next(l_maxRandom)

l_secondNewDataRowView.EndEdit()
```

Then the information is repainted, and now there are entries in all of the ListBoxes.

Clicking the button labeled Call AcceptChanges calls the `AcceptChanges` method and repaints the ListBox. The newly added rows now appear in the topmost ListBox because they are no longer considered "newly added" rows after `AcceptChanges` is called. There are no longer any newly added or deleted rows, so the bottom two ListBoxes are empty again.

Binding Data to a Control

Up to now all of the examples that have painted the contents of a `DataSet` have done so by iterating through the `DataTables.Rows` collection for each `DataTable` inside the `DataSet`. This is non-ideal for several reasons:

- The burden of responsibility to paint the data manually every time there is an update in the `DataSet` contents falls on the developer.
- The process of painting into a ListBox is error prone.
- It is unwieldy to write code that paints only those `DataRows` with certain RowState values.
- It is not easy to make the ListBox appealing to the eye when using it to display relational data.

The best solution to this problem is to use data binding, which connects a source of data to a consumer. The most popular source of data is the `DataView` because of its flexibility in filtering and sorting data automatically. The best-looking Winforms consumer of data is the `DataGrid` class, which is supported by the .NET Compact Framework in a limited way.

DETAILS ABOUT THE `DataGrid`

Chapter 3 details the features of the `DataGrid` that are missing from the .NET Compact Framework. In the present chapter, though, we focus on the nuts and bolts of using the `DataGrid` to display data.

Binding to a `DataGrid`

When a `DataSet` is bound to a `DataGrid`, the contents of the `DataSet` automatically appear on the `DataGrid` in a visually pleasing format. To bind a `DataSet` to a `DataGrid`, follow these steps:

1. Create a `DataView` through the techniques discussed earlier in this chapter.
2. Create a new instance of a `DataGrid`. If you are using the Smart Device Extensions IDE, all you need to do is drag an instance of the `DataGrid` from the toolbar to the canvas of your project.
3. Set the `DataGrid.DataSource` property to the `DataView` you created in step 1.

Although a brief overview of the `DataGrid` is provided here, it is treated in greater detail in Chapter 3. There you can learn how to alter the look, feel, and special behavior of the

DataGrid. The DataGrid is a very slimmed-down version compared to the desktop's, so there are many issues advanced programmers should be aware of when using the DataGrid on the .NET Compact Framework.

The following sample code demonstrates how to bind a DataGrid with a DataView. It comes from the example application DataGrid_With_DataView:

```
C#
// Assuming that m_DataSet was already set up…
m_sortAgeDataView = new DataView(m_DataSet.Tables[0]);
m_sortAgeDataView.Sort = "Age DESC, Name DESC";

// Bind the DataGrid to our DataView and it will
// automatically paint itself!

dataGrid1.DataSource = m_sortAgeDataView;

VB
m_sortAgeDataView = New DataView(m_DataSet.Tables(0))
m_sortAgeDataView.Sort = "Age DESC, Name DESC"

' Bind the DataGrid to our DataView and it will
' automatically paint itself!
Me.DataGrid1.DataSource = m_sortAgeDataView
```

Using Data Binding in a Sample Application

The sample application named DataGrid_With_DataView demonstrates binding a DataView to a DataGrid in a complete stand-alone application. The project files for this application are located in the folder SampleApplications\Chapter6\. There is a C# and a Visual Basic version.

DataGrid_With_DataView sets up a DataView using the same technique as was used in the DataView_Sort_And_Filter sample application. DataGrid_With_DataView uses only one DataView, which is sorted in descending order by age and name. Then, with just one line of code, the DataView binds to a DataGrid, and it displays automatically.

Comparing the Compact DataSet with the Desktop DataSet

The DataSet offering for the .NET Compact Framework has some notable omissions compared with the DataSet in the desktop .NET Framework. Specifically, developers are likely to notice these differences:

- The Compact Framework DataSet has no ability to serialize itself to anything other than XML. That is, it does not implement the ISerializable interface in a way that does anything useful. In a related point, the DataSet is the only object in the .NET Compact Framework capable of serializing itself in any way.
- The Clone() method is present, but using it on large DataSets can cause performance problems.
- Many overloads for supported methods have been cut.
- The .NET Compact Framework is capable of reading Diffgrams, but it is assumed that the DataSet is empty when a Diffgram is read. There is no support for the notion of updating a populated DataSet with the information contained in a Diffgram. The reason for this is that the DataSet on the .NET Compact Framework is meant for use as a local, small cache of data. Trying to use a DataSet to hold a large database in memory and updating it by passing Diffgrams to it would yield completely unacceptable performance.

In Brief

- The DataSet plays a central role in ADO.NET on the .NET Compact Framework. It is an in-memory cache of relational data.
- Internally, the DataSet holds collections of DataTables, DataRows, and DataColumns to store relational data. To directly access or alter the relational data in a DataSet, developers must use these collections.
- The UniqueConstraint class lets the DataSet automatically enforce uniqueness among the columns in a DataTable.
- Columns can be set to be automatically incremented when a new row is added to a DataTable.
- Column values can be automatically computed by using an Expression.
- The DataRelation class is used to express parent-child relationships between two DataTables.

- The `ForeignKeyConstraint` can be used to enforce a foreign key relationship between two tables. When changes are made to the data, the developer can choose what happens. The changes can cascade, default or null values can be assigned, or an exception can be thrown.
- Data in a `DataTable` cannot be sorted. Instead, developers can create a `DataView` that presents a sorted view of the rows in a `DataTable`. The sort key can be a single column, or developers can use additional columns as secondary sort keys.
- The `DataView` is a data-bindable object. Developers can attach a `DataView` to the `DataGrid` GUI object and get a tablular view of data. This is the cleanest way to present a `DataSet` on the .NET Compact Framework, and it only takes a few lines of code.

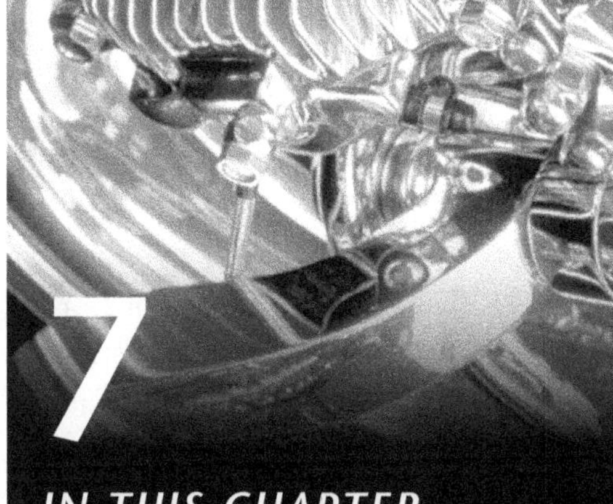

Programming with Microsoft SQL Server CE

7

IN THIS CHAPTER

Investigating the Features Supported by Microsoft SQL Server 2000 Windows CE Edition

Microsoft Structured Query Language (SQL, pronounced "se-quel" or "S-Q-L") Server 2000 Windows CE Edition (SQL Server CE) is a very tiny implementation of Microsoft's SQL Server 2000 database engine. Despite its size, SQL Server CE provides a fair amount of data storage and retrieval functionality.

SQL Server CE supports databases as large as 2GB. Binary Large Objects (BLOBS) are also supported. BLOBS can be up to 1GB in size. SQL Server CE supports a subset of the Data Definition Language (DDL) statements (see Table 7.2 for a list of supported DDL statements) and a subset of the Data Manipulation Language (DML) statements (see Table 7.3 for a list of supported DML statements). There is also support for single and multicolumn indexes, primary keys, constraints, and column defaults. Also, there is support for fast, forward-only, scrollable cursors, which are great for building applications that must perform efficiently. SQL Server CE also provides support for autonomous transactions and nested autonomous transactions. Finally, your database can also be password secured as well as encrypted with 128-bit file level encryption.

When developing a SQL Server CE application, you will need to add two assembly references to your project as well as three using (C#) or Imports (VB) statements to source code. The SQL Server CE manager exists in the System.Data.SqlServerCe assembly. You will also need to add a reference to the System.Data.Common assembly. As for the using/Imports statements, add the following lines of code to your source:

```
C#
using System.Data;
using System.Data.Common;
using System.Data.SqlServerCe;

VB
Imports System.Data
Imports System.Data.Common
Imports System.Data.SqlServerCe
```

Creating a Microsoft SQL Server CE Database

There are two options for creating a SQL Server CE database. One option is to use the SQL Server CE Query Analyzer to graphically create and design a SQL Server CE database. To learn more about the Query Analyzer, see Microsoft SQL Server CE Books Online.

You can also create a SQL Server CE database programmatically by using the SQL Server CE Data Provider classes defined in the System.Data.SqlServerCE namespace. When creating a database programmatically, you only need to interact with one SQL Server CE Data Provider class, System.Data.SqlServerCe.SqlCeEngine. The SqlCeEngine class provides programmable access to the engine-specific features in SQL Server CE. The SqlCeEngine provides two major pieces of functionality: the ability to create a new database and the ability to compact an existing database. To learn more about SQL Server CE database compaction, see the "Maintaining a SQL Server CE Database" Shop Talk.

SHOP TALK

MAINTAINING A SQL SERVER CE DATABASE

As with most data stored by an application, the internal structure of a SQL Server CE's database can, over time and repeated use, become fragmented. Needless to say, this fragmentation results in a waste of space and eventual degradation in your application's performance. To remedy this fragmentation, SQL Server CE provides the CompactDatabase method, exposed as the Compact method on the SqlCeEngine class. The Compact method performs the following tasks:

- Reorganizes a table's pages to reside in adjacent database pages, which improves performance by reducing table fragmentation

SHOP TALK

- Reorders the rows of the table when there is a unique or primary key constraint present, which increases search/insert performance
- Reclaims used space created by object and record deletions
- Resets incrementing identity columns so that the next values will be one more than the highest value in current rows
- Regenerates the table statistic, which may help improve the query optimization process
- Attempts to repair errors in the database
- Might modify database properties

The `Compact` method writes a new database to the file system. Because a new version of the database is being created, you can change the database's password, encryption properties, and locale identifier.

Before attempting to compact a database, it must be closed. Sufficient storage space must exist for both the original and compacted versions of the database. The source database path and destination databases path must not be the same. If any of these conditions are not met, the `SqlCeEngine` will throw a `SqlException`. Listing 7.1 demonstrates how to compact a SQL Server CE database:

LISTING 7.1 Using the `SqlCeEngine`'s `Compact` method

```
C#
string srcDBName = "testdb.sdf";
string destDBName = "testdb.sdf.tmp";
string newPassword = "Password = 123Testing";

SqlCeEngine engine = new SqlCeEngine("Data Source = " + srcDBName);

try {
  // Compact the database and add password protection
  engine.Compact("Data Source = " + destDBName + ";" + newPassword);
  engine.Dispose();

  File.Delete(srcDBName);
  File.Move(destDBName, srcDBName);
}
catch (SqlCeException e)
{
  // Handle the exception
}
finally
{
  // Clean up the SqlCeEngine object
  engine.Dispose();
}
```

SHOP TALK

```
VB
Dim srcDBName as string
Dim destDBName as string
Dim newPassword as string
srcDBName = "testdb.sdf"
destDBName = "testdb.sdf.tmp"
newPassword = "Password = 123Testing"

Dim engine as new SqlCeEngine("Data Source = " & srcDBName)

Try
   // Compact the database and add password protection
  engine.Compact("Data Source = " & destDBName & ";" & newPassword)
  engine.Dispose()

  File.Delete(srcDBName)
  File.Move(destDBName, srcDBName)
Catch e As SqlCeException
  // Handle the exception
Finally
  // Clean up the SqlCeEngine object
  engine.Dispose()
End Try
```

Creating a SQL Server CE database programmatically is very simple. There are only three steps:

1. The best practice is to ensure that the SQL Server CE database file (`.sdf`) does not exist before the database is created. The existing database will be deleted when creating the new database. So, caution must be taken when creating a new database.
2. An instance `SqlCeEngine` class must be instantiated and initialized with a connection string that describes how the engine should connect to the database. Table 7.1 describes the complete list of connection string properties.
3. Call the `CreateDataBase` method on the `SqlCeEngine` instance. Listing 7.2, demonstrates how to create a new SQL Server CE database.

LISTING 7.2 Creating a new SQL Server CE database

```
C#
public void CreateNewDatabase() {
  if(File.Exists("tempdb.sdf")
    File.Delete("tempdb.sdf");
```

```
  string connStr = "Data Source = tempdb.sdf; Password = testing123"

  using(SqlCeEngine engine = new SqlCeEngine(connStr)) {
    engine.CreateDatabase();
  }
}
```

```
VB
Sub Main()
  If File.Exists("tempdb.sdf") Then
    File.Delete("tempdb.sdf")
  End If

  Dim connStr As String
  connStr = "DataData Source = tempdb.sdf; Password = testing123"

  Dim engine As SqlCeEngine
  engine = New SqlCeEngine(connStr)
  engine.CreateDatabase()
  engine.Dispose()
End Sub
```

It is important to note that the `SqlCeEngine` object supports the `IDisposable` interface, and its resources can be cleaned up by calling the `Dispose` method. For C# users, wrapping the use of the `SqlCeEngine` instance in the `using` statement is the easiest and most reliable way to ensure that the `SqlCeEngine` is cleaned up after.

TABLE 7.1
SQL Server CE Command Line Parameters

PARAMETER	REQUIRED OR OPTIONAL	DESCRIPTION
Provider	Optional	The name of the data source provider. If the provider is not specified, `Microsoft.sqlserver.oledb.ce.2.0` is assumed.
`Data Source`	Required	The name of the database. By convention, you should name the SQL Server CE databases with the file extension `.sdf`.
`Locale Identifier`	Optional	The database locale. This property specifies the collation order for string comparisons in the database. The default database locale is Latin1_General (0 × 00000409).
`SSCE:Database Password`	Optional	The database password. This parameter must be specified if the database was created with a password.
`SSCE:Encrypt Database`	Optional	The property that specifies whether a newly created database should be encrypted. This property is only meaningful when creating a database. The created database is encrypted when this Boolean parameter is true and a database password is specified.

Adding Structure to a Microsoft SQL Server CE Database

After creating a SQL Server CE database, the next step is to add tables to the database. This can be done graphically by using SQL Server CE Query Analyzer or programmatically by using the SQL Server CE Data Provider classes. To learn more about the Query Analyzer, see Microsoft SQL Server CE Books Online.

To programmatically create database tables, you will need to connect to the database by using the `SqlCeConnection` class and issue DDL commands by using the `SqlCeCommand` class. Before we jump into how this is done, we will need to discuss the capabilities of SQL Server CE.

SQL Server CE supports a subset of the DDL. Table 7.2 describes the supported DDL statements.

TABLE 7.2
DDL Statements Supported by SQL Server CE

DDL STATEMENT	FUNCTION
`CREATE DATABASE`	Creates a new database and the file used to store the database.
`CREATE TABLE`	Creates a new table. Primary keys, unique and foreign keys, and defaults can be specified with this command.
`ALTER TABLE`	Modifies a table definition by altering, adding, or dropping columns and constraints.
`CREATE INDEX`	Creates an index on a given table.
`DROP INDEX`	Removes one or more indexes from the current database.
`DROP TABLE`	Removes a table definition and all data, indexes, and constraints for that table.

For more information on these commands, see the Microsoft SQL Server CE Books Online.

SQL Server CE also supports a subset of data types. Table 7.3 describes the data types supported by SQL Server CE.

TABLE 7.3
Data Types Supported by SQL Server CE

DATA TYPE	DESCRIPTION
`Bigint`	Integer (whole number) data from -2^{63} (–9,223,372,036,854,775,808) through $2^{63} - 1$ (9,223,372,036,854,775,807).
`Integer`	Integer (whole number) data from -2^{31} (–2,147,483,648) through $2^{31} - 1$ (2,147,483,647).
`Smallint`	Integer data from –32,768 to 32,767.
`Tinyint`	Integer data from 0 to 255.

TABLE 7.3
Continued

DATA TYPE	DESCRIPTION
`Bit`	Integer data with either a 1 or 0 value.
`numeric (p, s)`	Fixed-precision and scale-numeric data from $-10^{38} + 1$ through $10^{38} - 1$. *p* specifies precision and can vary between 1 and 38. *s* specifies scale and can vary between 0 and *p*.
`Money`	Monetary data values from $-2^{63}/10,000$ through $(2^{63} - 1)/10,000$ (–922,337,203,685,477.5808 through 922,337,203,685,477.5807 units).
`Float`	Floating-point number data from –1.79E+308 through 1.79E+308.
`Real`	Floating precision number data from –3.40E+38 through 3.40E+38.
`Datetime`	Date and time data from January 1, 1753, to December 31, 9999, with an accuracy of one three-hundredth second, or 3.33 milliseconds. Values are rounded to increments of .000, .003, or .007 milliseconds.
`nchar(`*n*`)`	Fixed-length Unicode data with a maximum length of 255 characters. Default length = 1.
`nvarchar(`*n*`)`	Variable-length Unicode data with a length of 1 to 255 characters. Default length = 1.
`ntext`	Variable-length Unicode data with a maximum length of $(2^{30} - 2) / 2$ (536,870,911) characters.
`binary(`*n*`)`	Fixed-length binary data with a maximum length of 510 bytes. Default length = 1.
`varbinary(`*n*`)`	Variable-length binary data with a maximum length of 510 bytes. Default length = 1.
`Image`	Variable-length binary data with a maximum length of $2^{30} - 1$ (1,073,741,823) bytes.
`uniqueidentifier`	A globally unique identifier (GUID).
`IDENTITY [(s, i)]`	This is a property of a data column, not a distinct data type. Only data columns of the integer data types can be used for identity columns. A table can have only one identity column. A seed and increment can be specified, and the column cannot be updated. *s* (seed) = starting value *i* (increment) = increment value
`ROWGUIDCOL`	This is a property of a data column, not a distinct data type. It is a column in a table that is defined by using the `uniqueidentifier` data type.

To learn more about theses data types, see the Microsoft SQL Server CE Books online.

Now, let's explore how to create the structure of a SQL Server database. The remaining samples in this chapter will rely on the tables we will now create. The sample database will contain two tables: the `Package` table and the `TrackingEntry` table. Tables 7.4 and 7.5 describe the columns and column types of the `Package` and `TrackingEntry` tables, respectively.

TABLE 7.4
Structure of the `Package` Table

COLUMN NAME	TYPE	SIZE
`ID`	`Int`	IDENTITY(1,1) PRIMARY KEY
`Code`	`Nvarchar`	12
`DestinationID`	`Nvarchar`	12

TABLE 7.5
Structure of the `TrackingEntry` Table

COLUMN NAME	TYPE	SIZE
`ID`	`Int`	IDENTITY(1,1) PRIMARY KEY
`PackageID`	`Int`	FOREIGN KEY
`LocationID`	`Nvarchar`	12
`ArrivalTime`	`Datetime`	
`DepartureTime`	`Datetime`	

The `Package` and `TrackingEntry` tables are part of a system that would help a shipping company track the location of a package. Each package has a package code (`Code`) and a destination ID (`DestinationID`). In practice it may be convenient to read the package code via a bar code scanner. The destination ID is a string that uniquely identifies the final shipping building the package leaves from before it is loaded onto a truck and delivered to the recipient.

As the package travels to its destination, it will make some stops along the way at intermediate shipping buildings. When a package comes into one of the intermediate shipping buildings, the package is scanned and an entry into the `TrackingEntry` table is created. This table tracks which shipping location it arrived at, what time it arrived, and what time it left the building. Listing 7.3, demonstrates how to create the `Package` and `TrackingEntry` tables.

LISTING 7.3 Creating the `Package` and `TrackingEntry` tables

```
C#
public static void CreateTrackingDatabase() {
  string connstr = @"Data Source=\My Documents\PTSystem.sdf";

  using(SqlCeConnection conn = new SqlCeConnection(connstr)) {
    conn.Open();

    // Create an the package table
    string ddlPackage =
```

```
        "CREATE TABLE Package( " +
        "ID int not null identity(1,1) PRIMARY KEY, " +
        "Code nvarchar(12) not null, " +
        "DestinationID nvarchar(12) not null)";
      RunDDLCommand(conn, ddlPackage);

      // Create the tracking entry table
      string ddlTrackingEntry =
        "CREATE TABLE TrackingEntry( " +
        "ID int not null identity(1,1), " +
        "PackageID int not null, " +
        "LocationID nvarchar(12) not null, " +
        "ArrivalTime datetime not null, " +
        "DepartureTime datetime null, " +
        "FOREIGN KEY (PackageID) REFERENCES Package(ID) )";
      RunDDLCommand(conn, ddlTrackingEntry);

      // Create an index on the tracking entry table
      string ddlArrivalTimeNdx =
        "CREATE INDEX ArrivalTime ON TrackingEntry(ArrivalTime )";
      RunDDLCommand(conn, ddlArrivalTimeNdx );
    }
}
```

```
VB
Sub CreateTrackingDatabase()
  Dim connstr As String
  connstr = "Data Source=\My Documents\PTSystem.sdf"

  Dim conn As SqlCeConnection
  conn = New SqlCeConnection(connstr)
  conn.Open()

  'Create an the package table
  Dim ddlPackage As String
  ddlPackage = "CREATE TABLE Package( " & _
    "ID int not null identity(1,1) PRIMARY KEY, " & _
    "Code nvarchar(12) not null, " & _
    "DestinationID nvarchar(12) not null)"
  RunDDLCommand(conn, ddlPackage)
```

```
  ' Create the tracking entry table
  Dim ddlTrackingEntry As String
  ddlTrackingEntry = "CREATE TABLE TrackingEntry( " & _
    "ID int not null identity(1,1), " & _
    "PackageID int not null, " & _
    "LocationID nvarchar(12) not null, " & _
    "ArrivalTime datetime not null, " & _
    "DepartureTime datetime null, " & _
    "FOREIGN KEY (PackageID) REFERENCES Package(ID) )"
  RunDDLCommand(conn, ddlTrackingEntry)

  ' Create an index on the tracking entry table
  Dim ddlArrivalTimeNdx As String
  ddlArrivalTimeNdx =
    "CREATE INDEX ArrivalTime ON TrackingEntry(ArrivalTime )"
  RunDDLCommand(conn, ddlArrivalTimeNdx)

  conn.Close()
End Sub
```

Let's take a look at the method. This method starts out by creating a connection to the SQL Server database by using the `SqlCeConnection` object. This object instance is created by using the connection string that grants access to the database. Notice that this connection is embedded in a `using` statement to ensure that the connection's resources are cleaned up after it is done being used. Next, the connection to the database is opened with a parameter-less call to the `SqlCeConnection.Open` method. We then create the `Package` table. Using a string that contains the `CREATE TABLE` SQL command. To learn about the `CREATE TABLE` command, see the Microsoft SQL Server CE Books Online. Next, we create the `TrackingEntry` table. This table contains a foreign key constraint on the `PackageID` column. The values inserted in the `PackageID` column must exist in the `ID` column of the `Package` table. In other words, a package that has not been entered into the `Package` table cannot have a `TrackingEntry` record. Finally, for performance reasons, we create an index consisting of an `ArrivalTime` column on the `TrackingEntry` table. This index will be used later in this chapter. To learn more about the `Create Index` SQL command, see the Microsoft SQL Server CE Books Online.

In the preceding sample, the `RunDDLCommand` method actually creates the different elements of the database. Listing 7.4 contains the code for the `RunDDLCommand` method.

LISTING 7.4 The implementation of the `RunDDLCommand` method

```
C#
public static void
RunDDLCommand(SqlCeConnection conn, string ddlCmdStr) {
  SqlCeCommand cmdDDL = null;
```

```
  try {
    cmdDDL = new SqlCeCommand(ddlCmdStr, conn);
    cmdDDL.CommandType = CommandType.Text;
    cmdDDL.ExecuteNonQuery();
  } catch(SqlCeException scee) {
    for(int curExNdx = 0; curExNdx < scee.Errors.Count; ++curExNdx) {
      MessageBox.Show("Error:"+scee.Errors[curExNdx].ToString()+"\n");
    }
  } finally {
    if( cmdDDL != null )
      cmdDDL.Dispose();
  }
}

VB
sub RunDDLCommand(conn as SqlCeConnection , ddlCmdStr as string )
  Dim cmdDDL As SqlCeCommand
  cmdDDL = Nothing

  Try
    cmdDDL = New SqlCeCommand(ddlCmdStr, conn)
    cmdDDL.CommandType = CommandType.Text
    cmdDDL.ExecuteNonQuery()
  Catch scee As SqlCeException
    Dim curExNdx As Int32
    For curExNdx = 0 To scee.Errors.Count
      MessageBox.Show("Error:" & scee.Errors(curExNdx).ToString())
    Next
  Finally
    If Not cmdDDL Is Nothing Then
      cmdDDL.Dispose()
    End If
  End Try
End Sub
```

The sole purpose of `RunDDLCommand` is to run a DDL command against a database by using a specified connection. First, a `SqlCeCommand` object is created by using the specified `SqlCeConnection` object and the specified command string that contains the DDL statement. We then specify how the command string should be interpreted by setting the `SqlCeCommand.CommandType` property. Because the command string is a just a SQL command string, we use `CommandType.Text`. The complete list of `CommandType` enumeration values are listed, along with a description, in Table 7.6. Finally, the `SqlCeCommand.ExecuteNonQuery` method is called to actually run the DDL statement against the database. The ExecuteNonQuery method should be used when running SQL commands that do not return a result other than the number of rows affected.

There are two other notable code blocks in the RunDDLCommand method. First, notice that we ensure that the command is properly disposed by wrapping the command execution code in a try/catch and calling SqlCeCommand.Dispose. The second notable block is in the catch block. We catch SQL errors that occur in the command execution code by catching all SqlCeException objects that are thrown. The SqlCeException contains an Errors property that is a list of errors that caused the SqlCeException to be thrown. The code in the catch block walks the list of errors and displays them to the user.

TABLE 7.6
The CommandType Enumeration Values

MEMBER NAME	DESCRIPTION
StoreProcedure	The name of a stored procedure. Note that SQL Server CE does not support stored procedures.
Text	A SQL text command (default).
TableDirect	When the CommandType property is set to TableDirect, the CommandText property should be set to the name of the table or tables to be accessed. All rows and columns of the named table or tables will be returned when you call one of the Execute methods. In order to access multiple tables, use a comma-delimited list, without spaces or padding, that contains the names of the tables to access. When the CommandText property names multiple tables, a join of the specified tables is returned.

Populating a Microsoft SQL Server CE Database

A SQL Server CE database can be populated by using SQL DML statements. SQL Server CE 2.0 supports a subset of the DML statement that is supported by SQL Server. The supported commands are listed in Table 7.7. See the Microsoft SQL Server CE 2.0 Books Online to learn more about the DML statements.

TABLE 7.7
DML Statements Supported by SQL Server CE

STATEMENT	FUNCTION
INSERT	Adds a new row to a table.
UPDATE	Modifies existing data in a table.
DELETE	Removes rows from a table.
SELECT	Retrieves rows from the database and allows the selection of one or many rows or columns from one or many tables. The SELECT statement supports inner and outer joins, and the Order By, Group By, and Having clauses.

SQL Server CE Query Analyzer can be used to execute DML statements against the database. The SqlCeCommand class can be used to programmatically execute these statements by using the SQL Server CE Data Provider.

To populate a SQL Sever CE database, you can run a series of INSERT commands against the database. Running an INSERT statement is very similar to running a CREATE TABLE DDL statement. The steps are basically the same.

1. Open a connection to the SQL Server CE database by using an instance of the SqlCeConnection class.
2. Create a SqlCeCommand object instance, and give it the INSERT command string to run against the server.
3. Set the command type, and execute the statement by using the ExecuteNonQuery method.

Listing 7.5 demonstrates how to insert a new package into the Package table.

LISTING 7.5 Inserting a new package into the Package table

```
C#
public static void
InsertNewPackage(string pckgCode, string destID) {
  String connstr = @"Data Source=\My Documents\PTSystem.sdf";

  using(SqlCeConnection conn = new SqlCeConnection(connStr)) {
    conn.Open();

    string dmlInsertPackage =
      "INSERT INTO Package(Code, DestinationID) " +
      "VALUES ('" + pckgCode + "', '" + destID + "')";
    SqlCeCommand cmdInsertPackage =
      new SqlCeCommand(conn, dmlInsertPackage);

    try {
      cmdInsertPackage = new SqlCeCommand(conn , dmlInsertPackage);
      cmdInsertPackage.CommandType = CommandType.Text;
      cmdInsertPackage.ExecuteNonQuery();
    } catch(SqlCeException scee) {
      for(int curNdx=0; curNdx<scee.Errors.Count; ++curNdx) {
        MessageBox.Show("Error:"+scee.Errors[curNdx].ToString()+"\n");
      }
    } finally {
      if(cmdInsertPackage != null)
        cmdInsertPackage.Dispose();
```

```
    }
  }
}

VB
sub InsertNewPackage(pckgCode as string , destID as string)
  Dim connstr As String
  connstr = "Data Source=\My Documents\PTSystem.sdf"

  Dim conn As SqlCeConnection
  conn = New SqlCeConnection(connstr)
  conn.Open()

  Dim dmlInsertPackage As String
  dmlInsertPackage = _
    "INSERT INTO Package(Code, DestinationID) " & _
    "VALUES ('" + pckgCode + "', '" + destID + "')"

  Dim cmdInsertPackage As SqlCeCommand
  cmdInsertPackage = New SqlCeCommand(dmlInsertPackage, conn)

  Try
    cmdInsertPackage = New SqlCeCommand(dmlInsertPackage, conn)
    cmdInsertPackage.CommandType = CommandType.Text
    cmdInsertPackage.ExecuteNonQuery()
  Catch scee As SqlCeException
    Dim curNdx As Int32
    For curNdx = 0 To scee.Errors.Count
      MessageBox.Show("Error:" & scee.Errors(curNdx).ToString())
    Next

    Finally
      If Not cmdInsertPackage Is Nothing Then
        cmdInsertPackage.Dispose()
      End If
    End Try

    conn.Close()
End Sub
```

Retrieving Data by Using SqlCeDataReader

Data can be retrieved from a SQL Server CE database by using the SqlCeDataReader class. The SqlCeDataReader class provides fast, forward-only access to rows of data.

The steps to retrieve data by using the SqlCeDataReader are similar to the steps required to run the INSERT SQL command:

1. We create an instance of the SqlCeConnection. The SqlCeDataReader will use this connection to retrieve the requested data rows.
2. A SqlCeCommand object should be created with the appropriate SELECT statement.
3. Set the command's command type, and call the SqlCeCommand.ExecuteReader method.

The ExecuteReader method executes the command text against the database by using the SqlCeConnection. A SqlCeDataReader that will provide access to the return data is then returned. The SqlCeConnection will be busy serving the SqlCeDataReader until the reader is closed.

The ExecuteReader method takes one parameter of the CommandBehavior type. The CommandBehavior type is an enumeration that the SqlCeCommand uses; it helps determine what results are returned and how the SqlCeConnection is left after the command executes. Table 7.8 lists all of the CommandBehavior member values and their descriptions.

TABLE 7.8
CommandBehavior Values

MEMBER NAME	DESCRIPTION
CloseConnection	The connection is closed after the data reader is closed.
Default	The query may return multiple result sets.
KeyInfo	The query returns column and primary key information. The query is executed without any locking on the selected rows.
SchemaOnly	The query returns column information only.
SequentialAccess	The query provides a way for the DataReader to handle rows containing columns of large binary values.
SingleResult	The query returns a single result set.
SingleRow	The query is expected to return a single row. It is possible to specify SingleRow when executing queries that return multiple result sets. In that case, multiple result sets are still returned, but each result set has a single row.

Once the SqlCeDataReader has been returned from the call to ExecuteReader, it is ready to retrieve data. The SqlCeDataReader.Read method advances the reader to the next record. Initially, the SqlCeDataReader is positioned directly before the first record. So, a call to Read is required before any data can be retrieved. The Read method will return true until the SqlCeDataReader reaches the end of result set. It then returns false.

Once you are positioned on a data row, you can use one of SqlCeDataReader's GetXXX methods to access the columns within each data row. For instance, the GetInt32 method retrieves an Int32 value from a column in the SqlCeDataReader's current row. The GetInt32 method takes a parameter of type int. This parameter represents the column's ordinal number. If the column's ordinal is unknown at design time, you can use the GetOrdinal method to look up the column's ordinal by column name. Listing 7.6 demonstrates how to retrieve all of the package information from the Package table.

LISTING 7.6 Retrieving all of the package information from the Package table

```
C#
public static void GetAllPackageInfo() {
  string pckgStr =
    "Package Data\nID: {0}\nCode: {1}\nDestination: {2}";
  string connstr = @"Data Source=\My Documents\PTSystem.sdf";

  using(SqlCeConnection conn = new SqlCeConnection(connstr)) {
    conn.Open();

    string dmlPackageInfo = "SELECT * FROM Package";
    SqlCeCommand cmdGetPackageInfo = null;
    SqlCeDataReader drPackageInfo = null;

    try {
      cmdGetPackageInfo = new SqlCeCommand(dmlPackageInfo, conn);
      cmdGetPackageInfo.CommandType = CommandType.Text;
      drPackageInfo =
        cmdGetPackageInfo.ExecuteReader(CommandBehavior.Default);

      while(drPackageInfo.Read()) {
        System.Windows.Forms.MessageBox.Show(
          string.Format(pckgStr,
          drPackageInfo.GetInt32(0),
          drPackageInfo.GetString(1),
          drPackageInfo.GetString(2)));
      }
    } catch(SqlCeException scee) {
      for(int curExNdx = 0; curExNdx < scee.Errors.Count; ++curExNdx) {
        System.Windows.Forms.MessageBox.Show(
          "Error:"+ scee.Errors[curExNdx].ToString()+"\n");
      }
    } finally {
      if( cmdGetPackageInfo != null )
```

```
        cmdGetPackageInfo.Dispose();

      if( drPackageInfo != null )
        drPackageInfo.Close();
    }
  }
}
```

VB

```
Sub InsertNewPackage(ByVal pckgCode As String, ByVal destID As String)
  Dim connstr As String
  connstr = "Data Source=\My Documents\PTSystem.sdf"

  Dim conn As SqlCeConnection
  conn = New SqlCeConnection(connstr)
  conn.Open()

  Dim dmlInsertPackage As String
  dmlInsertPackage = _
    "INSERT INTO Package(Code, DestinationID) " & _
    "VALUES ('" + pckgCode + "', '" + destID + "')"

  Dim cmdInsertPackage As SqlCeCommand
  cmdInsertPackage = New SqlCeCommand(dmlInsertPackage, conn)

  Try
    cmdInsertPackage = New SqlCeCommand(dmlInsertPackage, conn)
    cmdInsertPackage.CommandType = CommandType.Text
    cmdInsertPackage.ExecuteNonQuery()
  Catch scee As SqlCeException
    Dim curNdx As Int32
    For curNdx = 0 To scee.Errors.Count
      MessageBox.Show("Error:" & scee.Errors(curNdx).ToString())
    Next
  Finally
    If Not cmdInsertPackage Is Nothing Then
      cmdInsertPackage.Dispose()
    End If
  End Try

  conn.Close()
End Sub
```

SHOP TALK

OPTIMIZING SELECT QUERIES

When executing `SELECT` queries against a SQL Server CE database, you can get the fastest performance by using a `SqlCommand` with the `CommandType` property set to `CommandType.TableDirect`. `TableDirect` commands do not need to be compiled by the SQL Server CE Engine. Instead, a pointer into the table is created and maintained while the `SqlCeDataReader` moves through the results. This results in fast reads from the SQL Server CE database.

When creating a `TableDirect` command, the `SqlCeCommand`'s `CommantText` property must be set to the name of the table to be accessed. All rows and columns of the named table will be returned when you call one of the `Execute` methods. The following code demonstrates how to create `TableDirect` `SqlCeCommand` linked to the `Package` table.

```
C#
string connstr = @"Data Source=\My Documents\PTSystem.sdf";
SqlCeCommand cmdGetPackageInfo = null;
SqlCeDataReader drPackageInfo = null;

using(SqlCeConnection conn = new SqlCeConnection(connstr)) {
  conn.Open();

  cmdGetPackageInfo = new SqlCeCommand("Package", conn);
  cmdGetPackageInfo.CommandType = CommandType.TableDirect;
  drPackageInfo = cmdGetPackageInfo.ExecuteReader();

  // Use the reader to get the package data.
}

VB
Dim connstr As String
connstr = "Data Source=\My Documents\PTSystem.sdf"

Dim conn As SqlCeConnection
conn = New SqlCeConnection(connstr)
conn.Open()

Dim cmdGetPackageInfo As SqlCeCommand
Dim drPackageInfo As SqlCeDataReader
cmdGetPackageInfo = Nothing
drPackageInfo = Nothing

cmdGetPackageInfo = New SqlCeCommand("Package", conn)
cmdGetPackageInfo.CommandType = CommandType.TableDirect
```

SHOP TALK

```
drPackageInfo = _
  cmdGetPackageInfo.ExecuteReader(CommandBehavior.Default)

' Use the reader to get the package data
```

TableDirect queries can also be used to access multiple tables. The CommandText property should contain a comma-delimited list of table names without spaces or padding. When accessing multiple tables, a join of the specified tables is returned.

Using Parameterized SQL Commands

The SELECT statement used in Listing 7.5 was extremely simple. In practice your SELECT statements will most likely contain WHERE clauses that will help target specific data rows. For example, we can use the WHERE clause to select a specific package in the Package table. An example of the SELECT query follows:

```
SELECT * FROM Package WHERE ID = "0987654321"
```

This SELECT query retrieves the package whose ID column contains the value 0987654321. Because the ID column is the primary key, the record set will contain zero or one data row. You can specify the ID at runtime by using string.Format to parameterize the SELECT statement. A better way to accomplish command parameterization is to use the SqlCeParameter objects. First, the SELECT command changes to the following format:

```
SELECT * FROM Package WHERE ID = ?
```

Now, create a SqlCeCommand object with the preceding command. The SqlCeCommand object provides a Parameters property that is a collection of all parameters to the command. Add a parameter to the list that corresponds to the ID and then call the SqlCeCommand.Prepare method. The Prepare method compiles the statement against the database. This allows the command to be executed faster if it is used more than once. Finally, execute the command against the database as before. Listing 7.7 exemplifies the steps described in this section.

LISTING 7.7 Executing a parameterized SQL command

```
C#
public static void GetPackageInfo(int pckgID) {
  string pckgStr =
    "Package Data\nID: {0}\nCode: {1}\nDestination: {2}";
  string connstr = @"Data Source=\My Documents\PTSystem.sdf";

  using(SqlCeConnection conn = new SqlCeConnection(connstr)) {
    conn.Open();
```

```
    string dmlPackageInfo = "SELECT * FROM Package WHERE ID = ?";
    SqlCeCommand cmdGetPackageInfo = null;
    SqlCeDataReader drPackageInfo = null;

    try {
      cmdGetPackageInfo = new SqlCeCommand(dmlPackageInfo, conn);
      cmdGetPackageInfo.CommandType = CommandType.Text;
      cmdGetPackageInfo.Parameters.Add("ID", pckgID);
      cmdGetPackageInfo.Prepare();

      drPackageInfo =
        cmdGetPackageInfo.ExecuteReader(CommandBehavior.SingleRow);

      while(drPackageInfo.Read()) {
        System.Windows.Forms.MessageBox.Show(
          string.Format(pckgStr,
          drPackageInfo.GetInt32(0),
          drPackageInfo.GetString(1),
          drPackageInfo.GetString(2)));
      }
    } catch(SqlCeException scee) {
      for(int curExNdx = 0; curExNdx < scee.Errors.Count; ++curExNdx) {
        System.Windows.Forms.MessageBox.Show(
          "Error:"+ scee.Errors[curExNdx].ToString()+"\n");
      }
    } finally {
      if( cmdGetPackageInfo != null )
        cmdGetPackageInfo.Dispose();

      if( drPackageInfo != null )
        drPackageInfo.Close();
    }
  }
}
```

```
VB
Sub GetPackageInfo(ByVal pckgID As Int32)
  Dim connstr As String
  Dim pckgStr As String
  pckgStr = "Package Data\nID: {0}\nCode: {1}\nDestination: {2}"
  connstr = "Data Source=\My Documents\PTSystem.sdf"
```

```
  Dim conn As SqlCeConnection
  conn = New SqlCeConnection(connstr)
  conn.Open()

  Dim dmlPackageInfo As String
  Dim cmdGetPackageInfo As SqlCeCommand
  Dim drPackageInfo As SqlCeDataReader

  dmlPackageInfo = "SELECT * FROM Package WHERE ID = ?"
  cmdGetPackageInfo = Nothing
  drPackageInfo = Nothing

  Try
    cmdGetPackageInfo = New SqlCeCommand(dmlPackageInfo, conn)
    cmdGetPackageInfo.CommandType = CommandType.Text
    cmdGetPackageInfo.Parameters.Add("ID", pckgID)
    cmdGetPackageInfo.Prepare()

    drPackageInfo = _
      cmdGetPackageInfo.ExecuteReader(CommandBehavior.SingleRow)

    While (drPackageInfo.Read())
      System.Windows.Forms.MessageBox.Show( _
        String.Format(pckgStr, _
        drPackageInfo.GetInt32(0), _
        drPackageInfo.GetString(1), _
        drPackageInfo.GetString(2)))
    End While
  Catch scee As SqlCeException
    Dim curExNdx As Int32
    For curExNdx = 0 To scee.Errors.Count
      System.Windows.Forms.MessageBox.Show( _
        "Error:" & scee.Errors(curExNdx).ToString())
    Next
  Finally
    If Not cmdGetPackageInfo Is Nothing Then
      cmdGetPackageInfo.Dispose()
    End If

    If Not drPackageInfo Is Nothing Then
      drPackageInfo.Close()
    End If
  End Try
End Sub
```

Parameterized queries can be used in almost all SQL statements, DDL and DML. It is also possible to use more than one parameter in the query. For example, the following query can be used to SELECT and package with a parameterized code or destination ID.

```
SELECT * FROM Package WHERE Code = ? OR DestinationID = ?
```

When using the SELECT statement with multiple parameters, you must add the SqlCeParameters objects to the Parameters collection in the order that the question marks appear from left to right. The preceding example would require the Code parameter to be inserted into the Parameters collection *before* the DestinationID was inserted. After adding the parameters to the Parameters collection, they can no longer be referenced by name. So, you must ensure that they are inserted in the proper order.

Parameterized queries are a great way to increase the performance of a query that must be executed more then once. You must take care to change the value of the parameters after each call to prepare. Listing 7.8 demonstrates how this should be done.

LISTING 7.8 Executing a SQL command with multiple parameters

```
C#
public static void GetPackageInfo(int[] pckgID) {
  string pckgStr =
    "Package Data\nID: {0}\nCode: {1}\nDestination: {2}";
  string connstr = @"Data Source=\My Documents\PTSystem.sdf";

  using(SqlCeConnection conn = new SqlCeConnection(connstr)) {
    conn.Open();

    string dmlPackageInfo = "SELECT * FROM Package WHERE ID = ?";
    SqlCeCommand cmdGetPackageInfo = null;
    SqlCeDataReader drPackageInfo = null;

    try {
      cmdGetPackageInfo = new SqlCeCommand(dmlPackageInfo, conn);
      cmdGetPackageInfo.CommandType = CommandType.Text;
      cmdGetPackageInfo.Parameters.Add("ID", SqlDbType.Int);
      cmdGetPackageInfo.Prepare();

      for(int pckgNdx = 0; pckgNdx < pckgID.Length; ++pckgNdx) {
        cmdGetPackageInfo.Parameters[0].Value = pckgID[pckgNdx];
        try {
          drPackageInfo =
            cmdGetPackageInfo.ExecuteReader(CommandBehavior.SingleRow);
```

```
          while(drPackageInfo.Read()) {
            System.Windows.Forms.MessageBox.Show(
              string.Format(pckgStr,
              drPackageInfo.GetInt32(0),
              drPackageInfo.GetString(1),
              drPackageInfo.GetString(2)));
          }
        } catch(SqlCeException scee) {
          for(int curExNdx=0;curExNdx<scee.Errors.Count;++curExNdx) {
            System.Windows.Forms.MessageBox.Show(
              "Error:"+ scee.Errors[curExNdx].ToString()+"\n");
          }
        } finally {
          if( drPackageInfo != null )
            drPackageInfo.Close();
        }
      }
    } finally {
      if( cmdGetPackageInfo != null )
        cmdGetPackageInfo.Dispose();
    }
  }
}
```

VB

```
Sub GetPackageInfo(ByVal pckgID As Int32())
  Dim connstr As String
  Dim pckgStr As String
  pckgStr = "Package Data\nID: {0}\nCode: {1}\nDestination: {2}"
  connstr = "Data Source=\My Documents\PTSystem.sdf"

  Dim conn As SqlCeConnection
  conn = New SqlCeConnection(connstr)
  conn.Open()

  Dim dmlPackageInfo As String
  Dim cmdGetPackageInfo As SqlCeCommand
  Dim drPackageInfo As SqlCeDataReader

  dmlPackageInfo = "SELECT * FROM Package WHERE ID = ?"
  cmdGetPackageInfo = Nothing
  drPackageInfo = Nothing
```

```
    Try
      cmdGetPackageInfo = New SqlCeCommand(dmlPackageInfo, conn)
      cmdGetPackageInfo.CommandType = CommandType.Text
      cmdGetPackageInfo.Parameters.Add("ID", SqlDbType.Int)
      cmdGetPackageInfo.Prepare()

      Dim pckgNdx As Int32
      For pckgNdx = 0 To pckgID.Length
        cmdGetPackageInfo.Parameters(0).Value = pckgID(pckgNdx)

        Try
          drPackageInfo = _
            cmdGetPackageInfo.ExecuteReader(CommandBehavior.SingleRow)

          While (drPackageInfo.Read())
            System.Windows.Forms.MessageBox.Show( _
              String.Format(pckgStr, _
              drPackageInfo.GetInt32(0), _
              drPackageInfo.GetString(1), _
              drPackageInfo.GetString(2)))
            End While
        Catch scee As SqlCeException
          Dim curExNdx As Int32
          For curExNdx = 0 To scee.Errors.Count
            System.Windows.Forms.MessageBox.Show( _
              "Error:" & scee.Errors(curExNdx).ToString())
          Next
        Finally
          If Not drPackageInfo Is Nothing Then
            drPackageInfo.Close()
          End If
        End Try
      Next
    Finally
      If Not cmdGetPackageInfo Is Nothing Then
        cmdGetPackageInfo.Dispose()
      End If
    End Try

    Conn.Close()
  End Sub
```

Filling a DataSet by Using the SqlCeDataAdapter

The Compact Framework also provides the ability to load data directly from SQL Server CE into a DataSet. This is done by using the SqlCeDataAdapter to fill the DataSet. The SqlCeDataAdapter can fill the DataSet and then update the underlying database with any changes. In other words, the DataSet can manage all communication between your application and the SQL Server CE database.

The SqlCeDataAdapter manages the underlying database by running four different commands against the database when synchronization is required. These four commands are exposed as properties on the SqlCeDataAdapter. They are the SelectCommand property, the InsertCommand property, the UpdateCommand property, and the DeleteCommand property. All of the properties are of the SqlCeCommand type, and you can set them to any command you choose, or you can use the SqlCeCommandBuilder class to generate the commands.

The SelectCommand property is the SqlCeCommand object that defines the SQL command the SqlCeDataAdapter will use to retrieve data from the SQL Server CE database. The SqlCeDataAdapter will then use this data to populate the DataSet.

Precisely, the fill process consists of three steps:

1. Building the DataSet schema and mappings
2. Retrieving the data
3. Populating the DataSet

First, the SqlCeDataAdapter initializes the schema of the DataSet to match the schema of the data source. This means that DataTables are built to match the source database tables as well as building DataColumns that match the source database table columns. The relationships between DataSet and source database are known as *mappings* because they *map* DataSet objects to database objects. Next, the data is retrieved from the source database by using the SelectCommand property. Finally, the DataRows are created for the retrieved data, and the rows are inserted into the DataTables.

Before we investigate the code to fill a DataSet, let's look at a few of the properties of SqlCeDataAdapter that should be set before we attempt to populate the DataSet. These properties will determine how the SqlCeDataAdapter will fill the DataSet.

The MissingMappingAction property determines what action to take when a mapping is missing from a source table or a source column. Table 7.9 describes the values of the MissingMappingAction enumeration and how the SqlCeDataAdapter will handle the missing mapping.

TABLE 7.9
MissingMappingAction Values

VALUE	DESCRIPTION
Error	An InvalidOperationException is generated if the specified column mapping is missing.
Ignore	The column or table not having a mapping is ignored.
Passthrough (Default)	The source column or source table is created and added to the DataSet by using its source name.

The MissingSchemaAction property determines the action to take when existing DataSet schema does not match incoming data. Table 7.10 describes the values of the MissingSchemaAction enumeration and how the SqlCeDataAdapter will handle the missing schema objects.

TABLE 7.10
MissingSchemaAction Values

VALUE	DESCRIPTION
Add (Default)	Adds the missing columns to complete the schema
AddWithKey	Adds the missing columns and primary key information to complete the schema
Error	Generates an InvalidOperationException if the schema is incomplete
Ignore	Ignores the missing columns

By default, the SqlCeDataAdapter will use the names of the column as they are in the source database (MissingMappingAction.Passthrough), and the SqlCeDataAdapter will add any new columns to the DataSet (MissingSchemaAction.Add).

The code to fill a DataSet by using the SqlCeDataAdapter is very simple. Listing 7.9 demonstrates how to fill a DataSet with the contents of the sample Package table by using the SqlCeDataAdapter.

LISTING 7.9 Filling a DataSet with the contents of the sample Package table

```
C#
public static DataSet GetPackageDataSet() {
  string connstr = @"Data Source=\My Documents\PTSystem.sdf";

  using(SqlCeConnection conn = new SqlCeConnection(connstr)) {
    conn.Open();

    string dmlPackageInfo = "SELECT * FROM Package";
    SqlCeDataAdapter daPackages = new SqlCeDataAdapter();
    daPackages.MissingMappingAction = MissingMappingAction.Passthrough;
```

```
    daPackages.MissingSchemaAction = MissingSchemaAction.Add;
    daPackages.SelectCommand = new SqlCeCommand(dmlPackageInfo, conn);

    DataSet dsPackages = new DataSet();
    daPackages.Fill(dsPackages);

    return dsPackages;
}
```

VB

```
Function GetPackageDataSet() As DataSet
  Dim connstr As String
  connstr = "Data Source=\My Documents\PTSystem.sdf"

  Dim conn As SqlCeConnection
  conn = New SqlCeConnection(connstr)
  conn.Open()

  Dim dmlPackageInfo As String
  Dim daPackages As SqlCeDataAdapter
  dmlPackageInfo = "SELECT * FROM Package"
  daPackages = New SqlCeDataAdapter
  daPackages.MissingMappingAction = MissingMappingAction.Passthrough
  daPackages.MissingSchemaAction = MissingSchemaAction.Add
  daPackages.SelectCommand = New SqlCeCommand(dmlPackageInfo, conn)

  Dim dsPackages As DataSet
  dsPackages = New DataSet
  daPackages.Fill(dsPackages)

  conn.Close()
  Return dsPackages
End Function
```

First, create and open a connection to the SQL Server CE database. Then, create a `SqlCeDataAdapter` object and set the `MissingMappingAction` and `MissingSchemaAction` properties. Set the properties to the default. In practice this is unnecessary; it is done here just to demonstrate the use of the properties. Next, set the `SelectCommand` to a `SqlCeCommand` object that will select all of the data from the `Package` table. Finally, create a `DataSet` object and call `SqlCeDataAdapter.Fill` method to populate the `DataSet` with the `Package` data.

Handling the `FillError` Event

Errors may occur while the `SqlCeDataAdapter` is filling the `DataSet` with data. If an error does occur, the `SqlCeDataAdapter` will raise a `FillError` event. The `FillError` event provides a user with information that should be examined and used to decide if the fill operation should continue. The `FillError` event may be raised when errors occur that pertain to filling the `DataSet` object. For example, a piece of data could not be converted to the `DataColumn`'s Common Language Runtime type. A `FillError` will not be raised if the error is generated at the data source.

To handle this event, you must provide an implementation of the `FilllErrorEventHandler` delegate. This delegate receives a parameter of the `FillErrorEventArgs` type. The `FillErrorEventArgs` object contains properties that help decide whether the fill operation should continue. Table 7.11 lists the properties and their descriptions.

TABLE 7.11

`FillErrorEventArgs` Properties

PROPERTY	DESCRIPTION
`Continue`	Gets or sets a value indicating whether the fill should continue in light of the error
`DataTable`	Gets the `DataTable` being filled when the error occurred
`Errors`	Displays the errors that were generated
`VALUES`	Gets the values for the row being filled when the error occurred

When handling a `FillError` event, you should evaluate the properties of the `FillErrorEventArgs` parameter and then set the `Continue` property to signal whether the fill operation should continue. Listing 7.10 demonstrates how to handle the `FillError` event.

LISTING 7.10 Handling the `FillError` event

```
C#
protected static void
OnFillError(object sender, FillErrorEventArgs args) {
  if( typeof(System.OverflowException) == args.Errors.GetType() ) {
    // Handle the conversion precision loss

    // The fill should continue
    args.Continue = true;
  }
}
```

```
VB
Private Shared Sub _
OnFillError(ByVal sender As Object, ByVal args As FillErrorEventArgs)
  If TypeOf args.Errors Is System.OverflowException Then
    ' Handle the conversion precision loss

    ' The fill should continue
    args.Continue = True
  End If
End Sub
```

Updating the Microsoft SQL Server CE Database by Using the SqlCeDataAdapter

Once a DataSet has been populated by using the SqlCeDataAdapter, you can make changes to the data and then update the data source by using the same SqlCeDataAdapter. In order for the SqlCeDataAdapter to update the data source, you must specify three additional SqlCommand objects for the SqlCeDataAdapter's UpdateCommand, InsertCommand, and DeleteCommand properties.

These commands will need to be parameterized commands so that the SqlCeDataAdapter will be able to update individual rows in the data source tables. To learn more about parameterized SQL commands, see the section titled "Using Parameterized SQL Commands" in this chapter. Listing 7.11 demonstrates how to create a SqlCeDataAdapter that will handle filling and updating a DataSet that is mapped to the Package table.

LISTING 7.11 Using the SqlCeDataAdapter to update a database

```
C#
public static
SqlCeDataAdapter GetPackageDataAdapter(SqlCeConnection conn){
  string dmlPackageInfo = "SELECT * FROM Package";
  string dmlUpdatePackage="UPDATE Package " +
                         "SET CODE = ?, " +
                         "    DestinationID = ? " +
                         "WHERE ID = ?";
  string dmlInsertPackage="INSERT INTO " +
                         "Package(Code, DestinationID) " +
                         "VALUES (?, ?)";
  string dmlDeletePackage="DELETE FROM " +
                         "Package " +
                         "WHERE ID = ?";
```

```
    SqlCeDataAdapter daPackages = new SqlCeDataAdapter();
    daPackages.SelectCommand = new SqlCeCommand(dmlPackageInfo, conn);

    daPackages.UpdateCommand = new SqlCeCommand(dmlUpdatePackage, conn);
    daPackages.UpdateCommand.Parameters.Add("Code", SqlDbType.NVarChar);
    daPackages.UpdateCommand.Parameters.Add("DestinationID",
      SqlDbType.NVarChar);
    daPackages.UpdateCommand.Parameters.Add("ID", SqlDbType.Int);


    daPackages.InsertCommand = new SqlCeCommand(dmlInsertPackage, conn);
    daPackages.InsertCommand.Parameters.Add("Code", SqlDbType.NVarChar);
    daPackages.InsertCommand.Parameters.Add("DestinationID",
      SqlDbType.NVarChar);

    daPackages.DeleteCommand = new SqlCeCommand(dmlDeletePackage, conn);
    daPackages.DeleteCommand.Parameters.Add("ID", SqlDbType.Int);

    return daPackages;
  }

  VB
  Function _
  GetPackageDataAdapter(ByVal conn As SqlCeConnection) _
  As SqlCeDataAdapter
    Dim dmlPackageInfo As String
    Dim dmlUpdatePackage As String
    Dim dmlInsertPackage As String
    Dim dmlDeletePackage As String

    dmlPackageInfo = "SELECT * FROM Package"
    dmlUpdatePackage = "UPDATE Package " & _
                       "SET CODE = ?, " * _
                       "    DestinationID = ? " & _
                       "WHERE ID = ?"
    dmlInsertPackage = "INSERT INTO " & _
                       "Package(Code, DestinationID) " & _
                       "VALUES (?, ?)"
    dmlDeletePackage = "DELETE FROM " & _
                       "Package " & _
                       "WHERE ID = ?"
```

```
  Dim daPackages As SqlCeDataAdapter
  daPackages = New SqlCeDataAdapter
  daPackages.SelectCommand = New SqlCeCommand(dmlPackageInfo, conn)

  daPackages.UpdateCommand = New SqlCeCommand(dmlUpdatePackage, conn)
  daPackages.UpdateCommand.Parameters.Add("Code", SqlDbType.NVarChar)
  daPackages.UpdateCommand.Parameters.Add("DestinationID",
    SqlDbType.NVarChar)
  daPackages.UpdateCommand.Parameters.Add("ID", SqlDbType.Int)

  daPackages.InsertCommand = New SqlCeCommand(dmlInsertPackage, conn)
  daPackages.InsertCommand.Parameters.Add("Code", SqlDbType.NVarChar)
  daPackages.InsertCommand.Parameters.Add("DestinationID",
    SqlDbType.NVarChar)

  daPackages.DeleteCommand = New SqlCeCommand(dmlDeletePackage, conn)
  daPackages.DeleteCommand.Parameters.Add("ID", SqlDbType.Int)

  Return daPackages
End Function
```

The SqlCeDataAdapter updates the data source when you call its Update method. The Update method performs five steps when updating the data source.

1. The values to be updated are loaded from the DataRow object into the relevant command parameters.
2. The RowUpdating event is raised.
3. The relevant command is executed against the data source.
4. The RowUpdated event is raised.
5. The RowSet property of the DataRow is reset to RowState.Unchanged by calling the AcceptChanges method.

We will discuss the RowUpdating and RowUpdated events shortly, but let's take a look at some code that will update a DataSet. Listing 7.12 demonstrates how to update the Package table by using the SqlDataAdapter and a DataSet populated with the data from the Package table. This code uses the GetPackageDataAdapter method from Listing 7.11.

LISTING 7.12 Updating the `Package` table by using the `SqlDataAdapter`

```
C#
public static void UpdatePackageTable(DataSet dsPackages) {
  string connstr = @"Data Source=\My Documents\PTSystem.sdf";

  using(SqlCeConnection conn = new SqlCeConnection(connstr)) {
    conn.Open();

    SqlCeDataAdapter daPackages = GetPackageDataAdapter(conn);
    daPackages.Update(dsPackages);
  }
}

VB
Sub UpdatePackageTable(ByVal dsPackages As DataSet)
  Dim connstr As String
  connstr = "Data Source=\My Documents\PTSystem.sdf"

  Dim conn As SqlCeConnection
  conn = New SqlCeConnection(connstr)
  conn.Open()

  Dim daPackages As SqlCeDataAdapter
  daPackages = GetPackageDataAdapter(conn)
  daPackages.Update(dsPackages)
End Sub
```

Handling `SqlCeDataAdapter` Update Events

When you call the `Update` method on the `SqlCeDataAdapter`, there are two events raised: the `RowUpdating` event and the `RowUpdated` event. The `RowUpdating` event is raised before the `Update` command is executed against the data source. The `RowUpdated` event is raised after the `Update` command is executed against the data source.

The `RowUpdating` event can be handled by providing an implementation of the `SqlCeRowUpdatingEventHandler` delegate. The delegate receives a `SqlCeRowUpdatingEventArgs` object that contains data about the `Update` command that is about to be updated. Table 7.12 describes the properties of `SqlCeRowUpdatingEventArgs` and how they affect the update process.

TABLE 7.12

Properties of the SqlCeRowUpdatingEventArgs Class

PROPERTY	DESCRIPTION
Command	Gets or sets the SqlCommand to execute
Errors	Gets any errors generated when the command executes
Row	Gets the DataRow that is going to be sent through the Update command
StatementType	Gets the type of SQL statement that will be executed
Status	Gets the status of the Update command (see Table 7.13 for values)
TableMappings	Gets the DataTableMapping that will be sent through the Update command

When you receive a RowUpdating event, you should evaluate the properties of the SqlCeRowUpdatingEventArgs object and decide whether the update should continue. This is done by setting the value of the Status property. The Status property is of UpdateStatus. UpdataStatus is an enumeration, and its members are described in Table 7.13.

TABLE 7.13

Members of the UpdateStatus Enumeration

MEMBER	DESCRIPTION
Continue	The SqlCeDataAdapter should continue updating the DataRows.
ErrorsOccurred	The event handler reports that the update should be treated as an error. Examine the errors before deciding how to continue.
SkipAllRemainingRows	The remaining rows as well as the current row will not be updated.
SkipCurrentRow	The remaining rows will be updated, but the changes to the current row will be abandoned.

After a row is updated, the RowUpdated event is raised. You can handle this event by providing an implementation of the SqlCeRowUpdatedEventHandler delegate. This delegate receives an instance of the SqlCeRowUpdatedEventArgs as a parameter. Table 7.14 describes the properties of the SqlCeRowUpdatedEvent and their effects on the updating process.

TABLE 7.14

Properties of the SqlCeRowUpdatedEventArgs Class

PROPERTY	DESCRIPTION
Command	Gets or sets the SqlCommand that was executed
Errors	Gets any errors generated when the command was executed
RecordsAffected	Gets the number of rows changed, inserted, or deleted by the command
Row	Gets the DataRow that was sent through the Update command
StatementType	Gets the type of SQL statement that was executed
Status	Gets the status of the Update command (see Table 7.13 for values)
TableMappings	Gets the DataTableMapping that was sent through the Update command

Listing 7.13 gives examples of how to implement delegates that handle the `RowUpdating` and `RowUpdated` events. The listing uses the `GetPackageDataAdapter` method from Listing 7.11.

LISTING 7.13 Handling the `RowUpdating` and `RowUpdated` events

```
C#
protected static void OnRowUpdating(object sender,
  SqlCeRowUpdatingEventArgs args) {
  MessageBox.Show("RowUpdating Event Status: " + args.Status);
}

protected static void OnRowUpdated(object sender,
  SqlCeRowUpdatedEventArgs args) {
  MessageBox.Show("RowUpdated Event Status: " + args.Status);
}

public static void UpdatePackageTable(DataSet dsPackages) {
  string connstr = @"Data Source=\My Documents\PTSystem.sdf";

  using(SqlCeConnection conn = new SqlCeConnection(connstr)) {
    conn.Open();

    SqlCeDataAdapter daPackages = GetPackageDataAdapter(conn);
    daPackages.RowUpdating +=
      new SqlCeRowUpdatingEventHandler(OnRowUpdating);
    daPackages.RowUpdated +=
new SqlCeRowUpdatedEventHandler(OnRowUpdated);

    daPackages.Update(dsPackages);

    daPackages.RowUpdating -=
      new SqlCeRowUpdatingEventHandler(OnRowUpdating);
    daPackages.RowUpdated -=
      new SqlCeRowUpdatedEventHandler(OnRowUpdated);
  }
}

VB
Private Shared Sub _
OnRowUpdating(ByVal sender As Object, _
              ByVal args As SqlCeRowUpdatingEventArgs)
  MessageBox.Show("RowUpdating Event Status: " & args.Status)
End Sub
```

```
Private Shared Sub _
OnRowUpdated(ByVal sender As Object, _
             ByVal args As SqlCeRowUpdatedEventArgs)
  MessageBox.Show("RowUpdated Event Status: " & args.Status)
End Sub

Sub UpdatePackageTable(ByVal dsPackages As DataSet)
  Dim connstr As String
  connstr = "Data Source=\My Documents\PTSystem.sdf"

  Dim conn As SqlCeConnection
  conn = New SqlCeConnection(connstr)
  conn.Open()

  Dim daPackages As SqlCeDataAdapter
  daPackages = GetPackageDataAdapter(conn)

  AddHandler daPackages.RowUpdating, AddressOf OnRowUpdating
  AddHandler daPackages.RowUpdated, AddressOf OnRowUpdated

  daPackages.Update(dsPackages)

  RemoveHandler daPackages.RowUpdating, AddressOf OnRowUpdating
  RemoveHandler daPackages.RowUpdated, AddressOf OnRowUpdated

  conn.Close()
End Sub
```

Generating SqlCommand Objects with the SqlCeCommandBuilder

Looking back at Listing 7.11, you can see that several lines of code are required to set up the update, insert, and delete commands of the SqlCeDataAdapter. The SqlCeCommandBuilder can do this work for you, but the SqlCeDataAdapter must be handling a single source table.

First, you need to initialize the SqlCeDataAdapter and its SelectCommand property. You then create SqlCeCommandBuilder that passes the SqlCeDataAdapter as a parameter to the SqlCeCommandBuilder constructor. The SqlCeCommandBuilder will then create a command for SqlCeDataAdapter's UpdateCommand, InsertCommand, and DeleteCommand properties. Listing 7.14 demonstrates how to use the SqlCeCommandBuilder to build a SqlCeDataAdapter for the Package table.

LISTING 7.14 Using the `SqlCeCommandBuilder`

```
C#
public static
SqlCeDataAdapter GetPackageDataAdapter(SqlCeConnection conn){
  string dmlPackageInfo = "SELECT * FROM Package";
  SqlCeDataAdapter daPackages = new SqlCeDataAdapter();
  daPackages.SelectCommand = new SqlCeCommand(dmlPackageInfo, conn);

  SqlCeCommandBuilder cmdBldr = new SqlCeCommandBuilder(daPackages);
  MessageBox.Show(cmdBldr.GetUpdateCommand().CommandText);
  MessageBox.Show(cmdBldr.GetInsertCommand().CommandText);
  MessageBox.Show(cmdBldr.GetDeleteCommand().CommandText);

  return daPackages;
}

VB
Function _
GetPackageDataAdapter(ByVal conn As SqlCeConnection) _
As SqlCeDataAdapter
  Dim dmlPackageInfo As String
  Dim daPackages As SqlCeDataAdapter
  dmlPackageInfo = "SELECT * FROM Package"
  daPackages = New SqlCeDataAdapter
  daPackages.SelectCommand = New SqlCeCommand(dmlPackageInfo, conn)

  Dim cmdBldr As SqlCeCommandBuilder
  cmdBldr = New SqlCeCommandBuilder(daPackages)
  MessageBox.Show(cmdBldr.GetUpdateCommand().CommandText)
  MessageBox.Show(cmdBldr.GetInsertCommand().CommandText)
  MessageBox.Show(cmdBldr.GetDeleteCommand().CommandText)

  Return daPackages
End Function
```

In Brief

- SQL Server CE is a small implementation of Microsoft SQL Server that provides rich support for data storage and retrieval.
- The `SqlCeEngine` class provides access to core SQL Server CE functionality, such as creating, opening, and closing a SQL Server CE database.
- The `SqlCeDataReader` class provides fast, forward-only access to the data stored in a SQL Server CE database.
- The `SqlCeDataAdapter` class and the `DataSet` class provide simple, easy data synchronization between your application and the SQL Server CE database.
- The `SqlCeCommandBuilder` class can automatically generate `SqlCommand` objects for the `SqlCeDataAdapter`.

XML and the DataSet

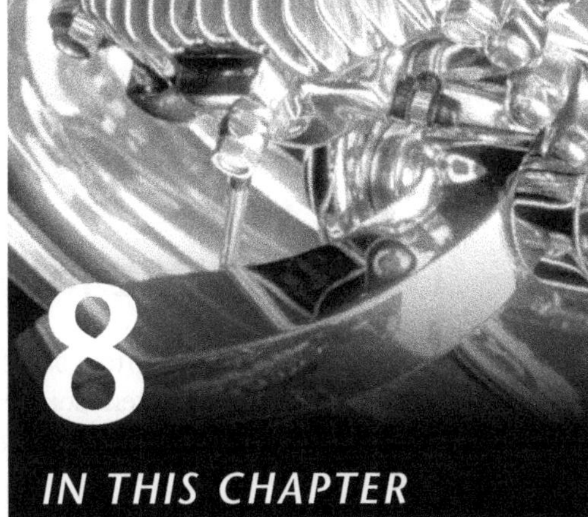

IN THIS CHAPTER

XML and the .NET Compact Framework DataSet

The DataSet provides a powerful means of manipulating caches of relational data in memory. A DataSet can be populated programmatically, as shown in Chapter 6, "ADO.NET on the .NET Compact Framework." It can also be populated by connecting to a SQL CE or desktop SQL server or by loading XML data that describes the contents of a DataSet into the DataSet. Because the DataSet can save itself as XML and load a DataSet description from XML, it is the only complex class in the .NET Compact Framework that is serializable by using only one or two lines of code.

There are two forms of XML data to and from which the .NET Compact Framework DataSet can be saved. The standard XML format can describe DataSets with multiple tables and express complex relational attributes of the DataSet. This chapter discusses how to use this format to describe a DataSet with all of the relational data features outlined in Chapter 6. Central to this discussion will be an overview of XML Schema, which describes the structure and layout of the DataSet.

The other format is the XML DiffGram. Support for this format is necessary to enable Web services, but the .NET Compact Framework DataSet supports the DiffGram format only to the extent as to allow a DataSet to be used as a parameter or return value in a Web service.

The techniques for loading and saving a DataSet to XML is nearly the same, regardless of the format use. Thus, we begin by thoroughly discussing the variety of means by which to load and save DataSets to XML.

Loading XML into a DataSet

XML can be loaded into a DataSet from a variety of sources, such as files or streams. Loading XML data from different sources often means using different method overloads in the DataSet. The specifics of loading XML data from a variety of sources are described in the following subsections.

Loading from a File

To load an XML file into a DataSet, use the DataSet.ReadXml(string) method and pass in the name of the file to load. For example:

C#

```
l_DataSet.ReadXml("\\MyDS.xml");
```

VB

```
l_DataSet.ReadXml("\MyDS.xml")
```

After the XML is loaded, the DataSet is populated as if the data had been entered programmatically. If the XML file does not describe a valid DataSet, then an exception is thrown. Typical scenarios that cause exceptions include these:

- The file to load does not exist.
- The file to load exists, but it cannot be opened due to a sharing violation.
- The data loaded into the DataSet has a constraint violation.
- The data loaded into the DataSet has a null value in a DataColumn that is forbidden to be null.
- The data loaded into the DataSet violates a ForeignKeyConstraint.

Loading Data from a Stream

You can load XML data coming from a stream by calling the DataSet.ReadXml(XmlReader, XmlReadMode) method. The XmlReader you pass in must be instantiated by somehow attaching it to an existing, valid stream. The stream, in turn, can be created by opening a file, using a memory buffer, or even accessing the stream from an infrared connection using the IrDAClient class, which is discussed in Chapter 5, "Network Connectivity with the .NET Compact Framework." The following subsections show specific examples of how to load XML data from a stream attached to a variety of sources:

Loading from a Stream Attached to a File

To load a stream attached to a file, create an `XmlTextReader` by passing in a `StreamReader` to the `XmlTextReader` constructor. Then pass the `XmlTextReader` into `DataSet.ReadXml()`. For example:

```
C#
System.Xml.XmlTextReader l_XmlTextReader = new System.Xml.XmlTextReader(new
        System.IO.StreamReader("\\DataSet.xml"));
l_DataSet.ReadXml(l_XmlTextReader, XmlReadMode.ReadSchema);
l_XmlTextReader.Close();
```

```
VB
Dim l_XmlTextReader as System.Xml.XmlTextReader = new System.Xml.XmlTextReader(new
        System.IO.StreamReader("\DataSet.xml");
l_DataSet.ReadXml(l_XmlTextReader, XmlReadMode.ReadSchema)
l_XmlTextReader.Close()
```

Loading from a Stream Attached to a Memory Buffer

To load XML from a stream attached to a memory buffer, follow these steps:

1. Acquire a reference to the byte array that will have a stream opened against it. You might receive such a byte array via a TCP connection from a remote party, as discussed in Chapter 5.
2. Instantiate a `MemoryStream`, passing the byte array from step 1 into the constructor.
3. Instantiate a `StreamReader` on the `MemoryStream` from step 2.
4. Instantiate an `XmlTextReader` on the `StreamReader` from step 3.
5. Call `DataSet.ReadXml()`, passing in the XmlTextReader of step 4 as an argument.

The following code demonstrates these steps. It is assumed that `l_ByteArray` already holds memory into which XML text was written, so that it can be opened with a `MemoryStream`.

```
C#
// DataSet was written to memory stream l_OutMemStream
// l_InBuff in an array of bytes
l_InBuff = l_OutMemStream.GetBuffer();
DataSet l_MemStreamDataSet = new DataSet();
System.IO.MemoryStream l_InMemStream = new System.IO.MemoryStream
        (l_InBuff, 0, 4000, false, true);
l_XmlTextReader = new System.Xml.XmlTextReader(l_InMemStream);
l_MemStreamDataSet.ReadXml(l_XmlTextReader, XmlReadMode.ReadSchema);
l_XmlTextReader.Close();
l_InMemStream.Close();
```

```
VB
' DataSet was written to memory stream l_OutMemStream
' l_InBuff in an array of bytes
l_InBuff = l_OutMemStream.GetBuffer()
dim l_MemStreamDataSet as DataSet = new DataSet()
dim l_InMemStream as System.IO.MemoryStream = new System.IO.MemoryStream
        (l_InBuff, 0, 4000, False,
        True)
l_XmlTextReader = new System.Xml.XmlTextReader(l_InMemStream)
l_MemStreamDataSet.ReadXml(l_XmlTextReader, XmlReadMode.ReadSchema)
l_XmlTextReader.Close()
l_InMemStream.Close()
```

Loading XML from a Stream Associated with an Infrared Connection

You can pass a DataSet to another device via the device's infrared ports by using the `IrDAClient` class, which is discussed in Chapter 5. The `IrDAClient` class exposes the `GetStream` method, through which you can acquire a stream and transmit the XML description of a DataSet. To do so, follow these steps:

1. Acquire an `IrDAClient` that is connected to another device. This process is described in detail in Chapter 5.
2. Acquire a handle to the underlying stream by calling `IrDAClient.GetStream()`.
3. Create a `StreamReader` by passing in the stream from step 2 into the `StreamReader` constructor.
4. Create an `XmlTextReader`. Pass the `StreamReader` of step 3 into the `XmlTextReader` constructor.
5. Pass the `XmlTextReader` into the `DataSet.ReadXml()` method.

The following sample code, borrowed from the XML_PhoneBook sample application, demonstrates these steps:

```
C#
m_PhoneBookDataSet = new DataSet();
StreamReader l_StreamReader = null;
XmlTextReader    l_XmlTextReader = null;
l_StreamReader = new StreamReader(this.m_IrDAClient.GetStream(),
        System.Text.Encoding.ASCII);

l_XmlTextReader = new XmlTextReader(l_StreamReader);
this.m_PhoneBookDataSet.ReadXml(l_XmlTextReader);
l_XmlTextReader.Close();
```

```
VB
Dim l_StreamReader As System.IO.StreamReader = Nothing
Dim l_XmlTextReader As System.Xml.XmlTextReader = Nothing
l_StreamReader = New System.IO.StreamReader(Me.m_IrDAClient.GetStream(),
        System.Text.Encoding.ASCII)

l_XmlTextReader = New System.Xml.XmlTextReader(l_StreamReader)
Me.m_PhoneBookDataSet.ReadXml(l_XmlTextReader)
l_XmlTextReader.Close()
```

Inferring Schema

The XML data describing a `DataSet` can include schema. Described in detail later this chapter, schema is a way to explicitly describe the structure of the relational data that the `DataSet` holds. If the `DataSet` loads XML that does not explicitly describe the data layout with schema, then the `DataSet` structure is inferred by examining the `DataRows` described in the XML. You can force the `DataSet` to infer the schema, even if there is a schema section in the XML, by using the method overload `DataSet.ReadXml(XmlReader, XmlReadMode.InferSchema)`.

Saving a DataSet as XML

A `DataSet` can be written as XML to variety of outputs, such as files or streams. Saving a `DataSet` as XML data to different output types often means using different method overloads in the `DataSet`. The specifics of saving a `DataSet` as XML data to various outputs are described in the following subsections.

Saving to a File

To save a `DataSet` in XML format to a file, call the `DataSet.WriteXml()` method, for example, like so:

```
C#
l_DataSet.WriteXml("\\MyDS.xml");

VB
l_DataSet.WriteXDml("\MyDS.xml")
```

`WriteXml()` throws an exception if there is a problem writing to the file. This is most typically caused by a sharing violation.

Saving Data to a Stream

You can save XML data to a stream by calling the `DataSet.WriteXml(XmlWriter, XmlWriteMode)` method. The `XmlWriter` you pass in must already be instantiated. That is, it must be constructed and an existing, valid stream must have been passed to the `XmlWriter` constructor. The stream in turn can be created by opening a file, using a memory buffer, or even accessing the stream from an infrared connection using the `IrDAClient` class. The following subsections show specific examples of how to write XML data to a stream attached to a variety of sources:

Saving to a Stream Attached to a File

To save to a stream attached to a file, follow these steps:

1. Create a new instance of a `System.IO.StreamWriter`. Pass in the name of the XML file to write to.
2. Acquire an `XmlTextWriter`. Pass in the `StreamWriter` of step 1 to the `XmlTextWriter` constructor.
3. Call `DataSet.WriteXml(XmlTextWriter, XmlWriteMode)`. Pass in the `XmlTextWriter` of step 2.

For example:

```
C#
l_XmlTextWriter = new System.Xml.XmlTextWriter(new System.IO.StreamWriter
        ("\\DataSet.xml"));
l_DataSet.WriteXml(l_XmlTextWriter, XmlWriteMode.WriteSchema);
l_XmlTextWriter.Close();
```

```
VB
l_XmlTextWriter = new System.Xml.XmlTextWriter(new System.IO.StreamWriter
        ("\DataSet.xml"))
l_DataSet.WriteXml(l_XmlTextWriter, XmlWriteMode.WriteSchema)
l_XmlTextWriter.Close()
```

Saving to a Stream Attached to a Memory Buffer

To save XML to a stream attached to a memory buffer, follow these steps:

1. Acquire a reference to the byte array that will have a stream opened against it. Make sure the buffer is large enough to accommodate the data you plan to write.
2. Instantiate a `MemoryStream`. Pass the byte array into the constructor.
3. Instantiate a `StreamWriter` on the memory stream from step 2.
4. Instantiate an `XmlTextWriter` on the `StreamWriter` from step 3.
5. Call `DataSet.WriteXml()`. Pass in the `XmlTextWriter` of step 4 as an argument.

The following code demonstrates these steps:

```
C#
// l_OutBuff is a byte array large enough to hold the XML
System.IO.MemoryStream l_OutMemStream = new System.IO.MemoryStream(l_OutBuff,
        0, 4000, true, true);
l_XmlTextWriter = new System.Xml.XmlTextWriter(l_OutMemStream,
        System.Text.Encoding.Default);
l_DataSet.WriteXml(l_XmlTextWriter, XmlWriteMode.IgnoreSchema);
l_XmlTextWriter.Close();
l_OutMemStream.Close();

VB
' l_OutBuff is a byte array large enough to hold the XML
Dim l_OutMemStream as System.IO.MemoryStream = new System.IO.MemoryStream(
        l_OutBuff, 0, 4000, true, true)
l_XmlTextWriter = new System.Xml.XmlTextWriter(l_OutMemStream,
        System.Text.Encoding.Default)
l_DataSet.WriteXml(l_XmlTextWriter, XmlWriteMode.IgnoreSchema)
l_XmlTextWriter.Close()
l_OutMemStream.Close()
```

Writing XML to a Stream Associated with an Infrared Connection

The `IrDAClient` class, discussed in Chapter 5, provides access to an underlying stream by calling the `GetStream()` method. You can use this feature to pass a `DataSet` to another device connected by an IR link. To do so, as the party who is writing out a `DataSet`, follow these steps:

1. Acquire an `IrDAClient` that is connected to another device. This process is described in detail in Chapter 5.
2. Acquire a handle to the underlying stream by calling `IrDAClient.GetStream()`.
3. Create a `StreamWriter` by passing in the stream from step 2 into the `StreamWriter` constructor.
4. Create an `XmlTextWriter`. Pass the `StreamWriter` of step 3 into the `XmlTextWriter` constructor.
5. Pass the `XmlTextWriter` into the `DataSet.WriteXml()` method.

The following code, borrowed from the XML_PhoneBook sample application, demonstrates these steps:

```
C#
StreamWriter l_StreamWriter = new StreamWriter(this.m_IrDAClient.GetStream(),
        System.Text.Encoding.ASCII);
```

```
System.Xml.XmlTextWriter l_XmlTextWriter = new System.Xml.XmlTextWriter
        (l_StreamWriter);
this.m_PhoneBookDataSet.WriteXml(l_XmlTextWriter, XmlWriteMode.WriteSchema);
l_XmlTextWriter.Flush();
l_XmlTextWriter.Close();

VB
Dim l_StreamWriter As System.IO.StreamWriter = New System.IO.StreamWriter
        (Me.m_IrDAClient.GetStream(), System.Text.Encoding.ASCII)
Dim l_XmlTextWriter As System.Xml.XmlTextWriter =
        New System.Xml.XmlTextWriter(l_StreamWriter)
Me.m_PhoneBookDataSet.WriteXml(l_XmlTextWriter, XmlWriteMode.WriteSchema)
l_XmlTextWriter.Flush()
l_XmlTextWriter.Close()
```

Writing Schema with the XML Data

You can control whether a DataSet writes schema information with the XML by using the DataSet.WriteXml(XmlWriter, XmlWriteMode) overload. Pass XmlWriteMode.WriteXmlSchema to force the DataSet to write schema information with the data or pass XmlWriteMode.IgnoreSchema to prevent the DataSet from writing schema information. XML Schema as it relates to DataSets is discussed in greater detail later in this chapter.

The XmlWriteMode has the following values:

- WriteSchema Write the schema when writing the DataSet as XML.
- IgnoreSchema Ignore the schema when writing the DataSet as XML.
- DiffGram Write the DataSet in XML DiffGram format. DiffGram format is discussed in greater detail later in this chapter.

If the DataSet that will ultimately load the XML data already knows what the data schema is, then there is no need to write the schema with the XML. It would be wasteful and redundant.

WHEN TO WRITE XML SCHEMA

If you don't know who will load your XML-serialized DataSet, then it is generally a good idea to write schema with your XML. It provides consumers of the XML data an exact description of the relational data structure. Even though both the .NET Desktop Framework and the .NET Compact Framework DataSet can infer schema, software trying to consume your XML but not based on .NET technology might not be able to infer schema.

Loading and Saving XML in a Sample Application

The XmlPhoneBook sample application demonstrates how to load and save `DataSets` in XML format. The XML can be stored on the local file system or through a stream associated with an `IrDAClient`. There is a C# and VB version of XmlPhoneBook in the folder `\SampleApplications\Chapter8`.

When the XmlPhoneBook application launches, the screen shows a `DataGrid` and a series of buttons. The `DataGrid` shows the contents of a `DataSet` maintained by the application. To set up a brand new `DataSet`, follow these steps:

1. Click the New PhoneBook button. The application responds by creating a new `DataSet`, but there are no records loaded.
2. Create a random record and insert it into the `DataSet` by clicking the Add Random button.
3. After the random record has been inserted into the `DataSet`, it appears in the `DataGrid`.

To save a `DataSet` to an XML file, click the XML File Save button. The application saves the `DataSet` in XML format to the file \PhoneBook.xml.

To load a previously saved `DataSet`, click the XML File Load button. The application attempts to populate the `DataSet` by loading the file \PhoneBook.xml.

XmlPhoneBook can also transmit a populated `DataSet` by using the infrared port. To do this, follow these steps:

1. Set up a populated `DataSet` in one device.
2. Line up the infrared ports on the devices.
3. Select the Host checkbox on only one of the devices. Click IR Send on one of the devices and IR Receive on the other.
4. The XmlPhoneBook application sets up an `IrDAClient`, as described in Chapter 5. It then acquires a handle to the stream associated with the `IrDAClient` by calling `IrDAClient.GetStream()`. Finally, the XML data is written to the stream by one device and read from the stream by the other.

There is neither a way to edit the records, nor to choose values for records to insert, nor to choose the filename for saving the `DataSet` as XML data. This is to keep the application simple and maintain a focus on loading and saving `DataSets` in XML format, rather than creating an exercise in user interface design.

Modeling Relational Data with XML Schemas

This section describes how to implement a variety of relational data concepts by using XML schemas. It is assumed that the reader has a firm understanding of how to load XML into a DataSet. The focus here is on the XML data itself and learn how to express relational data in a format that the .NET Compact Framework DataSet can handle. Learning to write XML from scratch can consume an entire large book in itself. This section proceeds by examining XML fragments that show off important DataSet features and commenting on the relevant points.

Examining XML with the XML_DataSetView Utility

The XML_DataSetView utility is located in the folder \SampleApplications\Chapter8. There are C# and Visual Basic versions.

The XML_DataSetView utility is included because it provides a convenient means of loading an XML file into a DataSet and viewing the tables. To use the XML_DataSetView utility, follow these steps:

1. Launch the application.
2. Enter the name of the XML file to load into the text box. Click the Load XML button. The file which you plan to load must be present before clicking the Load XML button. There are many ways to acquire an XML file to load. For example, you can create an XML file with the XmlPhoneBook sample application or the XML_Creator utility, or you can try writing one from scratch.
3. When you click "Load XML," the application responds by filling in the ListBox with the names of the tables in the loaded DataSet. The DataGrid shows the data in the table currently selected in the ListBox.

This sample application demonstrates no concepts that have not been previously covered. However, it is an enormously useful utility to have on hand to test XML data and schemas and determine how quickly your device can load them.

An important lesson you should take from this section is that trying to derive your own schema from scratch is not easy. Instead, there are simpler approaches you can take:

- Start with one of the following sample schemas shown and expand it to suit your needs.
- Use the schema editor in the Visual Studio IDE and test your creation using the XML_DataSetView utility.

- Create a DataSet that holds the relational database you want to model, and call WriteXmlSchema() to dump out the schema alone. You can also use one of the WriteXml() overloads that writes schema.

Examining a Simple Schema that Describes a Single Table

In the following first XML listing, we look at a single table that holds a variety of data types. The schema demonstrates how to declare a variety of column types, and the actual data following the schema demonstrates how the tables and columns of a very simple DataSet are laid out.

Let's start with some general observations. The entire DataSet, complete with schema, is declared by the name of the DataSet. In the following code, the name of the DataSet is NewDataSet, which is the default value for a DataSet that is created without a name. So, the entire DataSet XML description appears between the <NewDataSet> and </NewDataSet> tags.

The schema is denoted with the xs:schema keyword, and the end of the schema is marked with </xs:schema>.

The actual data rows for the tables in the DataSet appear after the schema. Each row in the table appears between two tags that bear the name of the table. This simple DataSet holds a single table named Demo Table, and there are two rows in that table. So, each row's data appears between the <Demo x0020 Table> and </Demo x0020 Table> tags. The x0020 represents the space in the name for the table.

The columns in the row are given values by referring to the column name. For example, the value for the column named AnInt16 is set to the value 1 with this code:

```
<AnInt16>1</AnInt16>
```

Column values are declared between the <Demo x0020 Table> and </Demo x0020 Table> tags. If the row has a column for which no value is given, then it receives a default value, which can be null. This is equivalent to programmatically adding a row to a DataSet without setting all of the column values. If there is a restriction on the value of a column, for example, not null, then missing a column value can cause an exception when the XML data is loaded.

The key observations about declaring data types in schema are outlined in Table 8.1.

TABLE 8.1

Declaring Data Types in Schema for XML DataSets

DATA TYPE	DECLARATION	COMMENT
String	xs:string	
Int16	xs:short	
Int32	xs:int	
Int64	xs:long	
UInt16	xs:unsignedShort	
UInt32	xs:unsignedInt	
UInt64	xs:unsignedLong	
Float	xs:float	This type is different from the decimal type because decimal values can represent higher precision without round-off error.
Decimal	xs:decimal	
Boolean	xs:boolean	
DateTime	xs:DateTime	
Char	xs:string	Char types are declared as string with a restriction that the length be exactly one. The following code sample shows the declaration for a Char type.

Sample code: Declaring a Char type with XML Schema

```
<xs:element name="AChar" minOccurs="0">
 <xs:simpleType>
  <xs:restriction base="xs:string">
   <xs:length value="1" />
  </xs:restriction>
 </xs:simpleType>
</xs:element>
```

The full text for the XML code describing a simple DataSet appears in Listing 8.1.

LISTING 8.1 XML for a simple DataSet

```
<NewDataSet>
 <xs:schema id="NewDataSet" xmlns=""
        xmlns:xs="http://www.w3.org/2001/XMLSchema"
        xmlns:msdata="urn:schemas-microsoft-com:xml-msdata">
  <xs:element name="NewDataSet" msdata:IsDataSet="true">
   <xs:complexType>
    <xs:choice maxOccurs="unbounded">
     <xs:element name="Demo_x0020_Table">
      <xs:complexType>
```

```
      <xs:sequence>
       <xs:element name="AString" type="xs:string" minOccurs="0" />
       <xs:element name="AnInt16" type="xs:short" minOccurs="0" />
       <xs:element name="AnInt32" type="xs:int" minOccurs="0" />
       <xs:element name="AnInt64" type="xs:long" minOccurs="0" />
       <xs:element name="AUInt16" type="xs:unsignedShort" minOccurs="0" />
       <xs:element name="AUInt32" type="xs:unsignedInt" minOccurs="0" />
       <xs:element name="AUInt64" type="xs:unsignedLong" minOccurs="0" />
       <xs:element name="AFloat" type="xs:float" minOccurs="0" />
       <xs:element name="ADecimal" type="xs:decimal" minOccurs="0" />
       <xs:element name="ABoolean" type="xs:boolean" minOccurs="0" />
       <xs:element name="AChar" minOccurs="0">
        <xs:simpleType>
         <xs:restriction base="xs:string">
          <xs:length value="1" />
         </xs:restriction>
        </xs:simpleType>
       </xs:element>
       <xs:element name="ADateTime" type="xs:dateTime" minOccurs="0"/>
      </xs:sequence>
     </xs:complexType>
    </xs:element>
   </xs:choice>
  </xs:complexType>
 </xs:element>
</xs:schema>
<Demo_x0020_Table>
 <AString>AString1</AString>
 <AnInt16>1</AnInt16>
 <AnInt32>2</AnInt32>
 <AnInt64>3</AnInt64>
 <AUInt16>4</AUInt16>
 <AUInt32>5</AUInt32>
 <AUInt64>6</AUInt64>
 <AFloat>7</AFloat>
 <ADecimal>8</ADecimal>
 <ABoolean>true</ABoolean>
 <AChar>c</AChar>
 <ADateTime>2001-06-16T00:00:00.0000000-07:00</ADateTime>
</Demo_x0020_Table>
<Demo_x0020_Table>
 <AString>AString2</AString>
```

```
    <AnInt16>10</AnInt16>
    <AnInt32>20</AnInt32>
    <AnInt64>30</AnInt64>
    <AUInt16>40</AUInt16>
    <AUInt32>50</AUInt32>
    <AUInt64>60</AUInt64>
    <AFloat>70</AFloat>
    <ADecimal>80</ADecimal>
    <ABoolean>false</ABoolean>
    <AChar>d</AChar>
    <ADateTime>2003-06-24T00:00:00.0000000-07:00</ADateTime>
  </Demo_x0020_Table>
</NewDataSet>
```

Autoincremented Fields and Unique Constraints

With the next chunk of XML code, we examine how to describe autoincremented fields and unique constraints.

A data column is declared as being autoincremented in the same line as its data type is declared. To declare a column autoincremented, use `msdata:AutoIncrement="true"`. Set the starting value with `msdata:AutoIncrementSeed` and the autoincrement step with `msdata:AutoIncrementStep`. For example, the following line declares the `ContactID` column to be autoincremented with a start value of 10 and a step value of 5:

```
<xs:element name="ContactID" msdata:AutoIncrement="true"
        msdata:AutoIncrementSeed="10"
        msdata:AutoIncrementStep="5" type="xs:int" minOccurs="0" />
```

To declare a column unique, use the `xs:unique` keyword. Declare the name of the constraint and indicate a selector and field for the constraint. The selector is the name of the `DataTable` in which the `DataColumn` resides. The field is the name of the `DataColumn`. These declarations appear in the schema after the `xs:complexType`, which specified the table name and columns in the table. For example, this XML snippet declares a unique constraint named `Constraint1`, in which the `PhoneNumber` DataColumn in the `Phone Contact DataTable` is declared unique:

BE CAREFUL WHEN IGNORING SCHEMA

Notice that information about such things as autoincremented fields, constraints, and so on is held in the schema section of the XML. The actual rows of data hold no such information. If you load a `DataSet` and ignore the schema that comes with it, the resulting populated `DataSet` will lack the constraints, autoincremented fields, and so on that the original creator of the `DataSet` intended.

```
<xs:unique name="Constraint1">
 <xs:selector xpath=".//Phone_x0020_Contacts" />
 <xs:field xpath="PhoneNumber" />
</xs:unique>
```

The entire XML representation of a `DataSet` with an autoincremented field and a unique constraint appears as follows in Listing 8.2.

LISTING 8.2 XML representation of a `DataSet` with an autoincremented field and a unique constraint

```
<NewDataSet>
 <xs:schema id="NewDataSet" xmlns=""
        xmlns:xs="http://www.w3.org/2001/XMLSchema"
        xmlns:msdata="urn:schemas-microsoft-com:xml-msdata">
  <xs:element name="NewDataSet" msdata:IsDataSet="true">
   <xs:complexType>
    <xs:choice maxOccurs="unbounded">
     <xs:element name="Phone_x0020_Contacts">
      <xs:complexType>
       <xs:sequence>
        <xs:element name="Name" type="xs:string" />
        <xs:element name="PhoneNumber" type="xs:string" minOccurs="0" />
        <xs:element name="ContactID" msdata:AutoIncrement="true"
               msdata:AutoIncrementSeed="10"
               msdata:AutoIncrementStep="5" type="xs:int" minOccurs="0" />
       </xs:sequence>
      </xs:complexType>
     </xs:element>
    </xs:choice>
   </xs:complexType>
   <xs:unique name="Constraint1">
    <xs:selector xpath=".//Phone_x0020_Contacts" />
    <xs:field xpath="PhoneNumber" />
   </xs:unique>
  </xs:element>
 </xs:schema>
 <Phone_x0020_Contacts>
  <Name>George Washington</Name>
  <PhoneNumber>340-1776</PhoneNumber>
  <ContactID>10</ContactID>
 </Phone_x0020_Contacts>
 <Phone_x0020_Contacts>
```

```
    <Name>Ben Franklin</Name>
    <PhoneNumber>336-3211</PhoneNumber>
    <ContactID>15</ContactID>
  </Phone_x0020_Contacts>
  <Phone_x0020_Contacts>
    <Name>Alexander Hamilton</Name>
    <PhoneNumber>756-3211</PhoneNumber>
    <ContactID>20</ContactID>
  </Phone_x0020_Contacts>
</NewDataSet>
```

Describing Multiple Tables, ForeignKeyConstraints, DataRelations, and Computed Fields with XML Schema

Declaring multiple tables is simply an extension of declaring a DataSet with a single table. In the schema section of the XML, tables are declared one after the other. All of the table declarations occur inside of an <xs:choice> node. Each table is started with an <xs:element> tag that also holds the name of the table.

DataRelations and ForeignKeyConstraints are described with <xs:keyref>. For example, this XML code, C# code, and Visual Basic code all describe the same thing:

```
XML
<xs:keyref name="MainToCholesterolFKConstraint" refer="Constraint1"
        msdata:ConstraintOnly="true" msdata:AcceptRejectRule="Cascade">
 <xs:selector xpath=".//BloodPressure" />
 <xs:field xpath="ContactID" />
</xs:keyref>

C#
ForeignKeyConstraint l_ForeignKC = new ForeignKeyConstraint
        ("MainToCholesterolFKConstraint",
        l_DataSet.Tables["PhoneContactsMainTable"].Columns["ContactID"],
        l_DataSet.Tables["BloodPressure"].Columns["ContactID"]);

VB
Dim l_ForeignKC as ForeignKeyConstraint = new ForeignKeyConstraint
        ("MainToCholesterolFKConstraint",
        l_DataSet.Tables("PhoneContactsMainTable").Columns("ContactID"),
        l_DataSet.Tables("BloodPressure").Columns("ContactID"));
```

Computed columns are described as part of the declaration for the column. When the column is first declared, the expression for computing the column is included. For example, the following XML code, C# code, and Visual Basic code all describe the same thing:

```
XML
<xs:element name="AverageReading" msdata:ReadOnly="true" msdata:Expression=
        "(Reading1 + Reading2 + Reading3) / 3"
        type="xs:decimal" minOccurs="0" />
```

```
C#
// Make the "AverageReading" column a computed column by using an expression
l_newTable.Columns["AverageReading"].Expression =
        "(Reading1 + Reading2 + Reading3) / 3";
```

```
VB
' Make the "AverageReading" column a computed column by using an expression
l_newTable.Columns("AverageReading").Expression =
        "(Reading1 + Reading2 + Reading3) / 3"
```

Listing 8.3 shows the XML code that completely describes a `DataSet`. It demonstrates how to describe multiple tables, `ForeignKeyConstraints`, and computed fields:

LISTING 8.3 XML for a completely described `DataSet`

```
<NewDataSet>
 <xs:schema id="NewDataSet" xmlns=""
        xmlns:xs="http://www.w3.org/2001/XMLSchema"
        xmlns:msdata="urn:schemas-microsoft-com:xml-msdata">
  <xs:element name="NewDataSet" msdata:IsDataSet="true">
   <xs:complexType>
    <xs:choice maxOccurs="unbounded">
     <xs:element name="PhoneContactsMainTable">
      <xs:complexType>
       <xs:sequence>
        <xs:element name="ContactID" msdata:AutoIncrement="true"
                msdata:AutoIncrementSeed="10"
                msdata:AutoIncrementStep="5" type="xs:int" minOccurs="0" />
        <xs:element name="FirstName" type="xs:string" minOccurs="0" />
        <xs:element name="LastName" type="xs:string" minOccurs="0" />
        <xs:element name="FullName" msdata:ReadOnly="true"
                msdata:Expression="FirstName + ' ' +
                LastName" type="xs:string" minOccurs="0" />
       </xs:sequence>
```

```
      </xs:complexType>
     </xs:element>
     <xs:element name="Cholesterol">
      <xs:complexType>
       <xs:sequence>
        <xs:element name="ContactID" type="xs:int" minOccurs="0" />
        <xs:element name="Reading1" type="xs:decimal" minOccurs="0" />
        <xs:element name="Reading2" type="xs:decimal" minOccurs="0" />
        <xs:element name="Reading3" type="xs:decimal" minOccurs="0" />
        <xs:element name="AverageReading" msdata:ReadOnly="true"
                msdata:Expression="(Reading1 +
                Reading2 + Reading3) / 3" type="xs:decimal" minOccurs="0" />
    </xs:sequence>
   </xs:complexType>
   </xs:element>
   <xs:element name="BloodPressure">
   <xs:complexType>
    <xs:sequence>
    <xs:element name="ContactID" type="xs:int" minOccurs="0" />
    <xs:element name="Reading1" type="xs:decimal" minOccurs="0" />
    <xs:element name="Reading2" type="xs:decimal" minOccurs="0" />
    <xs:element name="Reading3" type="xs:decimal" minOccurs="0" />
    <xs:element name="AverageReading" msdata:ReadOnly="true"
            msdata:Expression="(Reading1 +
            Reading2 + Reading3) / 3" type="xs:decimal" minOccurs="0" />
    </xs:sequence>
   </xs:complexType>
   </xs:element>
  </xs:choice>
  </xs:complexType>
  <xs:unique name="Constraint1">
  <xs:selector xpath=".//PhoneContactsMainTable" />
  <xs:field xpath="ContactID" />
  </xs:unique>
  <xs:keyref name="MainToCholesterolFKConstraint" refer="Constraint1"
          msdata:ConstraintOnly="true"
          msdata:AcceptRejectRule="Cascade">
  <xs:selector xpath=".//BloodPressure" /><xs:field xpath="ContactID" />
  </xs:keyref>
  </xs:element>
 </xs:schema>
 <PhoneContactsMainTable>
```

```
    <ContactID>10</ContactID>
    <FirstName>George</FirstName>
    <LastName>Washington</LastName>
    <FullName>George Washington</FullName>
  </PhoneContactsMainTable>
  <PhoneContactsMainTable>
    <ContactID>15</ContactID>
    <FirstName>Ben</FirstName>
    <LastName>Franklin</LastName>
    <FullName>Ben Franklin</FullName>
  </PhoneContactsMainTable>
  <PhoneContactsMainTable>
    <ContactID>20</ContactID>
    <FirstName>Alexander</FirstName>
    <LastName>Hamilton</LastName>
    <FullName>Alexander Hamilton</FullName>
  </PhoneContactsMainTable>
  <Cholesterol>
    <ContactID>10</ContactID>
    <Reading1>200</Reading1>
    <Reading2>300</Reading2>
    <Reading3>500</Reading3>
    <AverageReading>333.33333333333333333333333333</AverageReading>
  </Cholesterol>
</NewDataSet>
```

Creating XML Data with the XML_Creator Utility

The XML_Creator utility is included as a convenience for novice users who want to explore XML schema creation with `DataSets`. This utility is located in the `\SampleApplications\Chapter8` folder, and there are C# and Visual Basic versions.

XML_Creator does not demonstrate any new concepts related to the `DataSet`. Instead, it is a convenient framework by which to create `DataSets` with various relational data structures and save them in XML format. The XML_Creator was used to create all of the XML examples shown previously in this section.

DESIGNING XML CONSUMABLE BY THE .NET COMPACT FRAMEWORK

Using a utility like the XML_Creator to experiment with DataSets and the resulting XML is very useful because you know that the .NET Compact Framework DataSet will be able to consume the XML data that will be passed around. If you use a desktop application to design your XML DataSet representations, or if you do it by hand, always check it against the .NET Compact Framework's implementation of the DataSet.

To use the XML_Creator, launch it and then click the GO button. The application acquires a series of populated DataSets through a variety of method calls and saves each DataSet in XML format.

Loading and Saving Schema Alone

There are times when it is desirable to load or save only the schema portion of a DataSet. For example, you might have the rows of data ready to insert into a DataSet, but you want to make sure that your data conforms to a specific schema from someone else's DataSet. The other person can save their DataSet schema. Then you can take an empty DataSet, load the other person's schema, and then insert your own data. If your data violates the schema in some way, then an exception will be thrown as you enter it into the DataSet.

To load the schema from an XML representation of a DataSet and ignore any other data, call DataSet.ReadXmlSchema(). For example:

C#

```
l_DataSet.ReadXmlSchema("\\XML_DataSet.xml");
```

VB

```
l_DataSet.ReadXmlSchema("\XML_DataSet.xml")
```

To write only the schema information for a DataSet to XML, call DataSet.WriteXmlSchema(). For example:

C#

```
l_DataSet.WriteXmlSchema("\\DataSet_SchemaOnly.xml");
```

VB

```
l_DataSet.WriteXmlSchema("\DataSet_SchemaOnly.xml")
```

Writing a DataSet to XML without Schema Information

The process of loading schema information into a DataSet is computationally intensive and time consuming, especially on a device. An efficient way to deal with a DataSet is to load its schema just once when you first instantiate it. You can store the schema information in a separate XML file. Use ReadXmlSchema() to do this.

All subsequent writing of the actual tables can be written without the schema included. Since you will load this data back into another DataSet that already has a schema loaded, you can save yourself the time of having to load schema repeatedly. You can also save space and network bandwidth by writing the XML data without schema, if you know that everyone who will consume the XML already knows the schema.

> **READING SCHEMA IS EXPENSIVE**
>
> In situations where the amount of data in a DataSet is relatively small, but the schema is complex, reading the schema takes a significant portion of the time that it takes to read an XML document into the DataSet.

To write a DataSet to XML without writing schema information, call DataSet.WriteXml(XmlWriter, XmlWriteMode), and pass XmlWriteMode.IgnoreSchema as an argument. For example:

```
C#
l_XmlTextWriter = new System.Xml.XmlTextWriter(new System.IO.StreamWriter
        ("\\NO_SCHEMA_DS.xml"));
l_DataSet.WriteXml(l_XmlTextWriter, XmlWriteMode.IgnoreSchema);
l_XmlTextWriter.Close();

VB
l_XmlTextWriter = new System.Xml.XmlTextWriter(new System.IO.StreamWriter
        ("\NO_SCHEMA_DS.xml"))
l_DataSet.WriteXml(l_XmlTextWriter, XmlWriteMode.IgnoreSchema)
l_XmlTextWriter.Close()
```

Reading a DataSet from XML, Ignoring Schema Information

If you have a DataSet with a schema already loaded, you might want to read XML data without reading the schema associated with the XML. To read XML into a DataSet and ignore the schema information in the XML, call DataSet.ReadXml(XmlReader, XmlReadMode), and pass XmlReadMode.IgnoreSchema as an argument. For example:

```
C#
l_XmlTextReader = new System.Xml.XmlTextReader(new
        System.IO.StreamReader("\\MIGHT_HAVE_SCHEMA.xml.xml"));
l_DataSet.ReadXml(l_XmlTextReader, XmlReadMode.IgnoreSchema);
l_XmlTextReader.Close();

VB
l_XmlTextReader = new System.Xml.XmlTextReader(new
        System.IO.StreamReader("\MIGHT_HAVE_SCHEMA.xml.xml"))
l_DataSet.ReadXml(l_XmlTextReader, XmlReadMode.IgnoreSchema)
l_XmlTextReader.Close()
```

Inferring Schema

In some circumstances you may want to load XML data into an empty `DataSet` while ignoring the schema information in the XML. In this case, the `DataSet` is able to infer the schema by looking at the tables in the XML data. The `DataSet` creates its own schema to match that of the tables loaded, but there is no way to infer such things as autoincremented fields, constraints, and so on.

If the XML data loaded into an empty `DataSet` has no schema information, then the schema is inferred automatically. To explicitly ask that the schema be inferred, call `DataSet.ReadXml(XmlReader, XmlReadMode)`. Pass `XmlReadMode.InferSchema` as an argument. See these snippets, for example:

```
C#
l_XmlTextReader = new System.Xml.XmlTextReader(new
        System.IO.StreamReader
        ("\\DataSet_AutoInc_NotNull_UniqueContraint.xml"));
l_AnotherDS.ReadXml(l_XmlTextReader, XmlReadMode.InferSchema);
l_XmlTextReader.Close();

VB
l_XmlTextReader = new System.Xml.XmlTextReader(new
        System.IO.StreamReader
        ("\DataSet_AutoInc_NotNull_UniqueContraint.xml"))
l_AnotherDS.ReadXml(l_XmlTextReader, XmlReadMode.InferSchema)
l_XmlTextReader.Close()
```

Using XML to Design a DataSet

The Visual Studio IDE includes a feature called the Data Designer that is available only when working on projects that target the desktop .NET Framework. The Data Designer is a powerful graphical utility that makes it easy to design a DataSet's schema and populate it with data. The Data Designer can spit out code for you, which spares the developer from having to programmatically build DataSets and populate them. Unfortunately, this tool is not available when working with the .NET Compact Framework.

If you don't want to write code in C# or Visual Basic to set up your DataSet, you can instead describe your DataSet with XML schema. You also write XML to populate your DataSet with tables. The information previously presented in this chapter should give you enough knowledge to start with one of the samples included with this book and alter it to suit your needs. When you need a DataSet, simply create a DataSet, and call ReadXml() or ReadXmlSchema(), depending on whether you only want to load in schema.

Schema Editing with the IDE

If you want to use XML to edit data that will be loaded into a DataSet, you can use the XML editor in the Visual Studio IDE. The XML editor is a convenient, graphical way to edit XML data.

The easiest way to use the XML editor is to open an existing XML file that has the schema already set up the way you want it. If you create a new XML file in the Visual Studio IDE, by default it contains no elements, and the visual editor will not start. If you start with an existing XML file, then you can use the GUI to delete all of the old tables, set up a schema, and insert data as you wish.

Figure 8.1 shows the XML editor immediately after opening the file SampleDataSet.xml. This file is in the directory \SampleApplications\Chapter8. As you can see, when you first open an XML file, you see only the XML text.

To view the XML data in a graphical form, put the mouse over the editor, right-click, and select View Data. Figure 8.2 shows the result. All of the DataTables are on the left side. Click on a DataTable to select it, and you will be able to view the rows and columns on the right side. You can edit the values, delete rows, and insert rows by clicking on the bottom row and inserting your data.

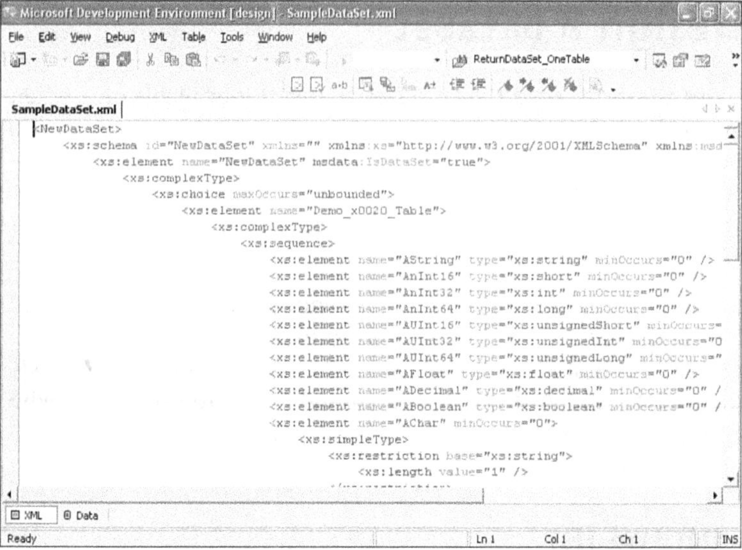

FIGURE 8.1 When SampleDataSet.xml is first opened, the XML editor shows the XML code.

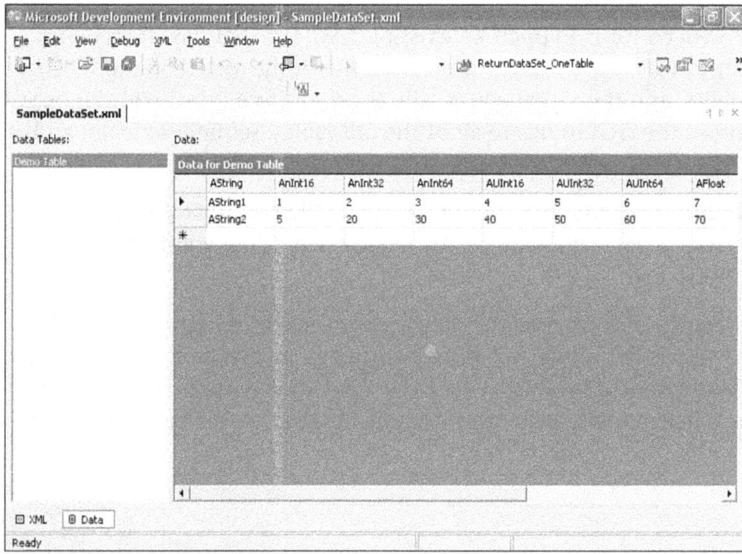

FIGURE 8.2 The XML editor can also show a data view.

If the XML you loaded is somehow invalid, then trying to view the XML data in graphical form will display an error indicating where in the XML the error lies. Thus, the XML editor is a convenient way to check the validity of `DataSet` schemas you design.

Saving and Loading DataSets as XML DiffGrams

The XML DiffGram is a format that makes it easy for two parties to refresh a shared data structure without having to transmit the entire data structure over the network. The desktop .NET Framework `DataSet` supports DiffGrams extensively. Thus, the desktop .NET Framework can use DiffGrams to support the following scenario:

- Machine A creates a very large `DataSet` of 2000 records, one that requires nearly 1MB to transmit over the network as XML.
- XML Data is flexible and human-readable. However, the cost is that XML data is not dense. Thus, even a modest amount of information can require a lot of text to be sent across the network. This is one of the reasons that the desktop .NET Framework `DataSet` extensively supports DiffGrams. It avoids having to retransmit redundant XML data.
- Machine A uses `DataSet.WriteXml()` to export its dataset into a string and transmits the data to Machine B.
- Machine B receives the XML and uses `DataSet.ReadXml()` to re-create a local copy of the `DataSet`.
- Machine B adds tree records, removes one record, and updates values in one record in the `DataSet`.
- Machine B sends a DiffGram back to Machine A. The DiffGram holds only five entries: three to describe the deleted records, one to describe the added record, and one to describe the updated record.

Machine B has used a DiffGram to describe updates to Machine A without having to transmit all of the redundant data back to Machine A.

The .NET Compact Framework's `DataSet` has very limited support for DiffGrams. It does not support the preceding scenario. Nonetheless, the scenario is very important to understand. If you try to write a .NET Compact Framework application that interacts with desktop applications, then you will run into trouble if the desktop applications try to use DiffGrams in the way described in the scenario.

The .NET Compact Framework DataSet supports DiffGrams under the following restrictions for reading and writing them:

- The incoming DiffGram must include the contents of an entire DataSet. The entire DataSet must be valid. That is, it may not violate any constraints. It is possible for a desktop application to send a fraction of a DataSet in a DiffGram. The records in the DiffGram might violate the DataSet's constraints if they are considered alone. When the .NET Compact Framework DataSet loads this DiffGram, it will consider the records as a stand-alone DataSet, discover that they are invalid, and throw an exception.
- The .NET Compact Framework DataSet loading the DiffGram must be empty immediately before loading the DiffGram. This can be done by calling DataSet.Clear() or using a newly instantiated DataSet.
- The incoming DiffGram may not have any data describing updated or deleted records.
- When a .NET Compact Framework DataSet writes a DiffGram, it writes the contents of the entire DataSet.
- There is no way to write only those records that have been added, updated, or deleted.

How to Read a Diffgram

The .NET Compact Framework DataSet can automatically infer that the XML data it reads is a DiffGram. If the file \ADSDiffgram.xml contains a valid DataSet in XML DiffGram format, then the following code will read it:

```
C#
System.Data.DataSet l_DataSet = new DataSet();
l_DataSet.ReadXml("\\ADSDiffgram.xml");
// l_DataSet automatically realizes that the data is in DiffGram format

VB
Dim l_DataSet as System.Data.DataSet = new System.Data.DataSet
l_DataSet.ReadXml("\ADSDiffgram.xml")
' l_DataSet automatically realizes that the data is in DiffGram format
```

When to Use DiffGrams

As you have noticed, DiffGram support in the .NET Compact Framework DataSet is quite limited. It is present because the level of support is good enough for the .NET Compact Framework to send and receive DataSets in Web service calls. Web services transmit DataSets by using the DiffGram representation.

When a DataSet is passed as a Web service argument, the DiffGram representation is embedded in a SOAP packet for you automatically. If you write a very advanced application that builds and cracks SOAP packets without using the built-in Web service support described in Chapter 9, "Using XML Web Services," then you may be forced to deal with DiffGrams. However, implementing such an application is beyond the scope of this book.

There are no other major advantages to using the DiffGram format when programming with the .NET Compact Framework DataSet. Thus, we will not discuss the DiffGram format in great detail.

In Brief

- The .NET Compact Framework DataSet supports a variety of methods to load and save the contents of a DataSet in XML format.
- The .NET Compact Framework DataSet supports the standard XML format extensively, but only supports DiffGrams well enough to enable passing DataSet in Web Services calls.
- XML data can be loaded into a DataSet in a variety of ways, such as from a file or a stream. There are multiple methods in the DataSet for loading XML data depending on the source of the data.
- A unique way to transmit a DataSet in XML format between two devices is by using the devices' infrared port. The IrDAClient class makes this possible.
- If XML describing a DataSet lacks schema information, the DataSet can infer the structure of the relational data, a process called *inferring schema.*
- The contents of a DataSet can be saved as XML in a variety of ways, targeting such outputs as files or streams. There are multiple methods in the DataSet for saving XML data depending on the destination for the data and the format of the XML written.
- When writing XML data from a DataSet, it is possible to write schema information with the data, to write only schema information but no data, or to write data only with no schema.
- XML Schema is used to describe the relational layout of the data in a DataSet. An XML Schema can be used to describe a variety of relational data constructs such as table relations, autoincrementing fields, and parent-child relationships.
- Many of the data types possible for a DataColumn can be also be described in an XML Schema.
- A DataSet can load the schema information from an XML document, ignoring the data values. Doing so causes the DataSet to set up its internal structure in such a way as to match the relational format specified in the schema.

- The Visual Studio XML editor is an easy way to manipulate and alter existing XML data that a DataSet can read.
- Although the DataSet can read and write DiffGrams, there are many restrictions on how it can do so. Specifically, the DataSet can only write all of its data at once to a DiffGram; it cannot write out only records which have been updated. Also, when a DataSet reads a DiffGram, the DataSet must initially be empty.

Using XML Web Services

IN THIS CHAPTER

Creating a Simple XML Web Service

Before a .NET Compact Framework XML Web service client can be created, there must exist an XML Web Service that the client can consume. In this chapter a simple Web service is created, and a .NET Compact Framework client is created to consume it. This XML Web service returns a famous quote, the name of the person who first spoke or wrote the quote, and the years during which the person lived.

The quotes will be stored in a Microsoft SQL Server database. When a request for a quote is made, the XML Web service will query for a random quote and return that quote to the client. You will need to set up this database before you can run this sample. See the `Readme.txt` file in the media for this chapter for instructions on setting up the database.

To create the XML Web service, start a new project in Visual Studio.NET. Use the ASP.NET Web Service template. Name the project `QuotableQuotesWebService`. An XML Web service named `Service1` will be created in the `Service1.asmx` file. Rename this XML Web service to `QuoteService` and the `.aspx` source file to `QuoteService.aspx`.

The XML Web service will expose one web method, `GetQuote`. This method returns the quote information. The quote information is first pulled from the Microsoft SQL Server database. There is a stored procedure in the `QuotableQuotes` database named `GetQuote` that we will use to query the quote information. Microsoft Visual Studio.NET will assist in writing code to interact with this stored procedure. First open the Server Explorer and locate the `GetQuote` stored procedure in the `QuotableQuotes` database. Drag the `GetQuote` stored procedure onto the XML Web service designer. This will create two objects: `sqlConnection1` and `sqlCommand1`. The `sqlConnection1` object is of type `SqlConnection` and represents the connection to the `QuotableQuotes` database. The `sqlCommand1` object is of type `SqlCommand` and represents the SQL command that will retrieve the quote information from the stored procedure. Rename the `sqlConnection1` and `sqlCommand1` to `quoteConnection` and `cmdGetQuote`, respectively.

Before implementing `GetQuote`, there needs to be a helper method to generate random quotes. The `cmdGetQuote` SqlCommand takes one parameter. This parameter is the unique ID of the quote record in the database. In the database each quote has a unique identifier of type `integer`. The identifiers sequentially increment by one, with the first quote having the ID of zero. The `QuotableQuote` XML Web service will return a random quote. To do this, the code must generate a random number between zero and the largest quote identifier in the database. The largest quote identifier must be retrieved from the database. There is a stored procedure named `GetLargestQuoteIdentifier` in the database to do just this. Locate the `GetLargestQuoteID` stored procedure, and drag it onto the designer. This will create another `SqlCommand` object. Rename this object to `cmdGetLargestID`. Listing 9.1 demonstrates how to retrieve the largest quote ID from the the database. This code should be placed in the `QuoteService` class.

LISTING 9.1

```
C#
public Int64 LargestID
{
  get
  {
    object largestID = cmdGetLargestID.ExecuteScalar();
    if(largestID== null || !(largestID is Int64))
      return -1;

    return (Int64)largestID;
  }
}

VB
Public ReadOnly Property LargestID() As Int64
  Get
    Dim largeID As Object
    largeID = Me.cmdGetLargestID.ExecuteScalar()
```

```
    If (largeID Is Nothing) Or Not (TypeOf largeID Is Int64) Then
      Return -1
    End If

    Return CLng(largeID)
  End Get
End Property
```

The preceding code retrieves that largest quote identifier from the `Quotes` table. First, the `cmdGetLargestID SqlCommand` object is used to get the largest quote identifier from the database. Then the value that is retrieved is checked for validity. Negative one (–1) is returned if the value is invalid.

After retrieving the largest quote identifer, a random quote ID can be generated. This is done with the `System.Random` class. The `System.Random` class represents a pseudo-random number generator. The `Next` method will be used to retrieve a random integer (`Int32`). The `Next` method can accept an integer (`Int32`) that represents the upper bound of the random number to be generated. In this sample the largest quote id will be passed as this parameter.

Tthe `GetQuote` method will return a custom data structure that contains the quote data. Listing 9.2 contains the `Quote` class that holds the quote data. The class should be placed in the `QuoteService.aspx` file inside of the namespace.

LISTING 9.2

```
C#
public class Quote
{
  public string String;
  public string Author;
  public string Date;
}

VB
Class Quote
    Public Str As String
    Public Author As String
    Public Data As String
End Class
```

Now that data structures are out of the way, the `GetQuote` method must be implemented. The `GetQuote` Web method needs to complete the following tasks:

1. Generate a random `Quote` ID
2. Get the quote data from the database
3. Fill the `Quote` data structure
4. Return the `Quote` data structure

The code in Listing 9.3 should replace the commented-out HelloWorld method in the `QuoteService.aspx` file.

LISTING 9.3

```
C#
[WebMethod]
public Quote GetQuote()
{
  quoteConnection.Open();

  try
  {
    Int64 largestID = LargestID;
    if(-1 == largestID)
      return null;

    Random rand = new Random(DateTime.Now.Millisecond);
    Int64 randomQuoteId = rand.Next((int)largestID);
    cmdGetQuote.Parameters["@id"] =
      new SqlParameter("@id", randomQuoteId);

    SqlDataReader reader = cmdGetQuote.ExecuteReader();
    if(!reader.Read())
      return null;

    Quote q = new Quote();
    q.String = reader.GetString(0);  // Get Quote String
    q.Author = reader.GetString(1);  // Get author's name
    q.Date = reader.GetString(2);    // Get the spoken date

    return q;
  }
  finally
  {
    quoteConnection.Close();
```

```
  }
}

VB
<WebMethod()> _
Public Function GetQuote() As Quote
  Me.quoteConnection.Open()

  Try
    Dim largeID As Long
    largeID = LargestID()

    If -1 = largeID Then
      Return Nothing
    End If

    Dim randomQuoteID As Int64
    Dim rand As New Random(DateTime.Now.Millisecond)

    randomQuoteID = rand.Next(CInt(largeID))
    Me.cmdGetQuote.Parameters("@id") = _
      New SqlParameter("@id", randomQuoteID)

    Dim reader As SqlDataReader
    reader = Me.cmdGetQuote.ExecuteReader()

    If Not reader.Read() Then
      Return Nothing
    End If

    Dim quote As New Quote
    quote.Str = reader.GetString(0)
    quote.Author = reader.GetString(1)
    quote.Data = reader.GetString(2)

    Return quote
  Finally
    Me.quoteConnection.Close()
  End Try
End Function
```

First the connection to the `QuotableQuotes` database is opened by calling the `Open` method on the `quoteConnection` object. Next a random number between zero and the largest quote identifier is generated by the Next method on the System.Random class. The ID is checked for validity. If the ID is valid, this number is set as the named parameter @id of the cmdGetQuote SqlCommand object. Next the ExecuteReader method of the SqlCommand object is called. This method executes the command against the Microsoft SQL Server database and returns a SqlDataReader object that provides access to the quote data. Then the SqlDataReader fills the Quote data structure. Finally, the Quote data structure is returned to the caller, and the finally block ensures the database connection is closed even in the case of an unexpected exception. Before the classes in the SqlClient namespace can be used, the System.Data.SqlClient namespace must be imported into the QuoteService.aspx code file.

There is one more detail that needs to be handled. By default, a new Web service is put in the http://tempura.org namespace. Microsoft recommends that each XML Web service have a unique XML namespace that identifies it. This allows client applications to distinguish it from other services on the Web. This can be done by applying the WebServiceAttribute attribute to the Web service class. Add the following line of code directly above the QuoteService class:

```
C#
[WebService(Namespace="http://netcfkickstart/QuoteService",
            Description="Provides access to famous quotes")]

VB
<WebService(Namespace:="http://netcfkickstart/QuoteService", _
            Description:="Provides access to famous quotes")> _
```

This attribute changes the namespace of the QuoteService as well as adds a brief description of the Web service.

With the WebServiceAttribute attribute applied to the Web service, the QuotableQoutes Web service can be compiled and tested. Press F5 to compile and debug the XML Web service. This brings up the QuoteService discovery Web page. The discovery Web page contains the string description from the WebServiceAttribute attribute. The page contains a link labeled Service Description. This link will display the formal WSDL file for the service. There is also a link with the text GetQuote. Clicking this link will bring up a Web page that allows testing of the GetQuote Web method.

The test page provides several pieces of information. Clicking the Invoke button on the test page will invoke the Web method and display the return data in Internet Explorer. The following XML is an example of the results from the GetQuote test page:

```
<?xml version="1.0" encoding="utf-8" ?>
<Quote xmlns:xsd="http://www.w3.org/2001/XMLSchema"
            xmlns:xsi="http://www.w3.org/2001/XMLSchema-instance"
            xmlns="http://netcfkickstart/QuoteService">
  <String>
```

```
"Once you eliminate the impossible, whatever remains, no matter how
improbable, must be the truth."
</String>
<Author>Sherlock Holmes</Author>
      <Date>1859-1930</Date>
</Quote>
```

Besides providing the ability to test the Web service, the test page provides three examples of how to format a request to the Web service and sample responses. These examples include formatting for `HTTP-POST`, `HTTP-GET`, and `SOAP`.

Understanding the .NET Framework Web Service Client

By using the HTTP protocol and a SOAP message, a Web service client can send requests and interpret responses from a Web server. Figure 9.1 depicts the lifetime of a Web method call from a client to the Web server.

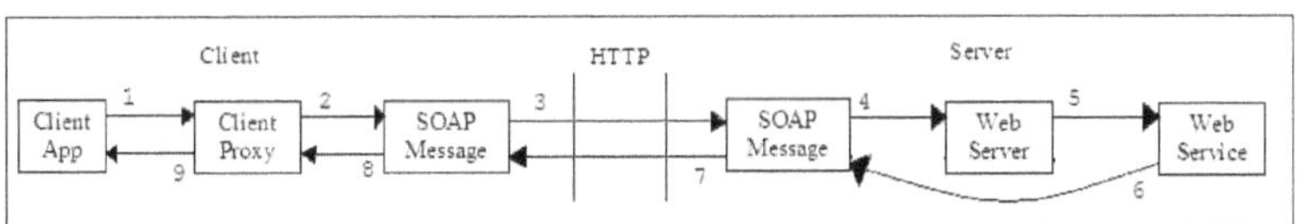

FIGURE 9.1 Lifetime of a Web method call on a Web service.

The following list describes nine steps in the lifetime of a Web method call.

1. The client application creates an instance of the proxy class. The proxy is a class that manages the communication between your application and the Web service. The client application calls a method on the proxy class that in turn invokes the Web service.
2. The proxy, using the client-side Web service architecture, serializes the XML Web service request and packages it in a SOAP message.
3. This SOAP message is then sent to the Web server over HTTP.
4. The server-side Web service architecture receives and deserializes the SOAP message.
5. Using the SOAP message, the class representing the Web service is created. Next, the arguments to the Web method are created from the SOAP message. Finally, the Web method is invoked with the specified arguments.
6. The server-side Web service architecture packages the out parameters and the return value into a SOAP message.
7. The SOAP message is sent back to the client over HTTP.

8. The client-side Web service architecture receives the SOAP message and deserializes the out parameters and return value. The values are returned to the proxy class.

9. The proxy class returns the out parameters and return value to the client application.

Creating a Client for the Simple XML Web Service

Now it is time to create a client for the `QuotableQuotes` XML Web service. Start out by creating a new Smart Device Application. Design the UI to match the UI in Figure 9.2. The code to call the XML Web service will be placed in the button-clicked handler on the button labeled `Get Quote`. First add a reference to the XML Web service into the Smart Device Application project.

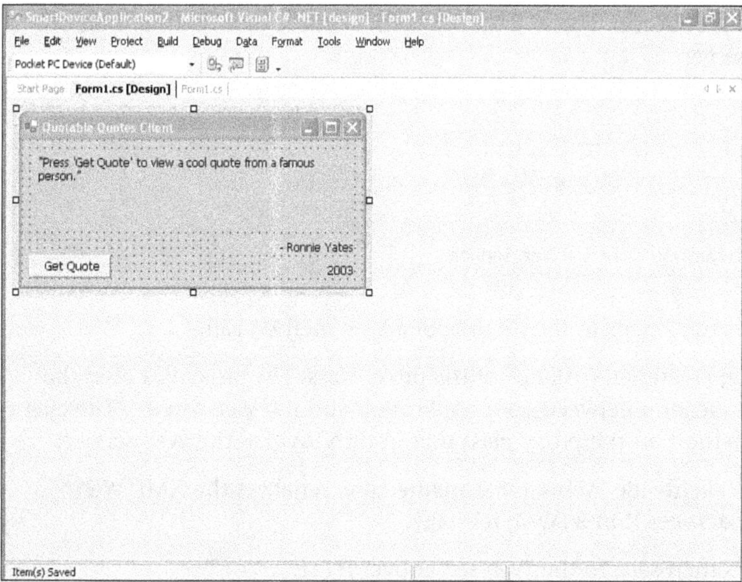

FIGURE 9.2 The user interface for the `QuotableQuotes` client application.

Adding a Web Reference to a Client Application

Now, a Web reference to service needs to be added to the client project. To do this, go to the Solution Explorer, right-click the `Reference` node, and select the `Add Web References...` menu item. This will bring up the Add Web Reference dialog (see Figure 9.3).

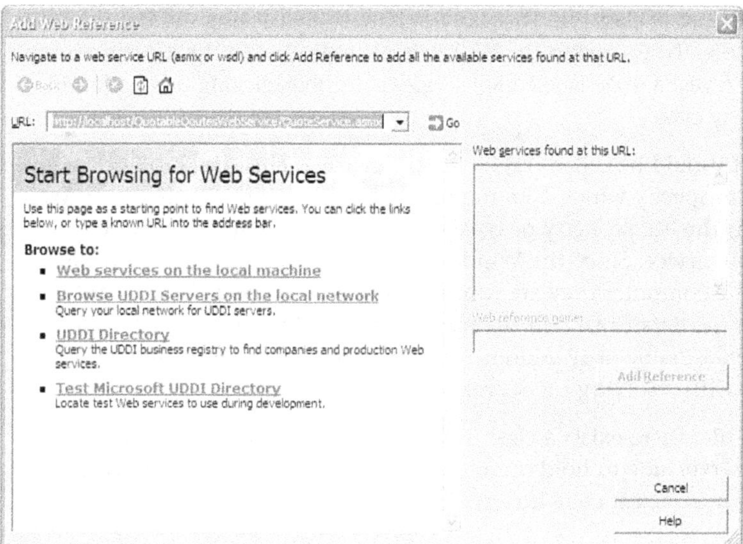

FIGURE 9.3 The Add Web Reference dialog.

The Add Web Reference dialog helps to locate online Web Services and use them in an application. The dialog displays four links. One link allows you to view all of the XML Web services on the local machine. The other three links browse three UDDI servers/directories. The "Browse UDDI Servers on the local machine" link allows you to browse the UDDI servers on the local network. The other two links are labeled UDDI Directory and the Test Microsoft UDDI Directory provide access to services that have registered with Microsoft. The "UDDI Directory" link browses the UDDI business registry to find companies and production Web services. The "Test Microsoft UDDI Directory" link locates test XML Web services that are meant to be used during development.

In this sample the `QuotableQuotes` XML Web service will not be registered with a UDDI. Instead of registering the `QuotableQuotes` XML Web service with a UDDI, the address to the `QuoteService.aspx` page will be entered into the Add Web Reference dialog. Enter the following URL, `http://localhost/QuotableQuotesWebService/QuoteService.asmx`, into the `Address` text box and then press `Enter` or click the `Go` button.

The dialog will then download the WSDL file for the QuoteService Web service. Once the download is complete, click the Add Reference button.

Viewing the Proxy Class

When the Add Reference button is clicked, Microsoft Visual Studio.NET generates a proxy class that will manage the interaction between the application and the `QuotableQuotes` Web service. In some cases the proxy class file may need to be viewed or modified, but by default the Solution

Explorer does not display the proxy class file. This can be changed by clicking the Show All Files button in the Solution Explorer. Now expand the Web References node and the Reference.map node underneath. This will reveal a node labeled Reference.cs. Double-clicking this node will display the code for the proxy class.

There are a few things that should be called out in the proxy class. First, the proxy class gives the client code the ability to specify which URL to use when contacting with the XML Web service. The proxy class has the `URL` property of type `String` that represents the URL to the `.aspx` page of the XML Web service. Since the Windows CE and Pocket PC emulators have a different IP address than the computer they are running, you will not be able to locate the XML Web service by using the default URL in the proxy that uses localhost as the server name. Instead, use the server's name or IP address. This book recommends using the IP address to avoid problems with resolving the server's name.

Second, in the proxy class file, there exists a class named `Quote`. This class corresponds to the `Quote` class created on the server-side to hold quote data. Listing 9.4 contains the declaration of the `Quote` class from the proxy class code file.

LISTING 9.4

```
C#
[System.Xml.Serialization.XmlTypeAttribute(
Namespace="http://netcfkickstart/QuoteService")]
public class Quote {
  /// <remarks/>
  public string String;

  /// <remarks/>
  public string Author;

  /// <remarks/>
  public string Date;
}

VB
<System.Xml.Serialization.XmlTypeAttribute _
  ([Namespace]:="http://netcfkickstart/QuoteService")> _ _
Public Class Quote

  '<remarks/>
  Public Str As String

  '<remarks/>
  Public Author As String
```

```
  '<remarks/>
  Public Data As String
End Class
```

The `System.Xml.Serialization.XmlTypeAttribute` attribute tells the client-side XML Web service architecture to use the `http://netcfkickstart/QuoteService` namespace when serializing an object of this type. The `Quote` class must be declared in the proxy so that the client and servers can use the same data structures when interacting. All custom data structures that are exposed via an XML Web service will be declared for use by the client in the proxy class code file.

Consuming the `QuotableQuotes` Web Service

Now that a reference is added to the client project and a proxy class has been generated, the XML Web service can be consumed by the client application. Open the designer for the client UI, and double-click the `Get Quote` button. This will bring up the code for the button-clicked handler. Before implementing this method, the namespace of the proxy class needs to be added to `Form1.cs`. The proxy class that was generated is placed in its own namespace under the namespace of the client. Add the following using directive under to `Form1.cs`:

```
C#
using QuotableQuotesClient.QuoteServiceWebReference;


VB
Imports QuotableQuotesClientVB.localhost
```

The handler simply needs to create an instance of the proxy class, invoke the `GetQuote` method, and display the quote data in the UI. Implement the handler by using the code in Listing 9.5.

LISTING 9.5

```
C#
private void btnGetQuote_Click(object sender, System.EventArgs e)
{
  QuoteService qs = new QuoteService();
  Quote quote = qs.GetQuote();

  if(null == quote)
  {
    MessageBox.Show("An error occurred retrieving a quote");
    Return;
  }

  UpdateQuoteUI(quote);
}
```

```
VB
Private Sub btnGetQuote_Click(ByVal sender As System.Object,
        ByVal e As System.EventArgs) Handles Button1.Click
   Dim qs As New QuoteService
   Dim quote As Quote

   quote = qs.GetQuote()
   If quote Is Nothing Then
      MessageBox.Show("An error occurred retrieving a quote")
    Return
  End If

  UpdateQuoteUI(quote)
End Sub
```

The UpdateQuoteUI is a simple helper method that extracts the data from a Quote object and updates the application's UI. Listing 9.6 contains the code for the UpdateQuoteUI method.

LISTING 9.6

```
C#
private void UpdateQuoteUI(Quote quote)
{
  lblQuote.Text = quote.String;
  lblAuthor.Text = "- " + quote.Author;
  lblDate.Text = ( quote.Date == "Unknown" ) ?
    string.Empty :
    quote.Date;
}

VB
Private Sub UpdateQuoteUI(ByVal quote As Quote)
  Me.Label1.Text = quote.Str
  Me.Label2.Text = "= " & quote.Author
  If quote.Data = "Unknown" Then
    Me.Label3.Text = String.Empty
  Else
    Me.Label3.Text = quote.Data
  End If
End Sub
```

Asynchronous Consumption of the Simple Web Service

So far the QuotableQuotes XML Web service has been used in a synchronous manner. The proxy object is created, and the `GetQuote` Web method is called. The code then blocks waiting for the `GetQuote` method to return. While this does get the job done, it is not always the desired behavior. For instance, imagine the client is invoking an XML Web service that batches process order requests. The client application would queue up a certain amount of orders and then send the orders to the server at some specified time, such as the end of the day or when the user clicks a Send Orders button. Assume it is a requirement that the application remains responsive and able to create new orders while the orders are being sent to the server. This scenario cries for asynchronous usage of XML Web services. The .NET Compact Framework provides this functionality while demanding very little code to be written by the developer.

The proxy class provides two methods for handling asynchronous Web XML service calls: `Begin`*`WebMethod`* and `End`*`WebMethod`*. In each case replace *`WebMethod`* with the name of the Web method. For example, the `QuoteService` proxy creates the `BeginGetQuote` and `EndGetQuote` methods.

The `Begin`*`WebMethod`* method takes two parameters in addition to the parameters the *`WebMethod`* takes. Since `GetQuote` does not accept any parameters, `BeginGetQuote` accepts only two parameters. The first addition parameter is of type `System.AsyncCallback`. This represents the method that will be called once the WebMethod has completed. The second addition parameter is of type object and can be anything that you want that represents the state of the WebMethod call.

To create an asynchronous client start by creating a stub callback function on the client. The method must be public and static. It must also have no return value and accept one parameter of type `System.IAsyncResult`. This parameter represents the result of the async call. It also allows access to the state object you passed to `Begin`*`WebMethod`*. Add the following method stub to the client:

```
C#
public static void GetQuoteCallBack(IAsyncResult ar)
{
  MessageBox.Show("GetQuote completed");
}

VB
Shared Sub GetQuoteCallBack(ByVal ar As IAsyncResult)
  MessageBox.Show("GetQuote completed")
End Sub
```

Now change the implementation of the button handler to call the XML Web service asynchronously. Just replace the current implementation of the button handler with the implementation from Listing 9.7.

LISTING 9.7

```
C#
private void btnGetQuote_Click(object sender, System.EventArgs e)
{
  QuoteService qs = new QuoteService();
  // Set the url of the proxy to the proper url of the web service

  AsyncCallback getQuoteCB = new AsyncCallback(
    QuotableQuotesClient.Form1.GetQuoteCallBack);

  object[] callBackState = {qs, this};
  qs.BeginGetQuote(getQuoteCB, callBackState);
}

VB
Private Sub Button1_Click(ByVal sender As System.Object,
        ByVal e As System.EventArgs) Handles Button1.Click
  Dim qs As New QuoteService
  Dim quote As Quote
  Dim getQuoteCB As New AsyncCallback(AddressOf Form1.GetQuoteCallBack)
  Dim callBackState(2) As Object

  callBackState(0) = qs
  callBackState(1) = Me

  qs.BeginGetQuote(getQuoteCB, callBackState)
End Sub
```

The preceding code first creates an instance of the Web service. An `AsycnCallback` object is created that represents a pointer `GetQuoteCallBack` method on the client. Finally, the Web service call is started by using the `BeginGetQuote` method. This method returns before the Web method call has completed. Note, an array is passed as the second parameter of the `BeginGetQuote` method. This array contains two elements: a reference to the Web service proxy and a reference to the client application. These objects will be used to retrieve the return value once the Web method completes.

The `GetQuoteCallBack` stub needs a proper implementation. In the callback, two actions need to be performed:

- Retrieving the quote data from the Web method call
- Filling the UI with the quote data

Listing 9.8 contains the complete implementation of the `GetQuoteCallBack` method.

LISTING 9.8

```
C#
public static void GetQuoteCallBack(IAsyncResult ar)
{
   object[] callBackState = (object[])ar.AsyncState;
  QuoteService qs = (QuoteService)callBackState[0];
  Form1 app = (Form1)callBackState[1];

  Quote quote = qs.EndGetQuote(ar);
  if(null == quote)
  {
    MessageBox.Show("No quote object received.");
    return;
  }

  app.UpdateQuoteUI(quote);
}

VB
Shared Sub GetQuoteCallBack(ByVal ar As IAsyncResult)
  Dim callBackState() = CType(ar.AsyncState, Object())
  Dim qs As New QuoteService
  Dim app As Form1
  Dim quote As Quote

  quote = qs.EndGetQuote(ar)
  If quote Is Nothing Then
    MessageBox.Show("No quote object received.")
    Return
  End If

  app = CType(callBackState(1), Form1)
  app.UpdateQuoteUI(quote)
End Sub
```

The array that was passed as the state object to the `BeginGetQuote` method provides access to the quote data. That array can be retrieved by accessing the AsyncState property on the IAsyncResult object. The reference to the proxy is the first element in the array. The proxy's `EndGetQuote` method returns the `Quote` object returned by the Web service. Finally, the `Quote` data structure is used to fill the application UI. Notice, the `GetQuoteCallBack` function is a static method. So, there is no way to call `UpdateQuoteUI` without an instance of the application. This is why an instance of the application was stored in the array passed to `BeginGetQuote` method. This instance is used to call `UpdateQuoteUI`.

Consuming a Web Service That Uses a DataSet

Thus far, our XML Web service has returned custom data structures. The .NET Compact Framework provides the ability to transfer even more complex data, such as the DataSet. Although the .NET Compact Framework does not support a typed DataSet, we will see that you can work around this by using a regular DataSet.

Sending and receiving a regular DataSet is done exactly like sending a custom data structure or a simple piece of data, such as a string. The Web method simply needs to accept or return a System.Data.DataSet object. In our first DataSet aware XML Web service, we send the complete Quotes table down to the client. This would allow the application to go offline and still be able to display quotes when the user presses GetQuote.

There is a stored procedure in the QuotableQuotes database called GetQuotes that returns every quote in the Quotes table. Go back to the XML Web service designer, and drag this procedure onto the designer. Rename the new command to cmdGetQuotes. We will expose the Web method GetQuotes on the QuoteService that will return a DataSet to the caller. The DataSet will be filled by using the SqlDataAdapter in conjunction with the cmdGetQuotes SqlCommand object we just added to the project. The following is the code:

```
C#
[WebMethod]
public DataSet GetQuotes()
{
  quoteConnection.Open();

  try
  {
    DataSet quotesDS = new DataSet();
    SqlDataAdapter quotesDa = new SqlDataAdapter(this.cmdGetQuotes);

    quotesDa.Fill(quotesDS);

    return quotesDS;
  }
  finally
  {
    quoteConnection.Close();
  }
}
```

```
VB
Public Function GetQuotes() As DataSet
  Me.quoteConnection.Open()
```

```
  Try
    Dim quotesDS As DataSet
    Dim quotesDa As SqlDataAdapter
    quotesDS = New DataSet
    quotesDa = New SqlDataAdapter(Me.cmdGetQuotes)

    quotesDa.Fill(quotesDS)
    Return quotesDS
  Finally
    Me.quoteConnection.Close()
  End Try
End Function
```

Compile and test this Web method in the same way the `GetQuote` Web method was tested. This time the result should be a long page of XML output to the browser. This is the XML representation of the `DataSet`. This XML is similar to the XML output from a call to `DataSet.WriteXml`. This is because the XML Web service architecture serializes a `DataSet` by calling the `DataSet.WriteXml` method.

Now, it is time to create a client that consumes the `DataSet` from the Web service. Start by creating a new Smart Device Application the same way you created the orginal `QuotableQuotesClient` application, right down to the same UI. Only button-clicked handler and the UI updating code will change. Once you create the application UI, add the following member variables to the `Form1` class:

```
C#
private DataSet quotesDataSet;
private int curQuoteRowNdx;
```

Now, double-click the Get Quotes button to bring up the button-clicked handler. Listing 9.9 contains the code for the handler.

LISTING 9.9

```
C#
private void btnQuote_Click(object sender, System.EventArgs e)
{
  if(null == quotesDataSet)
  {
    QuoteService qs = new QuoteService();
    // Set the proxy's url property to the correct url of the server

    quotesDataSet =  qs.GetQuotes();
    curQuoteRowNdx = 0;
  }
```

```
    if( quotesDataSet.Tables.Count <= 0 )
    {
      MessageBox.Show("Could not retreive the quotes dataset.");
      return;
    }

    if(curQuoteRowNdx >= quotesDataSet.Tables[0].Rows.Count)
      curQuoteRowNdx = 0;

    DataRow quote = quotesDataSet.Tables[0].Rows[curQuoteRowNdx];
    curQuoteRowNdx++;

    UpdateQuoteUI(quote, 0);
  }

  VB
  Private Sub Button1_Click(ByVal sender As System.Object,
          ByVal e As System.EventArgs) Handles Button1.Click
    If quotesDataSet Is Nothing Then
      Dim qs As New QuoteService
      ' Set the proxy's url property to the correct url of the server

      Me.quotesDataSet = qs.GetQuotes()
      Me.curQuoteRowNdx = 0
    End If

    If quotesDataSet.Tables.Count <= 0 Then
      MessageBox.Show("Could not retreive the quotes dataset.")
      Return
    End If

    If Me.curQuoteRowNdx >= quotesDataSet.Tables(0).Rows.Count Then
      Me.curQuoteRowNdx = 0
    End If

    Dim quote As DataRow
    quote = quotesDataSet.Tables(0).Rows(curQuoteRowNdx)
    curQuoteRowNdx = curQuoteRowNdx + 1

    UpdateQuoteUI(quote, 0)
  End Sub
```

The quotes DataSet is downloaded from the service only when the quotes DataSet is null. The DataSet is null the first time the Get Quotes button is clicked. On subsequent calls the quotes are cached in the local quotes DataSet, so there is no need to download them again.

If the quotes DataSet is null, then a QuoteService proxy object is created, the proxy's URL property is configured, and the GetQuotes Web method is called.

Next a check is performed to ensure that curQuoteRowNdx has stepped beyond the number of quotes in the DataSet. The curQuoteRowNdx specifies the index of the current quote in the DataSet. In this example, the application moves sequentially through the quotes every time the user clicks the Get Quotes button.

Then the corresponding DataRow is retrieved from the DataSet and UpdateQuoteUI is called, which will display the quote information to the user. The UpdateQuoteUI is now implemented to use a DataRow instead of a quote data structure. There is also an integer parameter named offset of type Int32. This offset is the column index to the first column of data. All of the other data can be found sequentially after the index. Listing 9.10 contains the new implementation of the UpdateQuoteUI method.

INCREASING THE PERFORMANCE OF WEB SERVICES THAT EXPOSE DATASET

When a XML Web service receives a DataSet, it calls the IXmlSerializable.ReadXml method to re-create the DataSet. This method can take a long time to execute depending on the amount of XML data. Improved performance can be seen if a custom object model is used instead of relying on the DataSet. Using the QuotesService as an example, a QuotesCollection data structure could be introduced into the XML Web service project. This collection would implement the ICollection interface and represent a collection of Quote objects. The GetQuotes method would return a QuotesCollection instead of a DataSet. The client would cycle through this collection instead of a DataTable of quote data. This will provide a significant performance increase, because the DataSet and its XML functionality would not be used at all.

LISTING 9.10

```
C#
private void UpdateQuoteUI(DataRow quote, int offset)
{
  lblQuote.Text = (string)quote[offset];
  lblAuthor.Text = "-" + quote[offset + 1];
  lblDate.Text = "Unknown" == (string)quote[offset + 2] ?
    string.Empty :
    (string)quote[offset + 2];
}

VB
Private Sub UpdateQuoteUI(ByVal quote As DataRow, ByVal offset As Int32)
  Label1.Text = CType(quote(offset), String)
  Label2.Text = "-" & CType(quote(offset + 1), String)
```

```
    If CType(quote(offset + 2), String) = "Unknown" Then
      Label3.Text = String.Empty
    Else
      Label3.Text = CType(quote(offset + 2), String)
    End If
End Sub
```

Compile and run the application. The application should act exactly as it did before, except the quotes will cycle instead of being displayed in a nondeterministic order.

Consuming Web Service That Exposes a Typed `DataSet`

The .NET Compact Framework does not support typed `DataSets`. If a Web service uses a typed `DataSet`, .NET Compact Framework clients will not be able to consume the Web service. Specifically, the client will fail when trying to compile the proxy into the application.

There are work-arounds for consuming a typed `DataSet`:

- Modify the XML Web service to use only regular `DataSets`.
- Modify the XML Web Service to expose only regular `DataSets`.
- Create a new XML Web service that consumes the original typed `DataSet` and converts it to a regular `DataSet`.

These first two options may seem similar, but they are quite different. The first option suggests that you remove all instances of the typed `DataSet` from the Web service and use only the regular `DataSet` while coding. The second option suggests that you modify the Web service to expose (send/receive) only regular `DataSets`. This means that the internal code can create and utilize the benefits of a typed `DataSet`, but the Web method uses only regular `DataSets` as parameters and return values.

It is simple to write an XML Web service to expose the regular `DataSet` but internally use a typed `DataSet`. Go back to the `QuoteableQuotes` XML Web service project, and go to the XML Web Service designer. Locate the `Quotes` table in the Server Explorer. It will be under `<Server Name>\SQL Servers\<Server Instance Name>\QuotableQuotes\Tables`. Now drag the `Quotes` table onto the designer. This creates a `SqlDataAdapter` that will create a typed `DataSet` for the `Quotes` table. Rename the `SqlDataAdapter` to `daQuotesDS`.

Now click the `SqlDataAdapter`, go to the `Data` menu, and select the Generate DataSet . . . menu item. This will bring up the Generate DataSet dialog. Change the name of the new `DataSet` to `QuotesDataSet`, and click OK. This will generate a typed `DataSet` for the `Quotes` table.

Create a Web method named `GetTypeQuotes`. The Web method will return a regular `DataSet`, but internally it will use the `QuotesDataSet` typed `DataSet` that was just created. Listing 9.11 contains a method that will convert the typed `DataSet` to a regular `DataSet` before returning it to the user.

LISTING 9.11

```
C#
[WebMethod]
public DataSet GetTypedQuotes()
{
  quoteConnection.Open();

  try
  {
    QuotesDataSet quotesDS = new QuotesDataSet();
    this.daQuotesDS.Fill(quotesDS);

    return quotesDS;
  }
  finally
  {
    quoteConnection.Close();
  }
}

VB
<WebMethod()> _
Public Function GeTypedQuotes() As DataSet
  Me.quoteConnection.Open()

  Try
    Dim quotesDS As New QuotesDataSet
    Me.daQuotesDS.Fill(quotesDS)

    Return quotesDS
  Finally
    Me.quoteConnection.Close()
  End Try
End Function
```

Notice there is no manual conversion going on here. The code relies on polymorphism. Since a typed `DataSet` is a subclass of the `DataSet` class, it supports the `WriteXml` method of the

IXmlSerializable interface. The XML written from the method will create the same data, even if it is read by a plain DataSet.

To finish this example the Quotable Quotes DataSet client need to be modified to use the GetTypedQuotes method. Only the button-clicked handler needs to be updated. The handler will now call GetTypeQuotes instead of GetQuotes. Also, the offset parameter of the UpdateQuoteUI method will be 1 instead of 0 in order to compensate for the ID column in the DataRow. Listing 9.12 contains the new code for this method.

LISTING 9.12

```
C#
private void btnQuote_Click(object sender, System.EventArgs e)
{
  if(null == quotesDataSet)
  {
    QuoteService qs = new QuoteService();
    // Set the proxy's url property to the correct url of the server

    quotesDataSet =  qs.GetTypedQuotes();
    curQuoteRowNdx = 0;
  }

  if(quotesDataSet.Tables.Count >= 0)
  {
    MessageBox.Show("Could not retreive the quotes dataset.");
    return;
  }

  if(curQuoteRowNdx >= quotesDataSet.Tables[0].Rows.Count)
    curQuoteRowNdx = 0;

  DataRow quote = quotesDataSet.Tables[0].Rows[curQuoteRowNdx];
  curQuoteRowNdx++;

  UpdateQuoteUI(quote, 1);
}

VB
Private Sub Button1_Click(ByVal sender As System.Object,
        ByVal e As System.EventArgs) Handles Button1.Click
  If quotesDataSet Is Nothing Then
    Dim qs As New QuoteService
```

```
    ' Set the proxy's url property to the correct url of the server

    Me.quotesDataSet = qs.GeTypedQuotes()
    Me.curQuoteRowNdx = 0
  End If

  If quotesDataSet.Tables.Count <= 0 Then
    MessageBox.Show("Could not retreive the quotes dataset.")
    Return
  End If

  If Me.curQuoteRowNdx >= quotesDataSet.Tables(0).Rows.Count Then
    Me.curQuoteRowNdx = 0
  End If

  Dim quote As DataRow
  quote = quotesDataSet.Tables(0).Rows(curQuoteRowNdx)
  curQuoteRowNdx = curQuoteRowNdx + 1

  UpdateQuoteUI(quote, 1)
End Sub
```

Compile and run this example and the application will run as it did before. The server side will be reaping the benefits of using typed `DataSet` while interoperating with the .Net Compact Framework client.

In Brief

- The .NET Compact Framework provides the ability for applications to consume XML Web services.
- The .NET Compact Framework provides the ability to consume simple data, such as strings, as well as complex data, such as user-defined data structures and `DataSets`.
- The .NET Compact Framework supports synchronous and asynchronous XML Web service calls.
- Although the .NET Compact Framework does not support typed `DataSets`, with a little work, XML Web service clients can consume them.

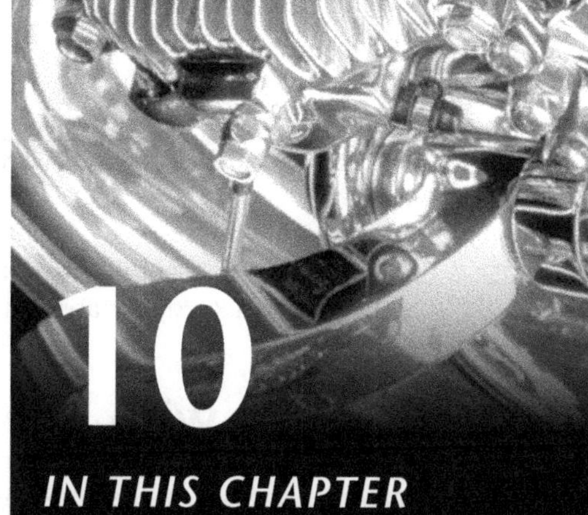

Manipulating XML with the XmlTextReader and the XmlTextWriter

IN THIS CHAPTER

XML Architecture Overview

When deciding which XML features to bring to the .NET Compact Framework's XML architecture, the designers chose to focus on code footprint size and performance instead of implementing the full set of W3C XML standards. The following lists the features that the .NET Compact Framework designers chose not to support:

- Document Type Definition Support
- XmlSchema
- XmlValidationReader
- XmlSerialization
- XPath

Even without these features the .NET Compact Framework provides robust, feature-rich support for manipulating XML data. Support for forward-only, noncached reading and writing is provided by `XmlTextReader` and `XmlTextWriter`, while the `XmlDocument` allows for the traversing and manipulating of XML data through a node tree data structure. The ability to extend the XML architecture is also available by extending XmlReader and XmlWriter abstract base classes.

Working with the XmlTextReader

The XmlTextReader is a fast, forward-only, read-only XML parser. Since the XmlTextReader is a forward-only reader, very little data is cached while parsing the XML stream. This allows for the XmlTextReader to parse a document while using a small footprint. The XmlTextReader does not provide any validation beyond checking the well-formedness of the XML. Since no validation is performed while reading the document, the XmlTextReader provides fast XML parsing.

The XmlTextReader throws an XmlException if it encounters an error parsing the XML stream. Although this allows you to catch the exception and continue parsing, this is not recommended. The state of the XmlTextReader is unpredictable after it has thrown an XmlException.

The desktop .NET Framework's implementation of the XmlTextReader checks the well-formedness of the DocumentType nodes but does not validate the document using the Document Type Definition (DTD). The .NET Compact Framework enforces an even stricter rule: Documents with DocumentType nodes cannot be processed by the XmlTextReader; the XmlTextReader will throw an XmlException if it encounters a DocumentType node while parsing.

Loading XML Data into an XmlTextReader

The first thing to do when parsing an XML document with an XmlTextReader is to load the stream into the reader. To do this, you simply need to create an instance of the XmlTextReader and tell it where it can find the document. The .NET Compact Framework provides several overloaded constructors for loading an XmlTextReader from different sources.

Loading an XML Stream from the Local File System

The easiest way to create an instance of the XmlTextReader is to call the constructor that takes a string representing the URI to where the XML document can be found, as shown in Listing 10.1.

LISTING 10.1

```
C#
XmlTextReader reader = new XmlTextReader("filename.xml");
VB
Dim reader As New XmlTextReader("filename.xml")
```

When passing a filename to this constructor, it is important to note that Windows CE does not have the concept of a current directory. This means that relative paths cannot be constructed without some interesting and tricky code. In fact, calling Directory.GetCurrentDirectory will raise a NotSupportedException. Calling the XmlTextReader constructor with a relative filepath will raise a FileNotFoundException unless the file exists in the root directory. Therefore, it is recommended that you always specify the fully qualified path of the file when creating an XmlTextReader.

Loading an XML File from a Remote Location

An XmlTextReader can also be created that is bound to an XML file that resides on a remote machine. This can be accomplished by passing the URL to the XmlTextReader constructor, as seen in Listing 10.2.

LISTING 10.2

```
C#
XmlTextReader xmlReader = new XmlTextReader("http://foo.com/bar.xml");

VB
Dim reader As New XmlTextReader("http://foo.com/bar.xml")
```

This constructor creates an XmlTextReader object that parses an XML file located at http://foo.com/bar.xml. The XmlTextReader will take care of creating a request to the server, opening a stream to the URL, and reading the XML data from the request. You need to interact with only the XmlTextReader.

In the real world, your XML data may not be anonymously accessible on your server. You may want to protect your data by applying some authentication scheme to the Web access. If your XML data is password-protected with Digest or Basic Authentication, the .NET Compact Framework can easily empower you to load an XmlTextReader with this protected data. Windows Authentication is not supported by the .NET Framework. Listing 10.3 exemplifies how to load an XmlTextReader from a password-protected Web server.

LISTING 10.3

```
C#
static void Main()
{
  NetworkCredential cred =
    new NetworkCredential("usrnm", "psswd", "domain");

  Stream s =
    GetDocumentStream("http://www/foo.com/bar.xml", cred,);
  XmlTextReader reader = new XmlTextReader(s);
```

```
    // Do something interesting with the XmlTextReader
  }

  static Stream GetDocumentStream(string address, ICredentials cred)
  {
    XmlUrlResolver xur = new XmlUrlResolver();
    Uri uri = new Uri(address);

    if(cred != null)
      xur.Credentials = cred;

    try
    {
      return (Stream)xur.GetEntity(uri, null,null);
    }
    catch(Exception e)
    {
      MessageBox.Show(e.ToString());
      throw e;
    }
  }

  VB
  Function GetDocumentStream
  (ByVal address As String, ByVal cred As ICredentials)
  As Stream
    Dim xur As New XmlUrlResolver
    Dim uri As New Uri(address)

    If (Not cred Is Nothing) Then
      xur.Credentials = cred
    End If

    Try
      Return xur.GetEntity(uri, Nothing, Nothing)
    Catch ex As Exception
      MessageBox.Show(ex.ToString())
    End Try
  End Function

  Sub Main()
```

```
  Dim url As String
  Dim s As Stream
  Dim cred As
    New NetworkCredential("usrnm", "psswd", "domain")
  url = "http://www.foo.com/bar.xml"
  s = GetDocumentStream("url", cred)
  Dim reader As New XmlTextReader(s)

  ' Do something interesting with the XmlTextReader
End Sub
```

The most important section of this code is in the `GetDocumentStream` method. This method takes two parameters: the URL address of the XML resource as a string and an `ICredentials` object that represents the credentials with which to request this XML resource. To acquire the remote resource stream, we use the `XmlURLResolver` that inherits from the `XmlResolver`. We will discuss the `XmlResolver` later in this chapter, but for now think of the `XmlResolver` as an object that can grab resources from remote locations. We set the `XmlURLResolver`'s `Credentials` property to the specified `ICredentials` object. The `XmlUrlResolver` uses this `ICredentials` object when the `GetEntity` method needs to authenticate with a remote server. The "Entity" in `GetEntity` does not refer to an XML character entity but to anything that resides on a remote server. The `GetEntity` method returns a `Stream` object attached to the remote resource. This stream is returned to the caller of `GetDocumentStream`. Remember that even XML documents that were loaded from a remote location cannot contain a `DOCTYPE` node.

Back in the `Main` method, we create a `NetworkCredentials` object that is used to hold our authentication credentials. Then we call the `GetDocumentStream` method, passing in the URL to the resource and the newly created `NetworkCredentials` object. The `GetDocumentStream` returns the `Stream` to the remote XML resource; we pass this stream to the `XmlTextReader` constructor. This binds the `XmlTextReader` to the XML data stream. Subsequent read operations will pull data from the remote stream.

Retrieving a XML document from a Web Server by using SSL is trivial. Use the same code from Listing 10.3, but instead of specifying `http://foo.com/bar.xml`, we would pass in `https://foo.com/bar.xml` if `foo.com` was using SSL. The `XmlTextReader` will handle the certificate authentication.

Using the XmlTextReader's Namespaces Property

The `Namespaces` property provides read and write access to a value indicating whether the reader does namespace support. This property is set to `true` by default, which means that namespaces are supported. If namespace support is turned off, namespace prefixes are not resolved to their respective namespace URIs. Instead, the prefix and local name are considered the local name of the element. This property can be changed only before any read operation has been executed. This means that the XmlTextReader's ReadState *property* must equal ReadState.Initial. An InvalidOperationException is thrown if the reader is in any other state. Listing 10.4 shows the difference between a namespace-enabled reader and a namespace-disabled reader.

LISTING 10.4

```
C#
reader.Namespaces = true;
reader.MoveToContent();
MessageBox.Show("Local Name: " + reader.LocalName);
MessageBox.Show("Prefix: " + reader.Prefix);
MessageBox.Show("Namespace: " + reader.NamespaceURI);
reader.Close();

VB
reader.Namespaces = true
reader.MoveToContent();
MessageBox.Show("Local Name: " & reader.LocalName);
MessageBox.Show("Prefix: " & reader.Prefix);
MessageBox.Show("Namespace: " & reader.NamespaceURI);
reader.Close();

Input Data
<q1:BusinessObjects xmlns:q1="http://test.org">
<q1:BOOK>
<q1:Title>Peter's Chair</q1:Title>
<q1:Author>Ezar Jack Keats</q1:Author>
<q1:ISBN>0-596-00058-8</q1:ISBN>
</q1:BOOK>
<q1:MAG>
<q1:Title>Muscle Fitness</q1:Title>
<q1:ISBN>0-989-00053-8</q1:ISBN>
</q1:MAG>
</q1:BusinessObjects>
```

Understanding the Normalization Property

The Normalization property provides read and write access to a value that controls whether the reader performs white-space and attribute value normalization. A complete description of normalization is beyond the scope of this book. To learn more about normalization, you can read the Extensible Markup Language (XML) 1.0 recommendation, section "Attribute-Value Normalization," located at `http://www.w3.org/TR/2000/REC-xml-20001006.html#AVNormalize`. For now, you can think of normalization as the process of combining adjacent white-space nodes into one white-space node and replacing entity references with their resolved values. Because the .NET Compact Framework's `XmlTextReader` will not consume DTD information, resolving anything other than general entities, `&#gt`, is not possible. Furthermore, because general entities are always resolved independently of the `Normalization` property, normalization on the .NET Compact Framework is white-space normalization.

The default value of this property is false, which means that attribute and white-space normalization is not performed. The `Normalization` property can be changed at any time during parsing, but the change will take effect only when a subsequent read operation is executed. One side effect of turning off normalization is that character range checking for numeric entities is also switched off. This allows the reader to accept character entities in the range of `�` and ``. With normalization enabled, the `XmlTextReader` would throw an `XmlException` when it encounters characters in this range. Listing 10.5 demonstrates how normalization property affects how the `XmlTextReader` parses XML data.

LISTING 10.5

```
C#
public static void TestNormalization(bool normOn)
{
  string data =
      @"<Root>
      <Element attr='     &lt;Testing Normalization&gt;
      New Line'/>
       </Root>";
      StringReader str = new StringReader(data);
      XmlTextReader reader = new XmlTextReader(str);
      reader.WhitespaceHandling=WhitespaceHandling.None;

      reader.MoveToContent();
      reader.ReadStartElement("Root");

      reader.Normalization=normOn;
      MessageBox.Show("Normalization On: " + normOn.ToString());
      MessageBox.Show("attr's Value: " + reader.GetAttribute("attr"));
        reader.Close();
}

public static void Main(string[] args)
{
```

```
        // Test with normalization off
        TestNormalization(false);
        // Test with normalization on
        TestNormalization(true);
}
```

VB

```
Sub TestNormalization(ByVal normOn As Boolean)
   Dim data As String
   data = "<Root>" & _
         "<Element attr='     &lt;Testing Normalization&gt; " & _
         "New Line'/></Root>"

  Dim str As New StringReader(data)
  Dim reader As New XmlTextReader(str)

  reader.WhitespaceHandling = WhitespaceHandling.None
  reader.MoveToContent()
  reader.ReadStartElement("Root")

  reader.Normalization = normOn
  MessageBox.Show("Normalization On: " & normOn.ToString())
  MessageBox.Show("attr's Value: " & reader.GetAttribute("attr"))
  reader.Close()
End Sub

Sub Main()
   ' Test with normalization off
    TestNormalization(False)
   ' Test with normalization on
    TestNormalization(True)
End Sub
```

Controlling White-Space Interpretation with the WhitespaceHandling Property

The WhitespaceHandling property provides read and write access to a value that determines how white space will be handled. The WhitespaceHandling property takes an enumeration value of type `System.Xml.WhitespaceHandling`. Table 10.1 shows the `WhitespaceHandling` values and how they affect the `XmlTextReader`.

TABLE 10.1

The `WhitespaceHandling` Enumeration Values and Meanings

VALUE	EFFECT ON `XmlTextReader`
`All`	Both `Whitespace` and `SignificantWhitespace` nodes are returned.
`None`	No `Whitespace` or `SignificantWhitespace` nodes are returned.
`Significant`	Only `SignificantWhitespace` is returned.

The `WhitespaceHandling` property can be changed at any time while parsing and will take effect on the next read operation. But attempting to change this property when the reader is closed would raise an `InvalidOperationException`. The default value of the `WhitespaceHandling` property is `WhitespaceHandling.All`.

`SignificantWhitespace` nodes are returned only when the node is within the `xml:space` attribute that is set to "preserve," because there is no DTD available. Listing 10.6 demonstrates how the `XmlTextReader` parses XML when the `WhitespaceHandling` property is set to different values.

LISTING 10.6

```
C#
public static void Main(string[] args)
{
      ReadXml(WhitespaceHandling.None);
      ReadXml(WhitespaceHandling.All);
      ReadXml(WhitespaceHandling.Significant);
}

public static void
ReadXml(WhitespaceHandling ws)
{
      XmlTextReader reader = new XmlTextReader("input.xml");
      reader.WhitespaceHandling = ws;

      while(reader.Read())
      {
        case XmlNodeType.Whitespace:
          MessageBox.Show("Whitespace");
          break;
        case XmlNodeType.SignificantWhitespace:
```

```
            MessageBox.Show("SignificantWhitespace");
            break;
          case XmlNodeType.Element:
            MessageBox.Show("Node " + reader.Name);
            break;
        }
        reader.Close();
}

VB
Sub Main()
  ReadXml(WhitespaceHandling.None)
  ReadXml(WhitespaceHandling.All)
  ReadXml(WhitespaceHandling.Significant)
End Sub

Sub ReadXml(ByVal ws As WhitespaceHandling)
   Dim reader As New XmlTextReader("input.xml")
   reader.WhitespaceHandling = ws

   While (reader.Read())
     If (reader.NodeType = XmlNodeType.Whitespace) Then
        MessageBox.Show("Whitespace")
     ElseIf (reader.NodeType = XmlNodeType.SignificantWhitespace)
     Then
        MessageBox.Show("SignificantWhitespace")
     ElseIf (reader.NodeType = XmlNodeType.Element) Then
        MessageBox.Show("Element")
     End If
   End While
   reader.Close()
End Sub
```

Using the XmlResolver Property

The XmlResolver property provides write-only access to the object used to resolve the location of the file being loaded. The XmlResolver is an abstract class that is used for resolving remote resources. By default, the XmlTextReader uses the XmlUrlResolver with no credentials.

On the desktop .NET Framework, the XmlResolver is also used to resolve DTD references, but this is not supported on the .NET Compact Framework. If a DocumentType node is encountered, the XmlTextReader will throw an XmlException.

This property can be changed at any time and takes effect the next time a read operation is executed. In most cases the XmlResolver will be used only to load the XmlTextReader. Because there will be no DTD information to resolve, there should never be any other reason to set the XmlResolver after it has been loaded. To save on memory usage, the XmlResolver property should be set to null after loading the document. If you don't need it to load the document, it can be set to null before the document is loaded. See Listing 10.3 for an example of how to use the XmlResolver property.

Reading XML Nodes

As you traverse the XML stream, the XmlTextReader will be a pointer to each node that is encountered. Once the XmlTextReader points at a valid node, there are four pieces of vital information that you can query for: Node Name, Node Namespace, Node Value, and Node Attributes. The XmlTextReader has accessors for each of these properties, but not all of them are applicable to every node type. The node value and node attributes properties are directly dependent on the type of the current node.

The node value property depends directly on the node type. Table 10.2 shows which node types have a value and the actual value that is returned for each. If the node type is not in the list, String.Empty is returned.

TABLE 10.2

XML Nodes with Values

NODE TYPE	VALUE
Attribute	The string value of the attribute
CDATA	The content of the CDATA section
Comment	The comment of the comment node
ProcessingInstruction	The entire content, not including the target
SignificantWhitespace	The white space within an xml:space = 'preserve' scope
Text	The content of the text node
Whitespace	The white space between markup
XmlDeclaration	The content of the declaration

The node attributes property depends on two things: the node type and whether the attributes are actually present on the node. Table 10.3 lists which nodes can contain attributes and which attributes are available on the node. Keep in mind that a node that has the ability to contain attributes is different from a node that actually contains attributes.

TABLE 10.3
XML Nodes That Can Contain Attributes

NODE TYPE	AVAILABLE ATTRIBUTE
Element	Any custom attribute
XmlDeclaration	Version, encoding, & standalone

The `XmlTextReader` also provides convenient methods to check whether a node has a value to return or if it has any attributes. The `HasValue` property returns true if the current node can have a value, but false if it is returned otherwise. Note that when `HasValue` is equal to true, it does not mean that the value is not the empty string. It means only that it is possible for the node to have a value. `HasAttributes` returns true if the current node has attributes, but returns false if it doesn't.

To facilitate checking the current node type, the `XmlTextReader` provides the `NodeType` property. The property returns an enumeration value of `System.Xml.XmlNodeType`. There is an enumeration value for each type of node that can be contained in an XML document. Some of these types, such as `XmlNodeType.DocumentType` or `XmlNodeType.Notation`, will never be encountered when using the .NET Compact Framework.

Reading XML Nodes with the `Read` Method

The `XmlTextReader` provides several methods to read through the XML stream. The `Read` method is the most basic of these methods. It iterates through each node in the document. When an `XmlTextReader` is first created and initialized, there is no information available, so the `Read` method must be called to read the very first node. This method returns true if the next node is read. It returns false if there are no more nodes to be read.

Listing 10.7 illustrates how to use the `Read` method to walk through the XML stream and print out data on the nodes that are found.

LISTING 10.7

```
C#
while(reader.Read())
{
  switch(reader.NodeType)
  {
  case XmlNodeType.Element:
```

```
    if(reader.IsEmptyElement)
      MessageBox.Show("<" + reader.Name + "/>");
    else
      MessageBox.Show("<" + reader.Name + ">");
      break;
  case XmlNodeType.EndElement:
    MessageBox.Show("</" + reader.Name + ">");
    break;
  case XmlNodeType.CDATA:
    MessageBox.Show("<![CDATA[" + reader.Value + "]]>");
    break;
  case XmlNodeType.Comment:
    MessageBox.Show("<!-- " + reader.Value + " -->");
    break;
  case XmlNodeType.Document:
    MessageBox.Show("Reading an XML document");
    break;
  case XmlNodeType.DocumentFragment:
    MessageBox.Show("Reading an XML document fragment");
    break;
  case XmlNodeType.ProcessingInstruction:
    MessageBox.Show("<? " +
              reader.Name + " " +
              reader.Value + "?>");
    break;
  case XmlNodeType.Text:
    MessageBox.Show("Text: " + reader.Value);
    break;
  case XmlNodeType.XmlDeclaration:
    MessageBox.Show("<?xml " + reader.Value + "?>");
    break;
  }
}
```

VB

```
While (reader.Read())
  If (reader.NodeType = XmlNodeType.Element) Then
    If (reader.IsEmptyElement) Then
      MessageBox.Show("<" & reader.Name & "/>")
    Else
      MessageBox.Show("<" & reader.Name & ">")
    End If
```

```
  ElseIf (reader.NodeType = XmlNodeType.EndElement) Then
    MessageBox.Show("</" & reader.Name & ">")
  ElseIf (reader.NodeType = XmlNodeType.CDATA) Then
    MessageBox.Show("<![CDATA[" & reader.Value & "]]>")
  ElseIf (reader.NodeType = XmlNodeType.Comment) Then
    MessageBox.Show("<!-- " & reader.Value & " -->")
  ElseIf (reader.NodeType = XmlNodeType.Document) Then
    MessageBox.Show("Reading an XML document")
  ElseIf (reader.NodeType = XmlNodeType.DocumentFragment) Then
    MessageBox.Show("Reading an XML document fragment")
  ElseIf (reader.NodeType = XmlNodeType.ProcessingInstruction) Then
    MessageBox.Show("<? " & _
                   reader.Name & " " &
                   reader.Value & "?>");
  ElseIf (reader.NodeType = XmlNodeType.Text) Then
    MessageBox.Show("Text: " & reader.Value)
  ElseIf (reader.NodeType = XmlNodeType.XmlDeclaration) Then
    Messagebox.Show("<?xml " & reader.Value & "?>");
  End If
End While
```

Reading the Start Element Tag

The ReadStartElement method is a helper method that checks whether the current node is a start element and advances the reader to the next node. Internally, this method first calls IsStartElement. If IsStartElement returns false, an XmlException is thrown. If IsStartElement returns true, the Read method is called. This will leave the XmlTextReader positioned on the content of the element. If the current node is a true empty element, <Empty/>, calling ReadStartElement will leave the XmlTextReader on the next node in the stream. If the current node is not a true empty element, <Empty></Empty>, the XmlTextReader is left on the end element, </Empty>.

You can optionally supply a name and/or namespace to this method. If this data is supplied, IsStartElement will check whether the current node has the matching name and/or namespace. If not, an XmlException is thrown.

The ReadStartElement should be used in conjunction with the ReadEndElement method. After you have called ReadStartElement and you have consumed any content of the element that exists, ReadEndElement should be called. Note that ReadEndElement will throw an XmlException if the current node is not an end element. Therefore, when ReadStartElement reads a true empty element, an XmlException will be thrown if the ReadEndElement is called next. To prevent this XmlException from being thrown, the XmlTextReader provides the IsEmptyElement property, which returns true if the current node is a true empty element, false otherwise.

The `ReadStartElement` method should be used when you want to move the reader past the start element and on to the content of the node. Listing 10.8 shows how to use `ReadStartElement`.

LISTING 10.8

```
C#
reader.Read();
reader.ReadStartElement("Exercise");

reader.ReadStartElement("Name");
string exName = reader.ReadString();
reader.ReadEndElement();

reader.ReadStartElement("BodyPart");
string bp = reader.ReadString();
reader.ReadEndElement();

reader.Close();

MessageBox.Show("The " + exName + " exercise works the " + bp);

VB
reader.Read()
reader.ReadStartElement("Exercise")

reader.ReadStartElement("Name")
Dim exName = reader.ReadString()
reader.ReadEndElement()

reader.ReadStartElement("BodyPart")
bp = reader.ReadString()
reader.ReadEndElement()

reader.Close()

MessageBox.Show("The " & exName & " exercise works the " & bp)
```

Reading Element Content as a String

The `ReadElementString` method is a helper method to read a text-only element. It first calls the `MoveToContent` method, discussed in the "Jumping to an Element's Content" section, to get the XmlTextReader positioned on the content of the current element. Then the method parses the content as a string value. The parsed string is returned, or if the element is empty, String.Empty is returned. This method will throw an XmlException in two cases: The current node is not a start element, or the element does not contain simple text. Simple text does not include markup such as child elements, comments, or processing instructions. If the node contains several text nodes or CDATA nodes, the text will be concatenated and returned to the user. ReadElementString will also consume the end element while reading the string content.

You can optionally supply a name and/or namespace to this method. If this data is supplied, MoveToContent will check whether the next node has the matching name and/or namespace URI. If not, an XmlException is thrown.

This method should be used when you need to pull all of the data from an element as a string. This method also alleviates the need to read the start and end elements with separate API calls. Listing 10.9 demonstrates how to use ReaderElementString.

LISTING 10.9

```
C#
reader.Read();
reader.ReadStartElement("Exercise");

string name = reader.ReadElementString();
string bodypart = reader.ReadElementString();

reader.Close();

MessageBox.Show("The " + name + " exercise works the " + bodypart);

VB
reader.Read()
reader.ReadStartElement("Exercise")

Dim name As String
Dim bodypart As String
name = reader.ReadElementString()
bodypart = reader.ReadElementString()

reader.Close()
MessageBox.Show("The " & name & " exercise works the " & bodypart)
```

Jumping to an Element's Content

The MoveToContent method is intended to get the reader to an element's content as fast and as reliably as possible. The method first checks whether the current node is a content node. A *content node* is an element, an end element, an entity reference, an end entity, or non–white space text. If the node is not a content node, this method will keep reading nodes until it finds the next content node in the XML stream or it reaches the end of the file. While searching for the next content node, MoveToContent will skip over DocumentType nodes, ProcessingInstruction nodes, Whitespace nodes, and SignificantWhitespace nodes. If the current node is an attribute of a content node, this method will move the reader back to the element that owns the attribute.

Conveniently, the MoveToContent method will return the System.Xml.XmlNodeType of the content node it finds. XmlNodeType.None is returned if the XmlTextReader has reached the end of the file. The MoveToContent method is extremely helpful when you want to skip over all non-content nodes and find a content node(s) of a specific type(s). Listing 10.10 illustrates how to use MoveToContent to perform such a task effectively.

LISTING 10.10

```
C#
while( XmlNodeType.None != reader.MoveToContent())
{
  if(XmlNodeType.Element == reader.NodeType
     && reader.Name == "book")
  {
    MessageBox.Show(reader.ReadElementString());
  }
}

VB
While (XmlNodeType.None <> reader.MoveToContent())
  If (XmlNodeType.Element = reader.NodeType _
      And reader.Name = "book")
  Then
    MessageBox.Show(reader.ReadElementString())
  End If
 End While
```

The code in Listing 10.10 will walk through the input XML stream and print out the text content of each node with the local name "book." Writing code using the MoveToContent method is quite robust. This code snippet will never break, no matter what new nodes or attributes are added to the input XML stream.

Reading XML Attributes

Now that you know how to move downward through a document, visiting nodes along the way, let's investigate moving horizontally through the reader, investigating attributes on elements. Since attributes appear adjacent to an element's start tag, I refer to moving through attributes as moving "horizontally" through the XML document.

Reading Attributes without Moving the Reader

One effective way to move through the attributes of an element is to set up a loop and simply read each one sequentially. The XmlTextReader provides properties and accessors to do just this. Listing 10.11 defines a method that searches the current element's attribute list looking for a specified value. The index of the attribute is returned if a matching value is found. String.Empty is returned otherwise.

LISTING 10.11

```
C#
public int
SearchAttributes(string value, XmlReader reader)
{
  if(!reader.HasAttributes)
    return -1;

  for(int ndx = 0;ndx<reader.AttributeCount;++ndx)
  {
    if(reader[ndx] == value)
      return ndx;
  }
  return -1;
}

VB
Function
SearchAttributes(ByVal value As String, ByVal reader As XmlReader)
As Integer
   Dim ndx As Integer
   If (Not reader.HasAttributes) Then
     Return -1
   End If
```

```
   For ndx = 0 To reader.AttributeCount
     If (reader(ndx) = value) Then
      Return ndx
    End If
  Next ndx
  Return -1
End Function
```

Listing 10.11 utilizes three very useful members of the XmlTextReader: HasAttributes, AttributeCount, and the Item indexer. The code uses HasAttributes to check for the presence of attributes. This property is discussed in the section titled "Reading XML Nodes" earlier in this chapter. The code first checks whether the current node even has attributes. If not, -1 is returned, signifying that no match was found. This property returns true if the current node has attributes. False is returned otherwise.

The *for* loop will sequentially step through each attribute. The upper bound of the loop is set to reader.AttributeCount. This property returns the number of attributes the element contains. This property is relevant only for XmlNodeType.Elements, XmlNodeType.DocumentType, and XmlNodeType.XmlDeclaration. Other nodes cannot have attributes.

Finally, the attribute value is compared to the specified value. If there is a match, the current index of the attribute is returned. The code uses an indexer to retrieve the value of the attribute. The indexer takes an integer representing the index of the attribute and returns the string value of the attribute at that index. The indexer will throw an ArgumentOutOfRangeException if the index is greater than or equal to the AttributeCount.

The item indexer property does not move the reader. Thus, in this example the reader remains on the current node throughout the method. Calling reader.Name after calling the indexer would return the current element's name and not the name of the attribute. In order to retrieve the attribute name, the code would have to move the XmlTextReader to attribute, using one of the horizontal move methods discussed in the next section, "Reading Attributes Using Horizontal Movement."

There are two other item indexers that will retrieve attribute values. The first indexer takes a single string parameter representing the qualified name of the attribute. The second indexer takes two string parameters. The first string parameter represents the local name of the attribute, and the second string parameter represents the namespace URI of the attributes. These indexers do not move the reader. They return String.Empty if the attribute is not found. These indexers are

READING THE ATTRIBUTES OF AN XML DECLARATION NODE

The XmlTextReader item indexers are also great for reading the attributes of an XML declaration, the only other node with attributes that will be encountered by the XmlTextReader. DocumentType nodes also have attributes, but the .NET Compact Framework's XmlTextReader does not allow documents with DTDs to be read. Listing 10.12 demonstrates reading the Standalone property of an XML declaration node.

very useful when the name of the attribute is known. They allow the reader to jump directly to the attribute and retrieve its value.

LISTING 10.12

```
C#
if(XmlNodeType.XmlDeclaration == reader.NodeType)
      MessageBox.Show("Standalone: " + reader["standalone"])

VB
If (XmlNodeType.XmlDeclaration = reader.NodeType) Then
  MessageBox.Show("Standalone: " & reader("standalone"))
End If
```

Reading Attributes by Using Horizontal Movement

In the previous section, we examined methods for reading the attributes of an element without actually moving the XmlTextReader. Now let us investigate the methods that read attributes by moving the reader.

The MoveToAttribute method allows you to move from an element to its attributes. This method functions much like the XmlTextReader item indexer, except, of course, that it moves the reader to the attribute. Since the reader is actually moved to the attribute, all of the methods and properties operate on the attribute and not on the container element. This means that calling the Name property after calling the MoveToAttribute method will return the name of the attribute and not the name of the element containing the attribute.

Like the XmlTextReader indexer, MoveToAttribute comes in three flavors. One overload takes a single integer parameter as the index of the attribute. And like its indexer counterpart, this method will throw a System.ArgumentOutOfRangeException if the index is less than zero or greater than or equal to AttributeCount. The next overload takes one string parameter as the qualified name of the attribute. The method returns true if an attribute with a matching name is found, false if otherwise. The reader is not moved if no attribute is found and false is returned. The final overload of MoveToAttribute takes two strings as parameters. The first represents the local name of the attribute and the second is the namespace URI of the attribute. Again, true is returned if the attribute was found and false if no attribute was found. Like its one parameter counterpart, the reader is not moved if no attribute is found.

Let's use the MoveToAttribute method to write another version of the SearchAttributes method from Listing 10.11. The version of SearchAttributes in Listing 10.13 will return the name of the matching attribute instead of the attribute's index.

LISTING 10.13

```
C#
public string
SearchAttributes(string value, XmlReader reader)
{
  if(!reader.HasAttributes)
    return string.Empty;

  for(int ndx = 0;ndx<reader.AttributeCount;++ndx)
  {
    reader.MoveToAttribute(ndx);
    if(reader.Value == value)
      return reader.Name;
  }
  return string.Empty;
}

VB
Function
SearchAttributes(ByVal value As String, ByVal reader As XmlReader)
As String
  Dim ndx As Integer

  If (Not reader.HasAttributes) Then
    Return String.Empty
  End If

  For ndx = 0 To reader.AttributeCount
    reader.MoveToAttribute(ndx)
    If (reader.Value = value) Then
      Return reader.Name
    End If
  Next ndx
  Return String.Empty
End Function
```

The code should look very familiar, except for the return type of the `SearchAttributes` method. We call `MoveToAttribute` with the index of the attribute we want to move to. Since the `XmlTextReader` is actually moved to the attribute, we can call `Name` to return the name of the attribute.

The `XmlTextReader` provides two additional methods for reading attributes that do not require knowing the index or the name of the attribute. The `MoveToFirstAttribute` method moves the reader from the current position to the first attribute on the current element. If there are no attributes, this method returns false and the reader does not move. If there is an attribute in the first position, true is returned and the reader is moved.

The `MoveToNextAttribute` method moves the reader to the next attribute on the element. If the current node is pointed to an element, this method is equivalent to `MoveToFirstAttribute`. If there is no next attribute, false is returned and the reader does not move. If a next attribute exists, the reader is moved and true is returned.

These two methods are very effective when navigating through an element's attributes. We revisit the `SearchAttributes` method from Listing 10.11 to demonstrate `MoveToFirstAttribute` and `MoveToNextAttribute`. The code in Listing 10.14 uses the two methods to loop through the attributes instead of using attribute indexes.

LISTING 10.14

```
C#
public string
SearchAttributes(string value, XmlReader reader)
{
  if(!reader.MoveToFirstAttribute)
    return string.Empty;

  do{
    if(reader.Value == value)
      return reader.Name;
  }while(reader.MoveToNextAttribute());
  return string.Empty;
}

VB
Function
SearchAttributes(ByVal value As String, ByVal reader As XmlReader)
As String
  If (Not reader.MoveToFirstAttribute) Then
    Return String.Empty
  End If

  Do
    If (reader.Value = value) Then
      Return reader.Name
    End If
  Loop While (reader.MoveToNextAttribute())
  Return String.Empty
```

Again, this code is similar to the last two versions. We call `MoveToFirstAttributes` to get the reader to go to the first attribute. The return value of the method is used to check whether an attribute exists at all. The *for* loop has been replaced with a *do* loop that calls `MoveToNextAttribute` until there are no more attributes, at which time the `MoveToNextAttributes` returns false.

This version of `SearchAttributes` seems to be fairly robust. As long as the reader is located on a node with attributes, the code should always find a match if one is present. There is one subtle side-effect of this current version. The astute reader (human) will notice that the reader (text) could be left anywhere on the current element. It would be nice if the reader were always left pointing to the element containing the attributes being searched. The .NET Compact Framework provides the `MoveToElement` method to perform the aforementioned task. This method returns true if the reader is positioned on an attribute. False is returned if the reader is not positioned on an attribute, in which case the reader is not moved. Listing 10.15 is yet another version of `SearchAttributes` that uses `MoveToElement` to guarantee that reader is left on the current element after the attributes are searched.

LISTING 10.15

```
C#
public string
SearchAttributes(string value, XmlReader reader)
{
    string ret = string.Empty;

    if(!reader.MoveToFirstAttribute)
      return ret;

    do{
      if(reader.Value == value)
      {
        ret = reader.Name;
        break;
      }
    }while(reader.MoveToNextAttribute());

    reader.MoveToElement();
    return ret;
}

VB
Function
SearchAttributes(ByVal value As String, ByVal reader As XmlReader)
```

```
As String
  Dim ret As String
  ret = String.Empty

  If (Not reader.MoveToFirstAttribute) Then
    Return ret
  End If

  Do
    If (reader.Value = value) Then
      ret = reader.Name
      Exit Do
    End If
  Loop While (reader.MoveToNextAttribute())

  reader.MoveToElement()
  Return ret
```

Understanding How to Perform Full-Content Reads

While parsing an XML document, the program may need to read the entire content of an element. For example, an XML element might contain a GIF image encoded using Base64 encoding. A program would need to locate the element and read all of the text from it and convert the text back to the image data. Or maybe an element contains the complete chapter of a book, and reading the whole text into one string is not practical for memory constraint reasons. Instead it would be nice to read the string in chunks, processing the data chunk by chunk.

The `XmlTextRead` provides methods to support these example scenarios and many more. The methods in the following list are referred to as *full-content read methods:*

- `ReadString`
- `ReadChars`
- `ReadBase64`
- `ReadBinHex`
- `ReadInnerXml`
- `ReadOuterXml`

We will now examine each method in detail. Let's start with reading the complete content as character data. The `XmlTextReader` provides two methods to facilitate this functionality: `ReadChars` and `ReadString`.

Reading All Content as String Data

The `ReadString` reads the content of an element or a text node as a string. This method behaves differently depending on whether the current node is an element or a text node. If the current node is an element, the method will attempt to concatenate all text, white space, significant white space, and `CDATA` nodes into one string. This string is returned to the user. This concatenation process stops when any markup is encountered. This includes mixed content as well as the end element. If the current node is a text node, the method will concatenate the same aforementioned nodes from the current text node to the element end tag or any markup that is encountered. It is also possible for the reader to be positioned on an attribute. In this case the reader will behave as if it were on the start element.

The return string can be either the concatenated string described previously or the empty string. The empty string can signify two things when returned from this method: It can mean that there is no more text to be read from the current node, or it can mean that the current node is not an element or a text node. Listing 10.16 demonstrates how to use the `ReadString` method.

LISTING 10.16

```
C#
string data =
  @"<Root>Text data followed by mark up.<Child/></Root>";

StringReader str = new StringReader(data);
XmlTextReader reader = new XmlTextReader(str);
reader.WhitespaceHandling=WhitespaceHandling.None;

reader.MoveToContent();
MessageBox.Show("Content of Root: " + reader.ReadString());


VB
Dim strData As String
strData = "<Root>Text data followed by mark up.<Child/></Root>"

Dim str = New StringReader(strData)
Dim reader = New XmlTextReader(str)
reader.WhitespaceHandling = WhitespaceHandling.None

reader.MoveToContent()
MessageBox.Show("Content of Root: " & reader.ReadString())
```

Reading String Data in Chunks

The ReadChars method reads the text content of an element into a character buffer. This method takes three parameters. The first parameter is the character buffer that the text will be copied into; an ArgumentNullException is thrown if this parameter is null. The second parameter is offset into the buffer in which the method should start writing the data; an ArgumentOutOfRangeException is thrown if the index is less then zero. The third parameter is the number of characters to copy into the buffer; an ArgumentOutOfRangeException is thrown if this parameter is less then zero. Also, an ArgumentException is thrown if the number of characters to copy is greater than the space available in the buffer starting from the specified offset index.

The ReadChars method returns the number of characters actually read from the XML stream. This number can be zero for one of two reasons. It can be zero if there are no more characters to be read from the element. It can also read zero if the reader is not positioned on an element.

This method should be used when an element contains a large amount of data or if allocating huge string objects would hurt performance and put a strain on memory usage. This method allows you to read the character data in smaller chunks, process the small chunks, and possibly reuse the same character buffer.

However, there are some characteristics particular to the ReadChars method that should be pointed out. The ReadChars method works only on element nodes. You will not be able to use the ReadChars method even if you are positioned on a text node. The text must be wrapped in an element node. The ReadChars method reads everything from the start tag to the end tag. This includes any markup or CDATA nodes. Well-formedness checking and normalization are turned off when using ReadChars. Also, this method eats the end tag when it finishes reading the content of the element. This is important to note because attempting to call ReadEndElement after reading the content of an element with ReadChars will raise an XmlException. Finally, reading attributes while using ReadChars is not possible. It is important that you examine all of the important attributes before calling ReadChars, because after the method is called, the attribute information will be lost.

Even with the characteristics particular to the ReadChars method, it is still very useful. It can be a very fast, efficient, and memory-conservative way to read text data from an XML element. Listing 10.17 demonstrates how to use the ReadChars method to buffer input from the XmlTextReader.

LISTING 10.17

```
C#
reader.WhitespaceHandling = WhitespaceHandling.None;
```

```
reader.MoveToContent();
reader.ReadStartElement("Hamlet");

int charsRead;
char[] buffer = new char[64];
while(0 != (charsRead = reader.ReadChars(buffer, 0, 64)))
{
  MessageBox.Show("Characters Read: " + charsRead);
  MessageBox.Show("Chars Read: " + new String(buffer));
}

VB
reader.WhitespaceHandling = WhitespaceHandling.None

reader.MoveToContent()
reader.ReadStartElement("Hamlet")

Dim charsRead As Integer
Dim buffer(64) As Char
While (Not 0 = (charsRead = reader.ReadChars(buffer, 0, 64)))
  MessageBox.Show("Characters Read: " & charsRead)
  MessageBox.Show("Chars Read: " & New String(buffer))
End While
```

Reading Binary-Encoded Data

The `XmlTextReader` provides two methods for decoding binary-encoded data: `ReadBase64` and `ReadBinHex`. Both methods read the text form of the binary data and decode the data into an array of bytes. Because the methods are so similar, we will discuss them in parallel. The methods both take three parameters. The first parameter is a byte buffer that will be used to copy out the decoded data; an `ArgumentNullException` is thrown if the buffer is null. The second parameter is offset into the buffer in which the method should start writing the data; an `ArgumentOutOfRangeException` is thrown if the index is less than zero. The third parameter is the number of bytes that should be copied into the buffer; an `ArgumentOutOfRangeException` is thrown if this parameter is less than zero or if the number of bytes to copy is greater than the space available in the buffer, starting from the specified offset index.

Both methods return the number of bytes that were actually read from the XML stream. Zero is returned when there is no more data to be obtained from the XML stream. Like with `ReadChars`, zero is also returned if the current node is not an element. Listing 10.18 demonstrates how to read binary encoded data with the `XmlTextReader`.

LISTING 10.18

```
C#
byte[] buffer = new byte[1024];
XmlTextReader reader = new XmlTextReader("input.xml");
reader.MoveToContent();
reader.ReadStartElement("Image");

int bytesRead;
while(reader.Name == "Base64" &&
      0 != (bytesRead = reader.ReadBase64(buffer, 0, 1024)))
{
  MessageBox.Show("Bytes Read: " + bytesRead);
}
while(reader.Name == "BinHex" &&
      0 != (bytesRead = reader.ReadBinHex(buffer, 0, 1024)))
{
  MessageBox.Show("Bytes Read: " + bytesRead);
}

reader.Close();

VB
Dim buffer(1024) As Byte
Dim reader As New XmlTextReader("input.xml")
reader.MoveToContent()
reader.ReadStartElement("Image")

Dim bytesRead As Integer
While (reader.Name = "Base64" And _
       Not 0 = (bytesRead = reader.ReadBase64(buffer, 0, 1024)))
  MessageBox.Show("Bytes Read: " & bytesRead.ToString())
End While

While (reader.Name = "BinHex" And _
       Not 0 = (bytesRead = reader.ReadBinHex(buffer, 0, 1024)))
  MessageBox.Show("Bytes Read: " & bytesRead.ToString())
End While
reader.Close()
```

Reading Element Content as Markup

There will be times when you will need to read all of the content from an element, whether it is plain text or mixed content, as one block of text. The `XmlTextReader` provides the methods `ReadInnerXml` and `ReadOuterXml`, which will read all content from an element and return it in string form.

These methods differ only in that `ReaderOuterXml` returns the start tag, content, and end tag of the current node, while `ReadInnerXml` returns only the content of the current node. The methods were designed to work against elements and attributes. If the reader is positioned on any other type of node, the empty string is returned. If the reader is positioned on an element, calling either method will move the reader to the next element after consuming the content of the current element, despite the fact that `ReadInnerXml` does not return the end tag. Table 10.4 describes what string will be returned from either method when positioned on the same node.

TABLE 10.4

Return Values of `ReadInnerXml` and `ReadOuterXml`

ELEMENT XML	POSITIONED ON	`ReadInnerXml`	`ReadOuterXml`
`<author>`	`<author>`	`<fn>Ronnie</fn>`	`<author>`
`<fn>Ronnie</fn>`		`<ln>Yates</ln>`	`<fn>Ronnie</fn>`
`<ln>Yates<ln>`			`</ln>Yates</ln>`
`</author>`			`</author>`

If the reader is positioned on an attribute, the `ReadInnerXml` will return the value of the attribute. If `ReadOuterXml` is called, the entire attribute as it appears in the XML stream will be returned to the user. Interestingly, in both cases the reader will not be moved to the next attribute. Instead, the reader will remain positioned on the same attribute. Table 10.5 illustrates the different values that are returned from `ReadInnerXml` and `ReadOuterXml` when the reader is positioned on an attribute.

TABLE 10.5

Return Values of `ReadInnerXml` and `ReadOuterXml`

ELEMENT XML	POSITIONED ON	`ReadInnerXml`	`ReadOuterXml`
<auth fn="Ronnie"/>	fn	Ronnie	fn="Ronnie"

These methods are designed to be used when you need to treat an element's content XML as a separate entity. For example, imagine that an application has a layer of business objects capable of saving and restoring their data in XML format. In this system there might be a central object serializer that would manage writing and reading the data out to a single XML file. This is not a difficult scenario to imagine, because the .NET Compact Framework does not provide an object serializer, such as the XmlSerializer or the BinarySerializer. Using ReadInnerXml or ReadOuterXml, the object serializer could hand off chunks of XML data to the separate business objects for them to parse and use for initialization.

Skipping over Nodes

Many times when reading an XML stream, you find that you are searching for one particular node or parent node in the document that meets a certain criterion. The XmlTextReader does not provide any methods that will allow you to find one particular node quickly. Therefore, searching an XML stream with the XmlTextReader can be both complicated and processor intensive. These issues can be alleviated if you design your XML data carefully. Listing 10.19 illustrates how to build a method that will search for a given error message in an XML document with the Skip method.

LISTING 10.19

```
C#
public string
FindErrorMessage(string filename, int catNumber, int errorNumber)
{
    XmlTextReader reader = new XmlTextReader(filename);
    reader.WhitespaceHandling = WhitespaceHandling.None;
    reader.Read();
    reader.ReadStartElement("Errors");

    int depth;
    while(!reader.EOF && reader.Name != "Errors")
    {
      if(reader["Number"] != catNumber.ToString())
      {
        reader.Skip();
        continue;
      }

      depth = reader.Depth;
      reader.Read();
      if(reader.Depth <= depth)
         return string.Empty;

      depth = reader.Depth;
      while(reader.Depth == depth)
      {
```

```
        if(reader["Number"] == errorNumber.ToString())
          return reader.ReadElementString();
        reader.Skip();
      }
      return string.Empty;
    }
    return string.Empty;
}
```

```
VB
Function _
FindErrorMessage (ByVal filename As String, _
                  ByVal catNumber As Integer, _
                  ByVal errorNumber As Integer) As String
  Dim reader = New XmlTextReader(filename)
  reader.WhitespaceHandling = WhitespaceHandling.None
  reader.Read()
  reader.ReadStartElement("Errors")
  Dim depth As Integer
  While (Not reader.EOF And Not reader.Name = "Errors")
    If (Not reader("Number") = catNumber.ToString()) Then
      reader.Skip()
    Else
      depth = reader.Depth
      reader.Read()
      If (reader.Depth <= depth) Then
        Return String.Empty
      End If

      depth = reader.Depth
      While (reader.Depth = depth)
        If (reader("Number") = errorNumber.ToString()) Then
          Return reader.ReadElementString()
        End If

        reader.Skip()
      End While
    End If

    Return String.Empty
  End While
```

```
    Return String.Empty
End Function

Input Data
<Errors>
    <Category Number="1">
        <Error Number="1">Stick A Fork In Me!</Error>
        <Error Number="2">…I'm Done!</Error>
        …
    </Category>
    <Category Number="2">
        <Error Number="1">Hmmm…Bad Mojo!</Error>
        <Error Number="2">This cannot be good!</Error>
        …
    </Category>
    <Category Number="3">
        <Error Number="1">File Not Found!</Error>
        <Error Number="2">What did you do!</Error>
        …
    </Category>
</Errors>
```

Closing the `XmlTextReader`

The `XmlTextReader` should be explicitly closed whenever possible. Explicitly closing it will free up any resources that may be in use while reading the stream. These could be file handles, Web response streams, sockets, or just plain string data. Letting the `XmlTextReader` go out of scope is not an acceptable way to free up these resources. Besides, explicitly closing the reader is a deterministic way to free up resources. That means there is no reason to guess when resources are available or to attempt to program around resources that are stuck in limbo.

Working with the `XmlTextWriter`

The `XmlTextWriter` provides a forward-only, read-only, noncached way of generating XML streams. The `XmlTextWriter` helps you create well-formed XML documents that conform to the W3C XML 1.0 recommendation and the namespaces in XML recommendation. The `XmlTextWriter` holds very little in memory other than the text to write to the file. It does, however, keep a stack of objects that represent the tags that have been written to the XML stream. The objects on the stack are used to resolve namespace prefixes and to close elements and attributes automatically, as well as to help ensure well-formedness of the document.

What Is Not Supported?

Just like with the `XmlTextReader`, document type nodes are not supported by the .NET Compact Framework's `XmlTextWriter`. Attempting to write a document type declaration using the `WriteDocumentType` method will raise a `NotSupportedException`. Those who create their own XML writers can handle DTDs and can inherit from `XmlWriter` and overwrite the `WriteDocumentType` method.

Creating the `XmlTextWriter`

To create the `XmlTextWriter`, we need to give it a stream to write to. `XmlTextWriter(string filename, Encoding encoding)` is the simplest constructor to use. You pass in the fully qualified name of the file to write to and the encoding to write the XML text with. The code in Listing 10.20 demonstrates how to create an `XmlTextWriter`.

LISTING 10.20

```
C#
XmlTextWriter writer = new XmlTextWriter("output.xml", Encoding.UTF8);

VB
Dim writer As New XmlTextWriter("output.xml", Encoding.UTF8);
```

Listing 10.20 creates an `XmlTextWriter` that will write to the file `output.xml` in the root directory of a Windows CE device by using UTF-8 character encoding.

The same filename restrictions that apply to loading an `XmlTextReader` apply to the `XmlTextWriter`. The fully qualified name of the file should be used to specify the output file.

Manipulating Namespace Support

The `Namespaces` property provides read and write access to a value indicating if the writer will perform namespace support. This property is set to true by default.

LISTING 10.21

```
C#
XmlTextWriter writer = new XmlTextWriter("namespaces.xml");
writer.Namespaces = true;
```

```
writer.WriteElementString("po",
                          "test",
                          "http://www.fake.com");

VB
Dim writer As New XmlTextWriter("namespaces.xml")
writer.Namespaces = True
writer.WriteElementString("po",
                          "test",
                          "http://www.fake.com")

Output
<po:test xmlns:po="http://www.fake.com" />
```

If namespace support is turned off, then any attempts to declare a namespace will raise `InvalidArgumentException`. The `Namespaces` property must be set before any write operation has been executed. In other words, the `WriteState` must be equal to `WriteState.Start`. An `InvalidOperationException` is thrown if the writer is in any other state.

Formatting the XmlTextWriter's Output

Before starting to write XML data, you should consider whether and how the XML should be formatted. Formatting can make your XML data appear more organized and more readable to the human eye.

Formatting is controlled by the `Formatting` property of the `XmlTextWriter`. The `Formatting` property accepts an enumeration of the type `System.Xml.Formatting`. The enumeration has two values: `Indented` and `None`. `None` is the default value, meaning that no formatting is performed by a newly created `XmlTextWriter`. Setting the `Formatting` property to `Formatting.Indented` means that child elements will be indented.

You can specify which character to use for indention as well as how many of those characters equal one indention. This is controlled by the `IndentChar` and `Indention` properties, respectively. The `IndentChar` property provides read and write access to the character used for indenting. The default value is the space character. To ensure valid XML, the property can be set to only a valid white space character: `0x9`, `0x10`, `0x13`, or `0x20`. The `Indention` property provides read and write access to how many `IndentChars` are written for each indentation. The default value is two. The `XmlTextWriter` will throw a `System.ArgumentException` if you attempt to change the `Indention` property to a negative value.

All three formatting properties can be set at any time, and they will be applied the next time an indention is inserted into the document. The `IndentChar` and `Indention` properties remain set throughout the life of the reader or until they are changed to new values. Setting the `Formatting` property to `Formatting.None` will *not* reset the `IndentChars` and `Indention` properties back to their defaults.

Turning formatting on will increase the size of your XML files and may cause performance degradation in the `XmlTextWriter`. When deciding whether to use formatting, you should determine if the XML needs to be human-readable. If not, it should not be formatted. You could use formatted XML while developing and debugging your application, but use unformatted XML for production code. The bottom line is that formatting XML is for humans. Formatting does not benefit XML parsers.

Writing an XML Declaration

Once the `XmlTextWriter` is created and initialized and the properties have been set to your liking, you can start writing XML data to the stream. The `XmlTextWriter` provides a rich API for writing nodes to the XML data stream. This section will look at the methods for writing the most important nodes: `XML Declarations`, `Elements`, and `Attributes`.

An XML declaration is not required in an XML document, but it can provide helpful information for parsers. The `XmlTextWriter` provides `WriteStartDocument` for writing the XML declaration to the XML stream. The API allows you to specify whether the document will be a standalone. Since `DocumentType` nodes are not supported by the .NET Compact Framework's `XmlTextWriter`, the document should always stand alone, but setting the standalone value to false will not raise an exception.

The `WriteStartDocument` must be the first write method executed after the constructor if you want to include an XML declaration. If this method is called after any other write method, an `InvalidOperationException` will be raised. If you choose not to write an XML declaration node, the writer will assume that you are writing an XML fragment node. This means that well-formedness rules that apply to the root node will not be checked. For example, you will be able to write more than one root node without raising an exception.

Listing 10.22 shows how to write a XML declaration to an XML stream. It also demonstrates calling `WriteEndDocument` to signal the end of the XML document to the writer.

LISTING 10.22

```
C#
XmlTextWriter writer =
  new XmlTextWriter("startdoc.xml", Encoding.UTF8);
writer.Formatting = Formatting.Indented;

writer.WriteStartDocument(true);
writer.WriteElementString("root", null);
writer.WriteEndDocument();
writer.Close();

VB
Dim writer As
  New XmlTextWriter("startdoc.xml", Encoding.UTF8)
writer.Formatting = Formatting.Indented

writer.WriteStartDocument(true)
writer.WriteElementString("root", null)
writer.WriteEndDocument()
writer.Close()

Output
<?xml version="1.0" encoding="utf-8" standalone="yes"?>
<root />
```

Writing XML Elements

Now that the XML declaration is written, it's time to start writing some real XML data. The most important node is the XML element. Elements can be written piece by piece or all at once.

Writing an element piece by piece is necessary when the element needs to contain attributes or child nodes. The `XmlTextReader` provides `WriteStartElement` and `WriteEndElement` to support this functionality. Calling `WriteStartElement` will write the specified start tag and associate it with an optional namespace and prefix. After the start tag is written, you can then write attributes on the element. You can also write child nodes. But you can no longer write an attribute on an element after a child node has been written.

Listing 10.23 demonstrates how to use `WriteElementString` and `WriteEndElement` to write an empty element.

WRITING FULL END ELEMENT TAGS

The `XmlTextWriter` will write true empty elements whenever you write an element with no content. If this is not the desired behavior, use `WriteFullEndElement` to make sure both the start tag and the end tag are written to the stream.

```
C#
// Make a true empty element
writer.WriteStartElement("True_Empty");
writer.WriteEndElement();

// Make an empty element with start and end tags
writer.WriteStartElement("Empty");
writer.WriteFullEndElement();
VB
' Make a true empty element
writer.WriteStartElement("True_Empty")
writer.WriteEndElement()

' Make an empty element with start and end tags
writer.WriteStartElement("Empty")
writer.WriteFullEndElement()

Output
<True_Empty />
<Empty>
</Empty>
```

LISTING 10.23

```
C#
XmlTextWriter writer =
  new XmlTextWriter("startelem.xml", Encoding.UTF8);
writer.Formatting = Formatting.Indented;

writer.WriteStartDocument();
writer.WriteStartElement("root");
writer.WriteEndElement();
writer.Close();

VB
Dim writer As
  New XmlTextWriter("startelem.xml", Encoding.UTF8)
```

```
writer.Formatting = Formatting.Indented

writer.WriteStartDocument()
writer.WriteStartElement("root")
writer.WriteEndElement()
writer.Close()

Output
<?xml version="1.0" encoding="utf-8"?>
<root />
```

An element that contains only text can be written by calling one, single API method, `WriteElementString`, which will write out the start tag, the text content, and the end tag. Unfortunately, attributes other than namespace declarations cannot be written. Listing 10.24 demonstrates how to use `WriteElementString`.

LISTING 10.24

```
C#
XmlTextWriter writer =
  new XmlTextWriter("elementstring.xml", Encoding.UTF8);
writer.Formatting = Formatting.Indented;

writer.WriteStartElement("StockQuote", "http://fakequote.com");
writer.WriteElementString
  ("Symbol", "http://fakequote.com", "MSFT");

writer.WriteElementString("Value",
                          "http://fakequote.com",
                          XmlConvert.ToString(123.32));
writer.WriteEndElement();
writer.Close();

VB
Dim writer As
  New XmlTextWriter("elementstring.xml", Encoding.UTF8)
writer.Formatting = Formatting.Indented

writer.WriteStartElement("StockQuote", "http://fakequote.com")
writer.WriteElementString _
  ("Symbol", "http://fakequote.com", "MSFT")

writer.WriteElementString("Value", _
```

```
                        "http://fakequote.com", _
                        XmlConvert.ToString(123.32))
writer.WriteEndElement()
writer.Close()

Output
<StockQuote xmlns="http://fakequote.com">
  <Symbol>MSFT</Symbol>
  <Value>123.32</Value>
</StockQuote>
```

Writing XML Attributes

In this section we examine how to write attributes on elements. We also look at declaring namespace prefixes and creating the `xml:space` and `xml:lang` attributes.

Just like elements, attributes can be constructed piece by piece or written all at once. Writing attributes piece by piece is similar to writing an element piece by piece. The `WriteStartAttribute` and `WriteEndAttribute` methods provide this functionality.

`WriteStartAttribute` writes the start of an attribute, "name=", or a start attribute with an optional user-defined namespace prefix, "prefix:name=". `WriteStartAttribute` needs to be followed by a call to some method that will write the content of the attribute. The `WriteEndAttribute` method should be called after writing the attribute value to signal that the attribute is completely written.

The `WriteStartAttribute` method will raise an `InvalidOperationException` if the writer's `WriteState` property is not equal to `WriteState.Element`. `WriteEndAttribute` will throw a `System.InvalidOperationException`, as well, if the writer's `WriteState` is not equal to `WriteState.Attribute`.

If a namespace URI and prefix is provided to the `WriteStartAttribute` method, a namespace prefix will be declared along with the attribute being written. It is important to remember that attributes do not pick up default namespaces. Therefore, attributes must be explicitly prefixed with a namespace prefix to be considered "in" a namespace. The code in Listing 10.25 demonstrates how to build an attribute piece by piece.

LISTING 10.25

```
C#
public static void Main()
{
  XmlTextWriter writer =
```

```
        new XmlTextWriter("startatt.xml", Encoding.UTF8);
    writer.Formatting = Formatting.Indented;

    writer.WriteStartElement("root");
    writer.WriteStartAttribute("po", "att1", "http://bogus");
    writer.WriteString("value");
    writer.WriteEndAttribute();
    writer.WriteEndElement();
    writer.Close();
}

VB
sub Main()
    Dim writer As
        New XmlTextWriter("startatt.xml", Encoding.UTF8)
    writer.Formatting = Formatting.Indented

    writer.WriteStartElement("root")
    writer.WriteStartAttribute("po", "att1", "http://bogus")
    writer.WriteString("value")
    writer.WriteEndAttribute()
    writer.WriteEndElement()
    writer.Close()
End Sub

Output
<root po:att1="value" xmlns:po="http://bogus" />
```

The XmlTextWriter also supplies the WriteAttributeString method to write an attribute with one method call. This method writes out the attribute, as well as any necessary namespace declarations. This method also performs a couple of checks to ensure that the attribute content will be well-formed. If the attribute value includes double or single quotes, they will be replaced with " and ', respectively. If you are writing an xml:space attribute, the writer verifies that the attribute value is valid. Valid values include the string's "preserve" and "default." Listing 10.26 demonstrates how to use WriteAttributeString.

LISTING 10.26

```
C#
public static void Main()
{
  XmlTextWriter writer =
    new XmlTextWriter("attstring.xml", Encoding.UTF8);
  writer.Formatting = Formatting.Indented;

  writer.WriteStartElement("root");
  writer.WriteAttributeString("att1", "http://bogus", "value1");
  writer.WriteEndElement();
  writer.Close();
}

VB
sub Main()
  Dim writer As
    New XmlTextWriter("attstring.xml", Encoding.UTF8)
  writer.Formatting = Formatting.Indented

  writer.WriteStartElement("root")
  writer.WriteAttributeString("att1", "http://bogus", "value1")
  writer.WriteEndElement()
  writer.Close();
End Sub

Output
<root d1p1:att1="value1" xmlns:d1p1="http://bogus" />
```

Writing `xml:space` and `xml:lang` Attributes

The `WriteAttributeString` method can also be used to write the special XML attributes `xml:space` and `xml:lang`. The `xml:space` attribute determines how white space should be handled within an element's scope. The `xml:space` attribute has only two possible values: `preserve` and `default`. `default` signals that the parse's white-space processing mode is acceptable for this element. `preserve` means that all the white space in the element's scope, significant or insignificant, should be preserved. When setting the value of the `xml:space` attribute, the `XmlTextWriter` will verifiy that the value is equal to one of the two strings. If it isn't, a `System.InvalidOperationException` is raised.

The `xml:lang` attribute specifies the language of the text content within the element's scope. This can be helpful when reading XML data that may be written in multiple languages. The

DECLARING NAMESPACES BY USING WriteAttributeString

WriteAttributeString actually has two usages. The first is overload of the WriteAttributeString method, which allows you to generate namespace declarations and redefine the default namespace. The code that follows demonstrates this functionality.

```
C#
public static void Main()
{
  XmlTextWriter writer =
    new XmlTextWriter("nsdecl.xml", Encoding.UTF8);
  writer.Formatting = Formatting.Indented;

  writer.WriteStartElement("root");

  // redefine the default namespace
  writer.WriteAttributeString("xmlns",
                              null,
                              "http://default");

  // define a namespace prefix "po"
  writer.WriteAttributeString("xmlns",
                              "po",
                              null,
                              "http://post_office");
     writer.WriteEndElement();
     writer.Close();
   }
   VB
   sub Main()
     Dim writer As
       New XmlTextWriter("nsdecl.xml", Encoding.UTF8)
     writer.Formatting = Formatting.Indented

     writer.WriteStartElement("root")

    ' redefine the default namespace
     writer.WriteAttributeString("xmlns", _
                                 nothing, _
                                 "http://default")

     ' define a namespace prefix "po"
     writer.WriteAttributeString("xmlns", _
```

```
                        "po", _
                        nothing, _
                        "http://post_office")
    writer.WriteEndElement()
    writer.Close()
  End Sub

  Output
  <root xmlns="http://default" xmlns:po="http://post_office" />
```

This code is worth a closer look. The default namespace for "root" and all of its child elements is redefined by calling `WriteAttributeString` and specifying `"xmlns"` as the local name of the attribute, `null` as the attribute namespace, and the namespace URI string as the attribute value.

The namespace prefix `"po"` is declared and associated with the namespace URI `http://post_office` by calling `WriteAttributeString` and specifying `"xmlns"` as the attribute prefix, po as the name of the attribute, `null` as the attribute namespace, and `http://post_office` as the attribute value.

`XmlTextWriter` makes no attempt to verify that the attribute value is valid. Listing 10.27 demonstrates how to set values for both special attributes.

LISTING 10.27

```
C#
public static void Main()
{
   XmlTextWriter writer =
     new XmlTextWriter("nsdecl.xml", Encoding.UTF8);
   writer.Formatting = Formatting.Indented;

   writer.WriteStartElement("root");

   // set the xml:space attribute to preserver
   writer.WriteAttributeString("xml",
                               "space",
                               null,
                               "preserve");

  // set the xml:lang attribute to lang:en
  writer.WriteAttributeString("xml",
                              "lang",
```

```
                                null,
                                "en");
    writer.WriteEndElement();
    writer.Close();
}

VB
sub Main()
  Dim writer As
    New XmlTextWriter("nsdecl.xml", Encoding.UTF8)
  writer.Formatting = Formatting.Indented

  writer.WriteStartElement("root")

  ' set the xml:space attribute to preserve
  writer.WriteAttributeString("xml", _
                              "space", _
                              nothing, _
                              "preserve")

  ' set the xml:lang attribute to lang:en
  writer.WriteAttributeString("xml", _
                              "lang", _
                              nothing, _
                              "en")
  writer.WriteEndElement()
  writer.Close()
End Sub

Output
<root xml:space="preserve" xml:lang="en" />
```

The `xml:space` attribute is set to `preserve` by calling `WriteAttributeString` and specifying `"xml"` as the prefix, `"space"` as the attribute's local name, `null` as the namespace, and `"preserve"` as the attribute value. The `xml:lang` attribute is set to `en`, almost exactly as we set `xml:space`, except that the attribute's local name was `"lang"` and the value was `"en"`.

Writing Element and Attribute Content

Writing elements and attributes piece by piece requires you to write the content separately from the start and end tags. Up to this point, this chapter has glossed over which methods are available for writing this content. This section will investigate these methods in more detail.

The XmlTextWriter provides the WriteString method to assist you in writing string data as content. To ensure that the resulting XML is still well-formed, the WriteString method will perform the following tasks before writing the string:

- The characters &, <, and > are replaced by &, <, and >, respectively.
- Character values in the range 0x–0x1F are replaced by numeric character entries. White space characters with this range are not converted.
- If WriteString is called to write the value of an attribute, double and single quotes are replaced by " and ', respectively.

It is important to note that the WriteString method considers null and the empty string to be equal. If the WriteString receives a null or String.Empty, no text content will be written out. For attributes, the value will be "".

The code in Listing 10.28 demonstrates WriteString used to build an element and its attribute.

LISTING 10.28

```
C#
public static void Main()
{
   XmlTextWriter writer =
     new XmlTextWriter("writestring.xml", Encoding.UTF8);
   writer.Formatting = Formatting.Indented;
   writer.WriteStartElement("root");

   writer.WriteStartElement("Element");
   writer.WriteStartAttribute("att1", "");
   writer.WriteString("< > & \" '");
   writer.WriteEndAttribute();
   writer.WriteStartAttribute("att2", "");
   writer.WriteString(null);
   writer.WriteEndAttribute();
   writer.WriteString("This is text < > &");
   writer.WriteEndElement();
   writer.WriteStartElement("Empty");
   writer.WriteString(null);
   writer.WriteEndElement();
:
```

```
    writer.WriteEndElement();
    writer.Close();
}

VB
sub Main()
    Dim writer As
      New XmlTextWriter("writestring.xml", Encoding.UTF8)
    writer.Formatting = Formatting.Indented
    writer.WriteStartElement("root")

    writer.WriteStartElement("Element")
    writer.WriteStartAttribute("att1", "")
    writer.WriteString("< > & \" '")
    writer.WriteEndAttribute()
    writer.WriteStartAttribute("att2", "")
    writer.WriteString(nothing)
    writer.WriteEndAttribute()
    writer.WriteString("This is text < > &")
    writer.WriteEndElement()
    writer.WriteStartElement("Empty")
    writer.WriteString(nothing)
    writer.WriteEndElement()

    writer.WriteEndElement()
    writer.Close()
End Sub

Output
<root>
  <Element att1="&lt; &gt; &amp; &quot; '" att2="">
    This is text &lt; &gt; &amp;
  </Element>
  <Empty />
</root>
```

If you have a large amount of text to write to the XmlTextWriter, it may be inefficient to load all of the text into one string and then write the string using WriteString. It would be more efficient to write a chunk of the text to a buffer and then write the buffer to the XmlTextWriter. Well, the XmlTextWriter supports this type of buffered writing with WriteChars.

The WriteChars method writes the content one buffer at a time. This method takes three arguments. The first argument is the array of characters to be written; an ArgumentNullException will be raised if you pass `null` as this parameter. The second argument is the position in the buffer indicating the start of the text to write. This parameter must not be less than zero. If it is, an `ArgumentOutOfRangeException` will be raised. The third argument is the number of characters to write; an `ArgumentOutOfRangeException` will again be thrown if this parameter is less than zero. Listing 10.29 demonstrates `WriteChars`.

LISTING 10.29

```
C#
char[] buffer = new char[64];
int bytesRead;

while(0 != (bytesRead = reader.Read(buffer, 0, 64)))
{
  writer.WriteChars(buffer, 0, bytesRead);
}

writer.WriteEndElement();
writer.Close();

VB
Dim buffer(64) As Char
Dim bytesRead As Integer

while(Not 0 = (bytesRead = reader.Read(buffer, 0, 64)))
  writer.WriteChars(buffer, 0, bytesRead)
End While

writer.WriteEndElement()
writer.Close()
```

This method is ideal for writing large amounts of data to the `XmlTextWriter`. This method uses less memory because the input is buffered. Performance is also increased because string manipulation, which can often cause unnecessary garbage, is bypassed by using the character array. This method is ideal for.NET Compact Framework, given the resource-constrained devices your code will be operating on.

Writing binary-encoded data is possible using the `WriteBase64` and `WriteBinHex` methods. The methods are very similar except for the obvious fact that they use different encoding algorithms. The algorithms take the same number of arguments, return the same return value (`void`), and throw the same exceptions in similar situations.

The methods take three arguments. The first argument is a byte array that contains the data to encode. Both methods will raise an `ArgumentNullException` if the byte buffer is null. The second

WRITING SURROGATE CHARACTERS WITH WriteChars

The WriteChars method requires special care when writing surrogate characters. If you split a surrogate character across multiple buffers, a System.ArgumentException will be thrown while writing the first half of the surrogate pair. When this exception is thrown, the last character written will be the character right before the first half of the surrogate pair. Therefore, after catching the exception, create a new buffer that includes the first half of the surrogate pair that caused the exception, followed by the second half of the surrogate pair. Then call WriteChars again, using the new buffer. This will ensure that the surrogate character is written to the stream correctly. The code that follows illustrates this concept:

```
C#
static void Main()
{
  XmlTextWriter writer =
    new XmlTextWriter("surrogate.xml", Encoding.UTF8);
  char [] charArray = new char[4];
  char low = Convert.ToChar(0xDC00);
  char high = Convert.ToChar(0xD800);

  writer.WriteStartElement("DocumentElement");
  charArray[0] = 'w';
  charArray[1] = 'x';
  charArray[2] = 'y';
  charArray[3] = high;

  try
  {
    writer. WriteChars(charArray, 0, charArray.Length);
  }
  catch (ArgumentException ex)
  {
    charArray[0] = high;
    charArray[1] = low;
    charArray[2] = 'z';
    writer.WriteChars(charArray, 0, 3);
  }

  writer.WriteEndElement();
  writer.Close();
}
```

attribute is the index into the buffer that indicates the start index of the bytes to write. The third argument is a number of bytes to write to the stream. Both methods will throw ArgumentOutOfRangeExceptions if either the index or the count of bytes is less than zero. An ArgumentException is also thrown if the byte array length minus the index is less than the number of bytes to write. Listing 10.30 demonstrates how to write binary encoded data with the XmlTextWriter.

LISTING 10.30

```
C#
public void WriteBinaryData(string filename)
{
  byte[] buffer = new byte[64];
  int bytesRead = 0;

  BinaryReader binReader =
    new BinaryReader(new FileStream(filename, FileMode.Open));
  XmlTextWriter xmlWriter =
    new XmlTextWriter("out.xml", Encoding.UTF8);
  xmlWriter.WriteStartElement("Encoded_ File");

  while(0 != (bytesRead = binReader.Read(buffer, 0, 64)))
  {
    xmlWriter.WriteBase64(buffer, 0, bytesRead);
  }

  xmlWriter.WriteEndElement();
  xmlWriter.Close();
}

VB
sub WriteBinaryData(filename as String)
  Dim buffer(64) As Char
  Dim bytesRead As integer

  Dim BinaryReader As
    New BinaryReader(New FileStream(filename, FileMode.Open)
       Dim xmlWriter As
    New XmlTextWriter("out.xml", Encoding.UTF8)
  xmlWriter.WriteStartElement("Encoded_File")
```

```
    while(Not 0 = (bytesRead = binReader.Read(buffer, 0, 64)))
      xmlWriter.WriteBase64(buffer, 0, bytesRead)
    End While

    xmlWriter.WriteEndElement()
    xmlWriter.Close()
  End Sub
```

These methods are ideal for writing any binary data to the XML stream. In fact, they are the only valid ways to write binary data to an XML stream using `XmlTextWriter`. A few examples of situations where you should use `WriteBase64` and `WriteBinHex` include embedding image data and sound bites. You should understand that embedding binary data in an XML stream will almost always result in a large XML stream, so embed wisely.

Converting .NET Compact Framework Data Types through `XmlConvert`

When writing encoded strings to an XML stream, `WriteString` and `WriteChars` encode any special characters so that the XML will be well formed. But what about writing .NET Compact Framework data types to the XML stream? It is not safe to call `ToString` on the data types, because `ToString` will not write the correct string. The data types must be written using the XML Schema (XSD) data type mapping defined by the XML Schema Part 2: DataTypes specification (`http://www.w3.org/TR/2001/REC-xmlschema-2-20010502/`).

The `XmlConvert` class provides static methods that will convert .NET Compact Framework data types to their correct XML Schema mapping. The class also supplies static methods to convert from a string to the .NET Compact Framework data type. It is important to use the `XmlConvert` class and not `System.Convert`, because `System.Convert` does not map Data Types to strings using XSD data type mapping. Also, if the data type was written using the `XmlConvert`, `XmlConvert` should also be used to read and create the data type. Listing 10.31 shows how to use `XmlConvert`.

LISTING 10.31

```
C#
public static void Main()
{
  XmlTextWriter writer =
    new XmlTextWriter("xmlconvert.xml", Encoding.UTF8);
  writer.Formatting = Formatting.Indented;
  writer.Indentation = 2;

  writer.WriteStartElement("root");
```

```
  writer.WriteElementString("boolean", XmlConvert.ToString(false));
  writer.WriteElementString("Single",
    XmlConvert.ToString(Single.PositiveInfinity));
  writer.WriteElementString("Double",
    XmlConvert.ToString(Double.NegativeInfinity));
  writer.WriteElementString("DateTime",
  XmlConvert.ToString(DateTime.Now));
  writer.WriteEndElement();
  writer.Close();

  XmlTextReader reader = new XmlTextReader("xmlconvert.xml");
  reader.WhitespaceHandling = WhitespaceHandling.None;

  reader.MoveToContent();
  reader.ReadStartElement("root");
  bool b = XmlConvert.ToBoolean(reader.ReadElementString());
  float s = XmlConvert.ToSingle(reader.ReadElementString());
  double d = XmlConvert.ToDouble(reader.ReadElementString());
  DateTime dt = XmlConvert.ToDateTime(reader.ReadElementString());
  reader.Close();

  MessageBox.Show("Boolean: " + b.ToString());
  MessageBox.Show("Single: " + s.ToString());
  MessageBox.Show("Double: " + d.ToString());
  MessageBox.Show("DateTime: " + dt.ToString());
}

VB
sub Main()
  Dim writer As _
    New XmlTextWriter("xmlconvert.xml", Encoding.UTF8)
  writer.Formatting = Formatting.Indented
  writer.Indentation = 2

  writer.WriteStartElement("root")
  writer.WriteElementString("boolean", XmlConvert.ToString(false))
  writer.WriteElementString("Single", _
    XmlConvert.ToString(Single.PositiveInfinity))
  writer.WriteElementString("Double", _
    XmlConvert.ToString(Double.NegativeInfinity))
```

```
    writer.WriteElementString("DateTime", _
      XmlConvert.ToString(DateTime.Now))
    writer.WriteEndElement()
    writer.Close()

    Dim reader = New XmlTextReader("xmlconvert.xml")
    reader.WhitespaceHandling = WhitespaceHandling.None

    reader.MoveToContent()
    reader.ReadStartElement("root")
    Dim b As Boolean
    b = XmlConvert.ToBoolean(reader.ReadElementString())
    Dim s As Float
    s = XmlConvert.ToSingle(reader.ReadElementString())
    Dim d As Double
    d = XmlConvert.ToDouble(reader.ReadElementString())
    Dime dt As DateTime
    XmlConvert.ToDateTime(reader.ReadElementString())
    reader.Close()

    MessageBox.Show("Boolean: " & b.ToString());
    MessageBox.Show("Single: " & s.ToString());
    MessageBox.Show("Double: " & d.ToString());
    MessageBox.Show("DateTime: ", dt.ToString());
  End Sub
```

Listing 10.31 writes a few strings to an XML file, using XmlConvert to convert the data types to strings. Then the code reads the values back into objects of their respective types, using XmlConvert to convert the strings to data types. Let's examine the most interesting lines. We write a Boolean value to the XML file. In this case the string "false" is written to the file. Next we write a single to the file. In this case the special value Single.PositiveInfinity is used, and "Inf" is written to the file. Then we write a double to the file. This time Double.NegativeInfinity is used, and "-Inf" is written to the file. Finally, we write the current date to the file. The format used to write the DateTime is "yyyy-MM-ddTHH:mm:ss." For example, with "1976-11-30T15:30:20," the year is 1976, the month is November, the day is the 30th, and the time is 3:30 pm and 20 seconds.

Now let's investigate reading the values from the XML file. First, we read the content of the first child of the root element as a string and convert the string to a Boolean using XmlConvert.ToBoolean. The XmlConvert.ToBoolean also handles the values one and zero, which correspond to true and false, respectively. Next we read the content of the second child element as a string, and using XmlConvert.ToSingle, we convert the string to float value. The next line of code does the same for the double value, but we call XmlConvert.ToDouble.

Remember that these strings correspond to the special negative and positive infinity values. In consequence, `XmlConvert` must be used to convert the strings; using `System.Convert` would result in a `FormatException`.

Finally, the `DateTime` string is converted back into a `DateTime` object using `XmlConvert.ToDateTime`. The string read from the XML file cannot be used to create a `DateTime` object directly. The `DateTime` string format used by `XmlConvert` is specific to the XSD data type mapping and is not supported by any of `DateTime`'s `ToString` methods. Trying to use `XmlConvert.ToDateTime` by using the string output of `DateTime.ToString` will result in a `FormatException`. However, the `DateTime.Parse` method *can* handle the string output from `XmlConvert.ToString(DateTime)`.

Table 10.6 shows which .NET Compact Framework data types and strings are returned using the XSD data type mapping. These types cannot be processed using `System.Convert`.

TABLE 10.6

Strings Returned from `XmlConvert.ToString()`

.NET COMPACT FRAMEWORK DATA TYPE	STRING RETURNED
`Boolean`	`true` or `false`
`Double.PositiveInfinity`	`INF`
`Double.NegativeInfinity`	`-INF`
`Single.PositiveInfinity`	`INF`
`Single.NegativeInfinity`	`-INF`
`DateTime`	`yyyy-MM-ddTHH:mm:ss`

The .NET Compact Framework data types require special handling when being written to or read from an XML stream. The .NET Compact Framework provides the `XmlConvert` class to handle these conversions. Do not be afraid to use these conversion methods liberally. They guarantee that your XML will be well-formed and will conform to W3C standards.

Writing Raw XML Markup

The `XmlTextWriter` provides a rich API for writing well-formed markup to an XML stream. There may be times when you need to strip away this rich API and write raw text to the XML stream without the `XmlTexWriter`'s attempting to encode the text. This could be helpful if your markup input is created by an external source and you just want to paste that markup into the XML stream.

The `WriteRaw` method provides this functionality. This method will not check the string for well-formedness. No encoding will be done, and no characters will be replaced with their respective character entities. The text is written "as is" to the XML stream. You can forget about formatting, too. Listing 10.32 demonstrates how to use `WriteRaw` to write raw markup to an XML stream.

LISTING 10.32

```
C#
public static void Main()
{
  string rawText = "<bad_xml att=\"bad chars ='\"<>\">\n\t" +
                   "This is bad content — & < > = ' \"\n" +
                   "</bad_xml>";
  XmlTextWriter writer =
    new XmlTextWriter("xmlconvert.xml", Encoding.UTF8);

  writer.Formatting = Formatting.Indented;
  writer.Indentation = 2;

  writer.WriteStartElement("root");
  writer.WriteRaw(rawText);
  writer.WriteEndElement();
  writer.Close();
}

VB
sub Main()
  Dim rawText As String
  rawText = "<bad_xml att=" & chr(34) & "bad chars="' & _
            chr(34) & "< >" & chr(34) & _
            "This is bad content - & < > = ' "& chr(34) &_
            "</bad_xml>"
  Dim writer As _
    New XmlTextWriter("xmlconvert.xml", Encoding.UTF8)

  writer.Formatting = Formatting.Indented
  writer.Indentation = 2

  writer.WriteStartElement("root")
  writer.WriteRaw(rawText)
  writer.WriteEndElement()
  writer.Close();
End Sub

Output
<root><bad_xml att="bad chars ='"<>">
      This is bad content — & < > = ' "
</bad_xml></root>
```

Writing Other Nodes

In most cases your XML will consist of only elements and attributes, but there are situations where you may need to add comments, CDATA, or other uncommon nodes to your XML document.

Writing Comments

Let's first look at writing comments. Comments can appear anywhere in the document after the prolog (everything before the root element) and even in the epilog (everything after the root element). Although the W3C XML specification states that a comment may appear anywhere in the document, even before the prolog, this is not true for documents created by the `XmlTextWriter`. Attempting to write a comment before writing an XML declaration will result in an `InvalidOperationException`. Listing 10.33 demonstrates how to add a comment to an XML document.

LISTING 10.33

```
C#
public static void Main()
{
  XmlTextWriter writer =
    new XmlTextWriter("xmlconvert.xml", Encoding.UTF8);
  writer.Formatting = Formatting.Indented;
  writer.Indentation = 2;

  // We can't write a comment here!!!
  writer.WriteStartDocument(true);
  writer.WriteComment("After prolog, before root");
  writer.WriteStartElement("root");
  writer.WriteComment("Inside root element");
  writer.WriteEndElement();
  writer.WriteComment("After the root element");
  writer.Close();
}
VB
sub  Main()
  Dim writer As
    New XmlTextWriter("xmlconvert.xml", Encoding.UTF8)
  writer.Formatting = Formatting.Indented
  writer.Indentation = 2
```

```
    ' We can't write a comment here!!!
    writer.WriteStartDocument(true)
    writer.WriteComment("After prolog, before root")
    writer.WriteStartElement("root")
    writer.WriteComment("Inside root element")
    writer.WriteEndElement()
    writer.WriteComment("After the root element")
    writer.Close()
End Sub

Output
<?xml version="1.0" encoding="utf-8" standalone="yes"?>
<!--After prolog, before root-->
<root>
  <!--Inside root element-->
</root>
<!--After the root element-->
```

Listing 10.33 is fairly simple. Each call to `WriteComment` accepts a string parameter. The resulting comment takes the form `<!--`*`string`*`-->`. An `ArgumentException` will be thrown if the string parameter will result in the comment's ending in `--->` or if the string contains a double hyphen (`--`).

Comments can be very useful. They can improve the readability of the XML document and improve the organization of your documents. That said, comments are primarily for humans. They do very little to help XML processors and will increase the size of your document. So, if speed and size are important to you, use comments sparingly.

Writing CDATA Sections

CDATA sections are used to escape blocks of text containing characters that would otherwise be recognized as markup. Left angle brackets and ampersands need not and cannot be avoided by using their character entity counterparts. Also, CDATA sections cannot be nested. CDATA sections take on the form `<![CDATA[`*`string`*`]]>`. The only text within a CDATA section that is recognized as markup is the CDATA section end tag, `]]>`. Therefore, attempting to write a string that contains `]]>` will raise an `InvalidOperationException`.

CDATA sections can appear anywhere character data may be used. This does not include the prolog or the epilog, only inside the root element where character data may appear.

CDATA is useful when text is produced from an external source and you want your document to include this data without parsing and without breaking the well-formedness of your document. Listing 10.34 demonstrates how to write a CDATA section by using the `WriteCData` method.

LISTING 10.34

```
C#
public static void Main()
{
  XmlTextWriter writer =
    new XmlTextWriter("xmlconvert.xml", Encoding.UTF8);
  writer.Formatting = Formatting.Indented;
  writer.Indentation = 2;

  // We can't write a CDATA section in the prolog!!!
  writer.WriteStartDocument(true);
  writer.WriteStartElement("root");
  writer.WriteCData("Inside root element");
  writer.WriteEndElement();
  // We can't write a CDATA section in the epilog!!!
  writer.Close();
}

VB
sub Main()
  Dim writer =
    New XmlTextWriter("xmlconvert.xml", Encoding.UTF8)
  writer.Formatting = Formatting.Indented
  writer.Indentation = 2

  ' We can't write a CDATA section in the prolog!!!
  writer.WriteStartDocument(True)
  writer.WriteStartElement("root")
  writer.WriteCData("Inside root element")
  writer.WriteEndElement()
  ' We can't write a CDATA section in the epilog!!!
  writer.Close()
End Sub
```

Writing Processing Instructions

XML provides the processing instruction as a means to pass information to an application that may parse the document. In the past, applications have abused comments and nonstandard tags to help out applications. For instance, HTML uses the META tag to tell search engines and other robots how the page should be indexed.

Instead of using nonstandard means to communicate with processing applications, you should write a processing instruction into your XML document. Processing instructions start with `<?` and end with `?>`. Directly after the `<?` is an XML name that specifies the target of the processing instruction. This could be the name of the application that should use the processing instruction, or it could be a unique name specifying the processing instruction. The only restriction on this target is that it must be an XML name. A string follows the target, and its meaning is application-specific.

Processing instructions are markup, but they are not elements, like comments. Processing instructions may appear anywhere in the XML document outside of a tag. This includes the sections before and after the prolog and inside the root element, as well as the sections before and after the epilog. Again, although it is legal for a processing instruction to appear before the prolog, attempting to call `WriteProcessingInstruction` before `WriteStartDocument` will raise an `InvalidOperationException`.

Listing 10.35 shows how to use the `WriteProcessingInstruction` method to create a common processing instruction, xml-stylesheet. This processing instruction attaches a stylesheet to an XML document.

LISTING 10.35

```
C#
static void Main(string[] args)
{
  XmlTextWriter writer =
    new XmlTextWriter("card.xml", Encoding.UTF8);
  writer.Formatting = Formatting.Indented;

  writer.WriteProcessingInstruction("xml-stylesheet",
    "href=\"card.css\" type=\"text/css\"");
  writer.WriteStartElement("BusinessCard");
  writer.WriteElementString("Name", "John Doe");
  writer.WriteElementString("Phone", "555-555-5555");
  writer.WriteElementString("Fax", "999-999-9999");
  writer.WriteElementString("E-Mail", "jdoe@mail.com");
  writer.WriteEndElement();
  writer.Close();
}

VB
sub Main()
  Dim writer As _
    New XmlTextWriter("card.xml", Encoding.UTF8)
  writer.Formatting = Formatting.Indented
```

```
    writer.WriteProcessingInstruction("xml-stylesheet", _
      "href=" & Chr(34) & "card.css" & Char(34) & _
      "type=" & Chr(34) & "text/css" & Chr(34))
    writer.WriteStartElement("BusinessCard")
    writer.WriteElementString("Name", "John Doe")
    writer.WriteElementString("Phone", "555-555-5555")
    writer.WriteElementString("Fax", "999-999-9999")
    writer.WriteElementString("E-Mail", "jdoe@mail.com")
    writer.WriteEndElement()
    writer.Close()
  End Sub
```

Writing White-Space Characters

There may be times when you need to format your document manually. The `WriteWhitespace` method can be used to write white space to your XML document. It is important to note that for the white space to be considered significant, the containing element should have the `xml:space` attribute value set to `preserve`. This is because significant white space is defined as any white space inside a mixed content model or any white space inside the scope of an `xml:space="preserve"` attribute. The `xml:space` attribute can appear on any element, but an element's content model can be marked as a mixed content only in the document's DTD. Since the `XmlTextReader` does not recognize DTD information, the only way to mark white space as significant is to put it in the scope of an `xml:space="preserve"` attribute. Listing 10.36 demonstrates how to utilize `WriteWhitespace` effectively.

LISTING 10.36

```
C#
static void Main()
{
  XmlTextWriter writer =
    new XmlTextWriter("ws.xml", Encoding.UTF8);

  writer.WriteStartDocument();
  writer.WriteStartElement("Root");
  writer.WriteAttributeString("xml", "space", "", "preserve");
  writer.WriteWhitespace("\t");
  writer.WriteElementString("bogus", "after white space");
  writer.WriteEndElement();
  writer.Close();
```

```
    XmlTextReader reader = new XmlTextReader("ws.xml");
    reader.WhitespaceHandling = WhitespaceHandling.Significant;

    while(reader.Read())
    {
      MessageBox.Show("Node Type: " + reader.NodeType);
      MessageBox.Show("Value: " + reader.Value);
    }

    reader.Close();
  }

  VB
  sub Main()
    Dim writer As _
      New XmlTextWriter("ws.xml", Encoding.UTF8)

    writer.WriteStartDocument()
    writer.WriteStartElement("Root")
    writer.WriteAttributeString("xml", "space", "", "preserve")
    writer.WriteWhitespace(Chr(9))
    writer.WriteElementString("bogus", "after white space")
    writer.WriteEndElement()
    writer.Close()

    XmlTextReader reader = new XmlTextReader("ws.xml")
    reader.WhitespaceHandling = WhitespaceHandling.Significant

    While(reader.Read())
      MessageBox.Show("Node Type: " & reader.NodeType)
      MessageBox.Show("Value: " & reader.Value)
    End While

    reader.Close()
  End Sub
```

In Brief

- The XmlTextReader is a fast, forward-only, read-only XML parser.
- The XmlTextWriter provides a forward-only, read-only, noncached way of generating XML streams.

- The `XmlTextReader` or the `XmlTextWriter` must first be created and initialized before it can be used.
- Part of this initialization includes setting the properties that will affect how it will handle its XML data.
- The `XmlTextReader` provides several methods for reading all XML node types.
- The `XmlTextWriter` provide methods for writing all XML node types.
- After using the `XmlTextReader` and `XmlTextWriter`, it is important to close the objects so their resources will be freed.

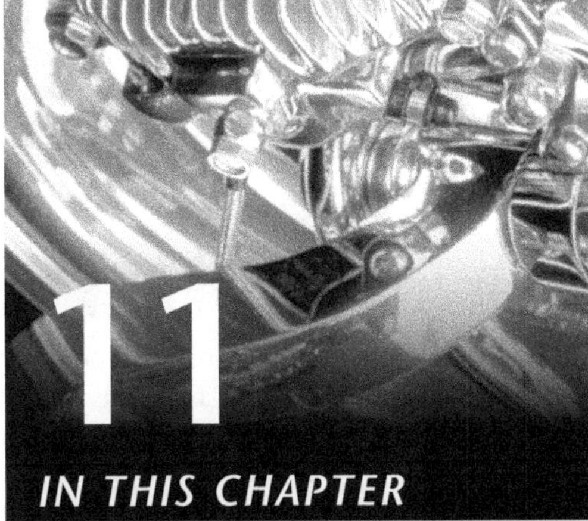

Working with the XML Document Object Model

IN THIS CHAPTER

Processing XML by Using the Document Object Model

So far we have discussed forward-only reading and writing of XML documents. The XML Document Object Model (XML DOM) creates an in-memory hierarchical view of the XML document that can be navigated in any direction. The XML DOM can be used to create, modify, and parse an XML document.

The XML DOM tree is made up of nodes. In the .NET Compact Framework, these nodes are represented by an abstract class named `XmlNode`. Each type of node is represented by an implementation of the `XmlNode class`. For instance, an element node is represented by the `XmlElement` class, which implements the `XmlNode` abstract class.

The `XmlDocument` class represents the very top of the XML DOM tree. The nodes underneath the XmlDocument node are arranged in a hierarchal structure made up mostly of parent-child relationships. Some relationships, such as element-attribute relationships, are considered container-contained relationships. The `XmlDocument` has a property named `DocumentElement` that represents the "root" element of the node tree. This node is incredibly useful for jumping directly to the root node and skipping over the prolog. The `XmlNode` class provides methods for navigating each of its child nodes.

Differences from the Desktop .NET Framework

The .NET Compact Framework's XML DOM implementation does not support schema validation or DTDs. This means that the only entities that will be resolved are the five standard XML entities: <, >, &, and '. Other entities will cause a System.NotSupportedException to be thrown while parsing or loading the document. Also, documents that contain DOCTYPE nodes cannot be loaded. Again, a System.NotSupportedException will be thrown while parsing or loading the document. Finally, element-to-element references using the ID, IDREF, and IDREFS attributes are not supported. Although they will not raise exceptions, you will not be able to use them for navigating from one element to another using the ID attribute. Each call to LoadXml in Listing 11.1 will raise a System.NotSupportedException because of unsupported DTD functionality.

LISTING 11.1

```
C#
// Raise an exception when reading the DOCTYPE node
XmlDocument doc = new XmlDocument();
doc.LoadXml("<!DOCTYPE person [ " +
            "<!ELEMENT first_name (#PCDATA)> " +
            "]>" +
            "<root>" +
            "</root>" );

// Raise an exception when reading an entity
doc = new XmlDocument();
doc.LoadXml("<root>&bogus;</root>" );

VB
'Raise an exception when reading the DOCTYPE node
Dim doc As New XmlDocument()
doc.LoadXml("<!DOCTYPE person [ " & _
            "<!ELEMENT first_name (#PCDATA)> " & _
            "]>" & _
            "<root>" & _
            "</root>" )

' Raise an exception when reading an entity
doc = new XmlDocument()
doc.LoadXml("<root>&bogus;</root>" )
```

Loading the XML DOM

If you want to load existing XML into an `XmlDocument`, there are two very similar methods that can be used: `Load` and `LoadXml`. The `Load` method can load an XML document from a `Stream`, `TextReader`, or `XmlReader`. The `LoadXml` method can load the XML only from a string. Listing 11.2 shows how to use both `Load` methods as well as create an `XmlDocument`.

LISTING 11.2

```
C#
// Load the document from a stream
XmlDocument doc = new XmlDocument();
doc.Load(new FileReader("bogus.xml", FileMode.Open));

// Load the document from a TextReader
doc = new XmlDocument();
doc.Load(new StreamReader("bogus.xml"));

// Load the document from a XmlReader
doc = new XmlDocument();
doc.Load(new XmlTextReader("bogus.xml");

// Load the document for an XML string
XmlDocument doc = new XmlDocument();
doc.LoadXml("<root><child/></root>");

VB
' Load the document from a stream
Dim doc As New XmlDocument()
doc.Load(New FileReader("bogus.xml", FileMode.Open))

' Load the document from a TextReader
doc = New XmlDocument()
doc.Load(New StreamReader("bogus.xml"))

' Load the document from a XmlReader
doc = New XmlDocument()
doc.Load(New XmlTextReader("bogus.xml")

' Load the document for an XML string
doc = New XmlDocument()
doc.LoadXml("<root><child/></root>")
```

Navigating the XML DOM

Now that the `XmlDocument` is loaded, you can navigate its tree structure. Usually, you'll want to skip the XML declaration and other elements in the prolog and start directly at the root node. This can be done by using the `DocumentElement` property. This property returns the root element of the node tree.

The `XmlDocument` can be thought of as a node tree with links both down and up the tree. This allows for navigation in all directions: parent to child, child to parent, and sibling to sibling. The `XmlNode` abstract base class provides several methods for navígating the tree structure.

Navigating Using `HasChildNodes` and `ChildNodes`

The `XmlNode` provides the `HasChildNodes` property as an easy way to check whether the node is linked to any child nodes. This property returns true if any child nodes are present and false otherwise. If child nodes are present, you can retrieve the list of child nodes by using the `ChildNodes` property. Listing 11.3 demonstrates how to use these two methods by implementing the recursive version of the depth-first search graph traversal algorithm.

LISTING 11.3

```
C#
void DepthFirst(XmlNode node)
{
  // Visit the node
  MessageBox.Show("Name: " + node.Name);

  if(!node.HasChildNodes)
  {
     foreach(XmlNode child in node.ChildNodes)
       DepthFirst(child);
  }
}

VB
sub DepthFirst( node as XmlNode )
  // Visit the node
  MessageBox.Show( "Name: " & node.Name )

  If Not node.HasChildNodes Then
    Dim i as Int32
    For i = 0 To node.ChildNodes.Length
      DepthFirst( node.ChildNodes(i) )
```

```
    Next i
  End If
End Sub
```

Navigating by Using Item Indexer Properties

The `XmlNode` class also provides an item indexer to retrieve its child nodes. The `XmlNode` indexer takes a string argument that represents the name of the child node to retrieve. Listing 11.4 illustrates how to use this method.

LISTING 11.4

```
C#
XmlDocument doc = new XmlDocument();
doc.LoadXml("<root><child1/><child2/></root>");
Console.WriteLine(doc.DocumentElement["child2"].Name);

VB
Dim doc = new XmlDocument()
doc.LoadXml("<root><child1/><child2/></root>")
 MessageBox.Show(doc.DocumentElement("child2").Name)
```

If you do not know the name of the child node you are searching for or you want to walk all of the child nodes by using a numeric indexer, it is much more convenient to call the `XmlNodeList`'s item indexer property. The `ChildNodes` property returns an `XmlNodeList` that holds all of the current node's children. Listing 11.5 rewrites the `DepthFirst` method from Listing 11.3, using the `XmlNodeList`'s numeric indexer.

LISTING 11.5

```
C#
static void DepthFirst(XmlNode node)
{
  // Visit the node
  MessageBox.Show("Name: " + node.Name);

  if(node.HasChildNodes)
  {
     for(int i = 0; i < node.ChildNodes.Count; ++i)
       DepthFirst(node.ChildNodes[i]);
  }
}
```

```
VB
sub DepthFirst(node As XmlNode)
  ' Visit the node
  MessageBox.Show("Name: " & node.Name)

  Dim i as Integer
  If(node.HasChildNodes)
    For i = 0 To node.ChildNodes.Count
      DepthFirst(node.ChildNodes(i))
    Next
  End If
End Sub
```

Navigating by Using `FirstChild` and `NextSibling`

The `XmlNode` class also provides properties for moving directly to the first and last child nodes. The `FirstChild` property returns the first child in a node's child list. A null reference is returned if the node has no children. `LastChild` returns the last child node in a node's child list. Again, a null reference is returned if the node has no children. If the node has only one child, then `FirstChild` and `LastChild` reference the same node.

The `XmlNode` class provides the ability to move to the node's siblings. `NextSibling` returns the node that immediately follows the current node or `null` if the node is the last in the sibling list. `PreviousSibling` returns the node that immediately precedes the current node or `null` if the node is the first in the sibling list.

Listing 11.6 shows how to implement the `DepthFirst` method from Listing 11.3, using the `FirstChild` and `NextSibling` methods.

LISTING 11.6

```
C#
void DepthFirst(XmlNode node)
{
  // Visit the node
  MessageBox.Show("NodeType: " + node.NodeType);

  XmlNode child = node.FirstChild;
  while(child != null)
  {
    DepthFirst(child);
    child = child.NextSibling;
  }
}
```

```
VB
sub DepthFirst(node As XmlNode)
  ' Visit the node
  MessageBox.Show("NodeType: " & node.NodeType)

  XmlNode child = node.FirstChild
  While(Not child  Is Nothing)
    DepthFirst(child)
    child = child.NextSibling
  End While
End Sub
```

Navigating by Using ParentNode and OwnerDocument

Two additional properties for navigating an XmlDocument tree are ParentNode and OwnerDocument. ParentNode returns the parent of the current node. The root element's parent is the XmlDocument, and the XmlDocument's parent node is null. Any node in the prolog or epilog also returns the XmlDocument as the parent node. The OwnerDocument property returns the XmlDocument to which the current node belongs. Listing 11.7 displays the path from any given node back to the root node.

LISTING 11.7

```
C#
static void Path(XmlNode node)
{
  XmlNode curNode = node;
  while(true)
  {
    MessageBox.Show("Name: " + curNode.Name);

    if(curNode.ParentNode == node.OwnerDocument)
      break;

    curNode = curNode.ParentNode;
  }
}

VB
sub Path(node As XmlNode)
  XmlNode curNode = node
  While(true)
    MessageBox.Show("Name: " & curNode.Name)
```

```
    If(curNode.ParentNode = node.OwnerDocument)
      Exit While
    End If

    curNode = curNode.ParentNode
  End While
End Sub
```

Accessing Text from Elements

When working with an XmlDocument and its child elements, you may need to retrieve text data from between the start and end tags of an element. Using the XmlTextReader, you can call ReadElementString or ReadString to get text with an element. With the XmlElement object you have two options, as well. The easiest way to get the text content of an element is to use the InnerText property, which returns the concatenated text values of the node and all its child nodes. For example, calling InnerText on the DocumentElement of an XmlDocument loaded with <root>first<child>,second</child>,last</root> will return the string first,second,last. This method can also be used to set the InnerText of the current node. It is very important to know that setting the InnerText property of a node will replace all of the node's children with the parsed content of the given string.

Listing 11.8 shows how to use the InnerText property to read the content of an XmlElement.

LISTING 11.8

```
C#
XmlDocument doc = new XmlDocument();
doc.LoadXml("<Book>" +
            "<Title>Catcher in the Rye</Title>" +
            "<Price>25.98</Price>" +
            "</Book>");
MessageBox.Show("Price: " +
                doc.DocumentElement["Price"].InnerText);

VB
Dim doc As New XmlDocument()
doc.LoadXml("<Book>" & _
            "<Title>Catcher in the Rye</Title>" & +
            "<Price>25.98</Price>" & +
            "</Book>")
MessageBox.Show("Price: " & _
                doc.DocumentElement["Price"].InnerText)
```

The second way to access the text of an `XmlElement` is to retrieve the value of its child text nodes. If the element contains only string data and no markup, you can retrieve the `Value` property of the node's first child. Of course, if the element contains mixed content, you will have to either call the `InnerText` property or walk the child nodes or gather the text manually. Listing 11.9 illustrates using the `FirstChild` and `Value` properties to get the text values from a node.

LISTING 11.9

```
C#
XmlDocument doc = new XmlDocument();
doc.LoadXml("<Book>" +
            "<Title>Catcher in the Rye</Title>" +
            "<Price>25.98</Price>" +
            "</Book>");

MessageBox.Show("Price: " +
                doc.DocumentElement["Price"].FirstChild.Value);


VB
Dim doc As New XmlDocument
doc.LoadXml("<Book>" & _
            "<Title>Catcher in the Rye</Title>" & _
            "<Price>25.98</Price>" & _
            "</Book>")

MessageBox.Show("Price: " &_
                doc.DocumentElement["Price"].FirstChild.Value)
```

Accessing Markup from Nodes

There may be times when you need to access an element's contents in markup form. The `XmlNode` class provides two properties for retrieving a node's XML markup content: `OuterXml` and `InnerXml`. `OuterXml` returns the content of the current node, with the current node as the root element. The `InnerXml` property returns only the content of the current element. The XML returned from `InnerXml` should be considered an XML fragment. Listing 11.10 shows how to use `OuterXml` and `InnerXml` to display the markup content of the node.

LISTING 11.10

```
C#
XmlDocument doc = new XmlDocument();
Doc.LoadXml("<Book>" +
            "<Title>Catcher in the Rye</Title>" +
            "<Price>25.98</Price>" +
            "</Book>");

MessageBox.Show("OuterXml:" + doc.DocumentElement.OuterXml);

MessageBox.Show("InnerXml:" + doc.DocmentElement.InnerXml);

VB
Dim doc As New XmlDocument()
Doc.LoadXml("<Book>" & _
            "<Title>Catcher in the Rye</Title>" & _
            "<Price>25.98</Price>" & _
            "</Book>")

MessageBox.Show("OuterXml:" & doc.DocumentElement.OuterXml)
MessageBox.Show("InnerXml:" & doc.DocmentElement.InnerXml)
```

These two methods are useful when a node's XML content should be considered separately from the document. For example, consider several classes that can use XML to initialize their data. All the XML data is being stored in a single XML file. There is an object that walks through the data and hands XML fragments to the various objects for initialization.

The two classes in Listing 11.11 exemplify how this may be designed. The `AppSerializer` class contains a method named `ReadStockQuote`, which reads an XML document, finds the stock quote node, and hands the outer XML of the node to the `StockQuote` object for initialization. The `StockQuote` class has a method called `ReadXml` that initializes its data from the XML string.

LISTING 11.11

```
C#
class AppSerializer
{
  StockQuote ReadStockQuote(string filename)
  {
    XmlDocument doc = new XmlDocument();
    StreamReader xmlFile = new StreamReader(filename);
    doc.Load(xmlFile);
    xmlFile.Close();
```

```
    // find the stock quote object
    XmlNode quoteNode = doc.DocumentElement;

    // hand outer xml to the quote object
    StockQuote quoteObj = new StockQuote();
    quoteObj.ReadXml(quoteNode.OuterXml);

    return quoteObj;
  }
}

class StockQuote
{
  string Name;
  string Price;

  void ReadXml(string xmlString)
  {
    // read the xml from the string
  }
}
```

VB

```
Public Class AppSerializer
  Public Sub StockQuote ReadStockQuote(filename As String)
    Dim doc As New XmlDocument()
    Dim xmlFile As New StreamReader(filename)
    doc.Load(xmlFile)
    xmlFile.Close()

    ' find the stock quote object
    Dim quoteNode As doc.DocumentElement;

    ' hand outer xml to the quote object
    Dim quoteObj As New StockQuote()
    quoteObj.ReadXml(quoteNode.OuterXml)

    return quoteObj;
  End Sub
End Class
```

```
Public Class StockQuote
  Public Sub ReadXml(string xmlString)
   ' read the xml from the string
   End Sub
End Class
```

Accessing an Element's Attributes

In most cases, navigating an XmlDocument node tree involves moving through parent-child relationships, except when you are accessing an element's attributes. The element-attribute relationship is a container-contained relationship. Since the XmlNode is an abstract class that represents the common functionality of all XML nodes, it provides only minimal functionality for accessing attribute information. Since attributes are most meaningful on XML elements, the XmlElement class provides a richer API for retrieving attribute information. In this section we will investigate the methods available on both classes.

The XmlNode class exposes the Attributes property to access its attributes. The Attributes property returns an AttributeCollection object that represents all of the XmlAttributes of an XmlElement. This includes namespace declarations, as well. You can iterate through the attribute nodes in two ways. You can iterate using the foreach statement or a for loop in combination with the Count and Item properties. Listing 11.12 illustrates each technique.

LISTING 11.12

```
C#
XmlDocument doc = new XmlDocument();
doc.LoadXml("<root att1=\"val1\" att2=\"val2\" />");

XmlNode node = doc.DocumentElement;

foreach(XmlAttribute att in node.Attributes)
{
  MessageBox.Show(att.Name + ":" + att.Value);
}

for(int ndx = 0; ndx < node.Attributes.Count; ++ndx)
{
  XmlAttribute att = node.Attributes[ndx];
  MessageBox.Show(att.Name + ":" + att.Value);
}
```

```
VB
Dim doc As New XmlDocument
doc.LoadXml("<root att1=" & Chr(34) & "val1" & Chr(34) & _
            "att2=" & Chr(34) & "val2" & Chr(34) & "/>")

Dim node As XmlElement
node = doc.DocumentElement

For Each xmlAtt As XmlAttribute In node.Attributes
  MessageBox.Show(xmlAtt.Name & ":" & xmlAtt.Value)
Next

Dim i As Int32
For i = 0 To node.Attributes.Count
  Dim att As XmlAttribute
  att = node.Attributes(i)
  MessageBox.Show(att.Name & ":" & att.Value)
Next i
```

You can access the value of the attribute by using the `Value` property, and you can access the name of the attribute by using the `Name` property. The `Attributes` property can also be used to access specific attributes by name. This is illustrated in Listing 11.13.

LISTING 11.13

```
C#
XmlDocument doc = new XmlDocument();
doc.LoadXml("<root att1=\"val1\" att2=\"val2\" />");

XmlNode node = doc.DocumentElement;
XmlAttribute att = node.Attributes["att1"];
MessageBox.Show(att.Name + ":" + att.Value);

VB
Dim doc As New XmlDocument();
doc.LoadXml("<root att1="&Chr(34)&"val1"&Chr(34)& _
            "att2="&Chr(34)&"val2"&Chr(34)& "/>")

Dim node AS doc.DocumentElement
Dim att As node.Attributes("att1")
MessageBox.Show(att.Name & ":" & att.Value)
```

The XmlElement class provides a richer API for accessing XmlAttributes. In addition to the Attributes property, the XmlElement class provides the HasAttributes property to find out whether an XmlElement contains any XmlAttributes. There is also the HasAttribute property, which determines whether the XmlElement has an XmlAttribute with a particular name.

LISTING 11.14

```
C#
XmlDocument doc = New XmlDocument();
doc.LoadXml("<root att1=\"val1\" att2=\"val2\" />");

XmlElement elem = doc.DocumentElement;
MessageBox.Show("Contains Attributes: " + elem.HasAttributes);
MessageBox.Show("Contains att1: " + elem.HasAttribute("att1"));

VB
Dim doc As New XmlDocument()
doc.LoadXml("<root att1="&Chr(34)&"val1"&Chr(34)& _
            "att2="&Chr(34)&"val2"&Chr(34)& "/>")

Dim elem As doc.DocumentElement
MessageBox.Show("Contains Attributes: " & elem.HasAttributes)
MessageBox.Show("Contains att1: " & elem.HasAttribute("att1"))
```

The XmlElement provides the GetAttribute method, which returns the value of a specified attribute, and the GetAttributeNode method, which returns an XmlAttribute object representing the specified attributes. The GetAttribute method is very useful if you want to retrieve only the value of the attribute. This method will locate the attribute and pull out the value of the attribute. The GetAttributeNode method is also helpful if you want to get the XmlAttribute object without having to interact with an XmlAttributeCollection object. These methods are used in Listing 11.15.

LISTING 11.15

```
C#
XmlDocument doc = new XmlDocument();
doc.LoadXml("<root att1=\"val1\" att2=\"val2\" />");

XmlElement elem = doc.DocumentElement;
MessageBox.Show("Attribute Value: " + elem.GetAttribute("att1"));

XmlAttribute att = elem.GetAttributeNode("att2");
MessageBox.Show("Attribute Value: " + att.Value);
```

```
VB
Dim doc As New XmlDocument()
doc.LoadXml("<root att1="&Chr(34)&"val1"&Chr(34)& _
            "att2="&Chr(34)&"val2"&Chr(34)& "/>")

Dim elem As doc.DocumentElement
MessageBox.Show("Attribute Value: " & elem.GetAttribute("att1"))

Dim att As elem.GetAttributeNode("att2")
MessageBox.Show("Attribute Value: " & att.Value)
```

Accessing Other `XmlNodes`

Thus far we have examined elements, attributes, and document nodes, but the .NET Compact Framework's XML DOM architecture also provides `XmlNode` implementations for each XML node type defined in the `System.Xml.XmlNodeType` enumeration. The two most important pieces of information, name and value, have different meanings in the context of different nodes. Table 11.1 lists each `System.Xml.XmlNodeType` and the values of their corresponding `Name` and `Value` properties.

TABLE 11.1

XmlNodeTypes and Their Names and Values

NODETYPE	NAME	VALUE
`Attribute`	Attribute Name	Attribute Value
`CDATA`	N/A	`CDATA`'s content
`Comment`	#comment	Text between `<!--` and `-->`
`Document`	#document	N/A
`DocumentFragment`	#document-fragment	N/A
`DocumentType`	N/A	N/A
`Element`	Fully qualified element name	N/A
`EndElement`	Fully qualified element name	N/A
`EndEntity`	N/A	N/A
`Entity`	N/A	N/A
`EntityReference`	N/A	N/A
`None`	N/A	N/A
`Notation`	N/A	N/A
`ProcessingInstruction`	#processing-instruction	Process instruction text minus the target
`SignificantWhitespace`	N/A	The white space within markup

TABLE 11.1
Continued

NODETYPE	NAME	VALUE
`XmlDeclaration`	Xml	The version number and encoding string
`Whitespace`	#whitespace	The white space in element content
`Text`	#text	The text content of an element or attribute

Searching an XML Document Node Tree

If there is one area where the `XmlDocument` API is thin, it is in providing the ability to search an `XmlDocument`. There are several methods that the `XmlDocument` does not support since the .NET Compact Framework lacks support for `XPath` or DTDs. On the desktop .NET Framework, the methods `SelectNode` and `SelectSingleNode` accept `XPath` query strings as parameters.

Also, the `GetElementById` method, which searches the document for the element with the specified ID, is not supported, because there is no DTD support. This deserves an explanation. An element's ID is a string value of a specific attribute. An attribute is specified as the special ID attribute in the DTD. Since the .NET Compact Framework does not support DTDs, there is no way to figure out which attribute is the special ID attribute.

The `XmlDocument` and `XmlElement` classes provide the `GetElementsByTagName` method to allow searching the document for an element by its name. When called on an `XmlDocument` object, this method retrieves all `XmlElements` with a specified name within the entire document. When called on an `XmlElement`, it finds all elements below the current element. Listing 11.16 shows how to use `GetElementsByTagName`.

LISTING 11.16

```
C#
public static void Main()
{
  XmlDocument doc = new XmlDocument();
  StreamReader s = new StreamReader("input.xml", Encoding.UTF8);
  doc.Load(s);
  s.Close();

  XmlNodeList authors = doc.GetElementsByTagName("Author");
  foreach(XmlElement author in authors)
  {
    if(author.GetAttribute("Name") == "John Doe")
```

```
      {
        XmlNodeList books = doc.GetElementsByTagName("Book");
        foreach(XmlElement book in books)
        {
          MessageBox.Show(book.FirstChild.InnerText);
        }
      }
    }
  }
```

```
VB
Sub Main()
  Dim doc As New XmlDocument
  Dim s As New StreamReader("input.xml", Encoding.UTF8)

  doc.Load(s)
  s.Close()

  Dim authors As XmlNodeList
  authors = doc.GetElementsByTagName("Author")

  For Each author As XmlElement In authors
    If author.GetAttribute("Name") = "John Doe" Then
      Dim books As XmlNodeList
      books = doc.GetElementsByTagName("Book")

      For Each book As XmlElement In books
        MessageBox.Show(book.FirstChild.InnerText)
      Next
    End If
  Next
End Sub
```

Creating `XmlNodes`

An `XmlDocument` also provides a rich API for modifying the XML node tree. Before you can modify the XML node tree, you will need to create nodes to add to the node tree. All of the `XmlNode` implementations, with the exception of the `XmlDocument`, have private constructors, so they cannot be instantiated directly. `XmlNode` implementations can be created only via the `XmlDocument` object. These methods also link the `XmlNode` to the `XmlDocument`, but they do not add the `XmlNodes` to the `XmlDocument`. You must explicitly add the `XmlNode` to the `XmlDocument`.

PERFORMANCE OF XmlDocument SEARCHING

Given that the GetElementsByTagName may search elements at any depth, it is important to design your XML document and search methods carefully. Your XML should be designed so that XML elements with similar names are found at similar depths. Also, the GetElementsByTagName should be called on an XmlElement to prevent searching the whole document. Finally, try to narrow the search area as much as possible by finding the ancestor XmlElement closest to the target elements.

Creating XmlDeclarations

If you are creating a new XMLDocument from scratch, you may want to start by creating an XmlDeclaration node to add to the node tree. The XmlDocument's CreateXmlDeclaration method provides this functionality.

The CreateXmlDeclaration method takes three string parameters. The first parameter is the XML version number, which must always equal "1.0." If the string is any other value, an ArgumentException is thrown. The second parameter is the string value of the encoding attribute. The value must be a string supported by the System.Text.Encoding class; otherwise, you will not be able to save the document. You may pass in null or empty string if you do not want the encoding attribute to be written while saving the document. Also, if the XmlDocument is saving to either a TextWriter or an XmlTextWriter, the encoding value is discarded, and the encoding of the specified writer is used instead. The final parameter is the string value of the stand-alone attribute. This string must be either "yes" or "no;" otherwise, an ArgumentException is raised. Setting the stand-alone attribute's value to null or String.Empty signifies that the stand-alone attribute should not be written when the document is saved.

LISTING 11.17

```
C#
// <?xml version="1.0" encoding="ASCII" standalone="yes"?>
XmlDeclaration decl = XmlDeclaration("1.0", "ASCII", "yes");
// <?xml version="1.0" encoding="UTF8"?>
decl = XmlDeclaration("1.0", "UTF8", null);
// <?xml version="1.0"?>
decl = XmlDeclaration("1.0", null, null);

VB
' <?xml version="1.0" encoding="ASCII" standalone="yes"?>
Dim decl As New XmlDeclaration("1.0", "ASCII", "yes")
' <?xml version="1.0" encoding="UTF8"?>
decl = XmlDeclaration("1.0", "UTF8", Nothing)
' <?xml version="1.0"?>
decl = XmlDeclaration("1.0", null, Nothing)
```

Creating XmlElements

To create an XmlElement, the XmlDocument provides the CreateElement methods. The methods create an XmlElement node with the specified name and an optional namespace. This code illustrates how to create an element:

```
C#
XmlElement newElement = xmlDoc.CreateElement("Book");

VB
Dim newElement As xmlDoc.CreateElement("Book")
```

Creating XmlAttributes

To create an XmlAttribute, the XmlDocument provides the CreateAttribute method. This method creates an XmlAttribute node linked to the current XmlDocument. The value must be set after the XmlAttribute is created (see Listing 11.18).

LISTING 11.18

```
C#
XmlAttribute newAttribute = XmlDoc.CreateAttribute("att");
newAttribute.Value = "val";

VB
Dim newAttribute As XmlDoc.CreateAttribute("att")
newAttribute.Value = "val"
```

After creating the attribute, you must add it to an XmlElement. The XmlElement's SetAttributeNode method provides this functionality. The SetAttributeNode takes the newly created XmlAttribute and adds it to the XmlElement's list of attributes. The SetAttributeNode method has an overload that takes two parameters: one string representing the name of the attribute, and another string representing the namespace URI of the attribute. This overload returns a new XmlAttribute that is already attached to the XmlElement and the XmlElement's owner document. The XmlAttribute will not have a value, so you must set the Value property after calling this method. Listing 11.19 demonstrates how to add a new XmlAttribute to an XmlElement by using SetAttributeNode.

LISTING 11.19

```
C#
XmlAttribute newAttribute = element.SetAttributeNode("att", null);
```

```
newAttribute.Value = "val";

VB
Dim newAttribute As element.SetAttributeNode("att", Nothing)
newAttribute.Value = "val"
```

An alternate way of adding a method is the `XmlElement`'s `AddAttribute` method. This method is very convenient. It creates an `XmlAttribute`, adds the `XmlAttribute` to the `XmlElement`, sets the `XmlAttribute`'s owner document, and then sets the `XmlAttribute`'s value (see following code snippet).

```
C#
element.SetAttribute("title", "Going Back to Cali");

VB
element.SetAttribute("title", "Going Back to Cali");
```

Creating Other Node Types

Of course, the `XmlDocument` class provides methods for creating every node type. Table 11.2 lists the additional `System.Xml.XmlNodeTypes` and their corresponding creation methods.

TABLE 11.2
Other XmlNodeTypes and Their Creation Methods

NODETYPE	CREATION METHOD
`CDATA`	`CreateCDataSection(string data)`
`Comment`	`CreateComment(string data)`
`DocumentFragment`	`CreateDocumentFragment()`
`EntityReference`	`CreateEntityReference(string name)`
`ProcessingInstruction`	`CreateProcessingInstruction(string target, string data)`
`Whitespace`	`CreateWhitespace(string ws)`
`Text`	`CreateTextNode(string text)`
`SignificantWhitespace`	`CreateSignificantWhitespace(string ws)`

The `CreateEntityReference` method must be handled with care. The method will accept any legal name that can be used to create an `EntityReference` node. This entity reference will be written out when the document is saved. The catch is that unless the name parameter is equal to one of the general entities (`gt`, `lt`, `apos`, `quot`, `amp`), a `NotSupportedException` will be raised when the document is loaded back into an `XmlDocument`. Listing 11.20 will raise an exception at line 9 because of this.

LISTING 11.20

```
C#
XmlDocument doc = new XmlDocument();
XmlElement ele = doc.CreateElement("", "root", "");
doc.AppendChild(ele);
ele.AppendChild(doc.CreateEntityReference("entity"));
doc.Save("blah.xml");

XmlDocument doc2 = new XmlDocument();
doc.Load(new StreamReader("blah.xml"));
string val = doc.DocumentElement.FirstChild.Value;

VB
Dim doc As New XmlDocument()
Dim ele As doc.CreateElement("", "root", "")
doc.AppendChild(ele)
ele.AppendChild(doc.CreateEntityReference("entity"))
doc.Save("blah.xml")

Dim doc2 As New XmlDocument();
doc.Load(New StreamReader("blah.xml"))
Dim val As doc.DocumentElement.FirstChild.Value;
```

Inserting XmlNodes

It is important to remember that the XmlNode creation method creates an XmlNode only in the context of the current document. It does not, however, add the XmlNodes to the document. You must explicitly insert the nodes into the document tree. The XmlNode class provides several methods for adding nodes to the document tree.

Before we begin discussing the insertion methods, let's talk about which nodes can contain other nodes. Not all nodes are container nodes, and not all container nodes can contain every node type. For example, an XmlAttribute node is a container node that can contain only XmlText nodes. It cannot contain an XmlProcessingInstruction node. An InvalidOperationException is raised when a specified node cannot be inserted as a valid child of a container node because the specified node has the wrong type. Table 11.3 is a chart that describes which nodes are containers and which nodes they can contain. The classes down the left side of the chart are the container nodes. The classes across the top of the graph are the classes that can be contained. An "X" in the intersection of a row and a column signifies that the class on that row can contain the class in that column.

TABLE 11.3
Container Nodes and the Nodes They Can Contain

	XmlNode	XmlDocument	XmlAttribute	XmlElement	XmlText	XmlComment	XmlCDataSection	XmlProcessingInstruction	XmlDeclaration	XmlDocumentFragment	XmlEntityReference	XmlWhitespace	XmlSignificantWhitespace
XmlDocument				X		X		X	X			X	X
XmlAttribute					X						X		
XmlElement				X	X	X	X	X			X	X	X
XmlText													
XmlComment													
XmlCDataSection													
XmlProcessingInstruction													
XmlDeclaration													
XmlDocumentFragment				X	X	X	X	X			X	X	X
XmlEntityReference					X								
XmlWhitespace													
XmlSignificantWhitespace													

Using the AppendChild and PrependChild Methods

The two most commonly used node-insertion methods are the `AppendChild` method, which appends the specified node as the last child of the current node, and `PrependChild`, which prepends the specified node as the first child of the current node. Listing 11.21 creates two elements, adds one to the beginning of the root node's list, and adds another to the end of the root node's list.

LISTING 11.21

```
C#
XmlDocument doc = new XmlDocument();
doc.LoadXml("<root/>");
XmlElement firstChild = doc.CreateElement("", "First", "");
```

```
XmlElement lastChild = doc.CreateElement("", "Last", "");

doc.DocumentElement.PrependChild(firstChild);
doc.DocumentElement.AppendChild(lastChild);
doc.Save("out.xml");

VB
Dim doc As new XmlDocument()
doc.LoadXml("<root/>")
Dim firstChild As XmlElement
firstChild = doc.CreateElement("", "First", "")
Dim lastChild As XmlElement
lastChild = doc.CreateElement("", "Last", "")

doc.DocumentElement.PrependChild(firstChild)
doc.DocumentElement.AppendChild(lastChild)
1doc.Save("out.xml")
```

These methods return the node that was added to the document tree. It is possible for two exceptions to be thrown when using these methods. As discussed before, an InvalidOperation will be thrown if the specified node cannot be inserted as a valid child, and an ArgumentException will be raised if the node is created from a document different from the one that created this node. In this case you can use the XmlDocument.ImportNode method to import the node into the current document. The ImportNode method returns an XmlNode that can then be inserted into the document.

Working with the InsertBefore and InsertAfter Methods

If you need to insert a node in the middle of a container node's child list, you cannot use AppendChild or PrependChild. Instead, you must use InsertBefore and InsertAfter. Both methods take two XmlNode objects as parameters. The first XmlNode object is the XmlNode to be inserted. The second XmlNode object is the reference node. When using InsertBefore, the new node will be inserted before the reference node, and it will be inserted after the reference node when calling InsertAfter. Listing 11.22 creates two elements and adds one before the reference node and one after the reference node.

LISTING 11.22

```
C#
XmlDocument doc = new XmlDocument();
```

```
doc.LoadXml("<root><first/><last/></root>");
XmlNode first = doc.DocumentElement.FirstChild;
XmlNode last = doc.DocumentElement.LastChild;

XmlElement secondChild = doc.CreateElement("", "second", "");
XmlElement thirdChild = doc.CreateElement("", "third", "");

doc.DocumentElement.InsertAfter(secondChild, first);
doc.DocumentElement.InsertBefore(thirdChild, last);
doc.Save("blah.xml");

VB
Dim doc As New XmlDocument()
doc.LoadXml("<root><first/><last/></root>")
Dim first As doc.DocumentElement.FirstChild
Dim last As doc.DocumentElement.LastChild

Dim secondChild As XmlElement
secondChild =  doc.CreateElement("", "second", "")
Dim thirdChild As XmlElement
thirdChild = doc.CreateElement("", "third", "")

doc.DocumentElement.InsertAfter(secondChild, first)
doc.DocumentElement.InsertBefore(thirdChild, last)
doc.Save("blah.xml")
```

Just like `AppendChild` and `PrependChild`, the `InsertBefore` and `InsertAfter` methods return the `XmlNode` that is added to the document tree. Also, there are two exceptions to beware of when using this method. Like with any insertion method, an `InvalidException` will be thrown if the specified node is not a valid child. An `ArgumentException` will be raised if the reference node is not a child of the current node or if the new child is created from a different document. Again, you can use `ImportNode` to import the node into the current document.

Inserting `XmlAttributes` into an `XmlDocument`

`XmlAttributes` can be inserted into a document by using the `SetAttribute` method described earlier in this chapter. Alternatively, you can add them using the insertion methods on the `XmlAttributeCollection` class. The insertion methods, similar to the `XmlNode` insertion methods, are `Append`, `Prepend`, `InsertBefore`, and `InsertAfter`.

`Append` and `Prepend` add an `XmlAttribute` to the end of the attribute list and to the beginning of the attribute list, respectively. The methods return the `XmlAttribute` that was inserted. If there exists an attribute in the collection with the same name, then the original attribute is removed from the collection, and the new attribute is added. Listing 11.23 illustrates how to use Append and Prepend.

LISTING 11.23

```
C#
XmlDocument doc = new XmlDocument();
doc.LoadXml("<root/>");
XmlAttribute first = doc.CreateAttribute("att1");
first.Value = "first";
XmlAttribute second= doc.CreateAttribute("att2");
second.Value = "second";

XmlAttributeCollection atts = doc.DocumentElement.Attributes;
atts.Prepend(first);
atts.Append(second);
doc.Save("blah.xml");

VB
Dim doc As New XmlDocument()
doc.LoadXml("<root/>");
Dim first As doc.CreateAttribute("att1")
first.Value = "first"
Dim second= doc.CreateAttribute("att2")
second.Value = "second"

Dim atts As doc.DocumentElement.Attributes
atts.Prepend(first)
atts.Append(second)
doc.Save("blah.xml")
```

InsertBefore and InsertAfter add an XmlAttribute before and after a specified reference attribute, respectively. If the reference node is null, then the InsertBefore method will insert the attribute at the end of the collection, whereas InsertAfter inserts the attribute at the beginning of the collection. If the reference attribute is not in the collection, then an ArgumentException is raised. As with Append and Prepend, when an existing attribute has the same name, it will be removed, and the new attribute will be inserted. Listing 11.24 shows how to use these two methods.

LISTING 11.24

```
C#
XmlDocument doc = new XmlDocument();
```

```
doc.LoadXml("<root att1=\"first\" att4=\"last\"/>");

XmlAttribute second = doc.CreateAttribute("att2");
second.Value = "2nd";

XmlAttribute third = doc.CreateAttribute("att3");
third.Value = "3rd";

XmlAttributeCollection atts = doc.DocumentElement.Attributes;
XmlAttribute first = atts[0];
XmlAttribute last = atts[1];

atts.InsertAfter(second, first);
atts.InsertBefore(third, last);
doc.Save("blah.xml");

VB
Dim doc As New XmlDocument()
doc.LoadXml("<root att1="&Chr(34)&"first"&Chr(34)& _
                   "att4="&Chr(34)&"last"&Chr(34)& "/>")

Dim second As doc.CreateAttribute("att2")
second.Value = "2nd"

Dim third A doc.CreateAttribute("att3")
third.Value = "3rd";

Dim atts As doc.DocumentElement.Attributes
Dim first As atts[0]
Dim last As atts[1]

atts.InsertAfter(second, first)
atts.InsertBefore(third, last)
doc.Save("blah.xml")
```

Replacing the `XmlNodes` within an `XmlDocument`

Using the insertion methods is not the only way to insert nodes into an `XmlDocument`. You can also insert new nodes by replacing nodes that already exist in the node tree by using the `ReplaceChild` method. The `ReplaceChild` method takes two parameters, much like the `InsertBefore` and `InsertAfter` methods, but instead of using the existing node as a reference point, the existing node is removed from the node tree, and the new node is inserted in its place. The method returns the node that has been replaced. Listing 11.25 demonstrates how to use the ReplaceChild method.

LISTING 11.25

```
C#
XmlDocument doc = new XmlDocument();
doc.LoadXml("<root att1=\"first\" att4=\"last\"/>");
XmlElement newRoot = doc.CreateElement("NewRoot");
doc.ReplaceChild(newRoot, doc.DocumentElement);
doc.Save("blah.xml");

VB
Dim doc As new XmlDocument()
doc.LoadXml("<root att1="&Chr(34)&"first"&Chr(34)& _
            "att4="&Chr(34)&"last"&Chr(34)& "/>")
Dim newRoot As doc.CreateElement("NewRoot")
doc.ReplaceChild(newRoot, doc.DocumentElement)
doc.Save("blah.xml")
```

The ReplaceChild method may throw two exceptions. It may throw an ArgumentException if the new node was created from a different document or if the existing node is not a child of the container node. Also, the InvalidOperationException may be raised if the new node is not a valid child of the container node or if the new node is an ancestor of the container node. This would create a cycle in the node tree, which is not allowed.

ATTRIBUTE REPLACEMENT USING THE INSERTION METHODS

To replace an attribute node by using the XmlAttributeCollection, you can simply use the insertion methods, passing an attribute with the same name as the attribute you would like to replace. The insertion method will remove the existing attribute and add the new attribute. The following code demonstrates removing an attribute through the Append method:

```
C#
XmlDocument doc = new XmlDocument();
doc.LoadXml("<root att1=\"first\" att4=\"last\"/>");

XmlAttribute newFirst = doc.CreateAttribute("att1");
newFirst.Value = "1st";

XmlAttributeCollection atts = doc.DocumentElement.Attributes;
atts.Append(newFirst);
doc.Save("blah.xml");
```

```
VB
Dim doc As New XmlDocument()
0doc.LoadXml("<root att1="&Chr(34)&"first"&Chr(34)& _
            "att4="&Chr(34)&"last"&Chr(34)& "/>")

Dim newFirst As doc.CreateAttribute("att1")
newFirst.Value = "1st"

Dim atts As doc.DocumentElement.Attributes
atts.Append(newFirst)
doc.Save("blah.xml")
```

Removing XmlNodes for an XmlDocument

In addition to inserting and replacing nodes, you may need to remove a node from the tree altogether. You can easily remove a node from the node tree by using the RemoveChild method on the XmlNode class.

The RemoveChild method takes one parameter: the node to remove. If this node is successfully removed, then it returns a reference to the removed node. If the node is not a child of the container node, then an ArgumentException is raised. Listing 11.26 shows how to use the RemoveChild method.

LISTING 11.26

```
C#
XmlDocument doc = new XmlDocument();
doc.LoadXml("<root><child/></root>");
doc.DocumentElement.RemoveChild(doc.DocumentElement.FirstChild);
doc.Save("out.xml");

VB
Dim doc As New XmlDocument()
doc.LoadXml("<root><child/></root>")
doc.DocumentElement.RemoveChild(doc.DocumentElement.FirstChild)
doc.Save("out.xml")
```

The XmlNode class also provides the RemoveAll method, which removes all of the child nodes and any existing attributes from the container node. Listing 11.27 shows how to use the RemoveAll method.

LISTING 11.27

```
C#
XmlDocument doc = new XmlDocument();
doc.LoadXml
("<root att=\"val\" ><child/><child att=\"val\"/></root>");
doc.DocumentElement.RemoveAll();
doc.Save("out.xml");

VB
Dim doc As New XmlDocument()
doc.LoadXml
("<root att="&Chr(34)&"val"&Chr(34)&"><child/>" & _
"<child att="&CHr(34)&"val"&Char(34)& "/></root>");
doc.DocumentElement.RemoveAll()
doc.Save("out.xml")
```

Removing XmlAttributes from their XmlElements

The API for removing attributes is much richer than that for removing other nodes. This is due to the ability to remove nodes through methods on the XmlAttributeCollection class as well as methods on the XmlElement class.

The XmlAttributeCollection class provides three methods for removing an XmlAttribute: Remove, RemoveAt, and RemoveAll. The Remove method removes a single specified attribute from the attribute collection. The RemoveAt method removes a single attribute corresponding to the specified index from the collection. The RemoveAll method removes all of the attributes in the collection. Listing 11.28 demonstrates how to use each method.

LISTING 11.28

```
C#
XmlDocument doc = new XmlDocument();
doc.LoadXml("<root><child att1=\"val\" " +
            "att2=\"val\"/><child att=\"val\" "+
            "att3=\"val\"/></root>");

XmlAttributeCollection atts =
  doc.DocumentElement.FirstChild.Attributes;
atts.Remove(atts[0]);
atts.RemoveAt(0);
```

```
atts = doc.DocumentElement.LastChild.Attributes;
atts.RemoveAll();

doc.Save("out.xml");

VB
Dim doc As New XmlDocument()
doc.LoadXml("<root><child att1="&Chr(34)&"val"&Chr(34)& _
            "att2="&Chr(34)&"val"&Chr(34)& "/>" & _
            "<child att="&Chr(34)&"val"&Chr(34)& _
            "att3="&Chr(34)&"val"&Chr(34)&"/>" & _
             "</root>");

Dim atts As doc.DocumentElement.FirstChild.Attributes
atts.Remove(atts(0))
atts.RemoveAt(0);

atts = doc.DocumentElement.LastChild.Attributes
atts.RemoveAll()

doc.Save("out.xml")
```

The XmlElement also provides three methods for removing attributes: RemoveAttributeNode, RemoveAttributeAt, RemoveAllAttributes. The RemoveAttributeNode is just like XmlAttributeCollection.Remove: It removes a single specified attribute from the element's attribute collection. There is also RemoveAttribute, which takes the name of the attribute, as opposed to the XmlAttribute object, and removes the attribute with the corresponding name. The RemoveAttributeAt method corresponds to XmlAttributeCollection's RemoveAt method: It removes a single attribute corresponding to the specified index from the element's attribute collection. Finally, the RemoveAllAttributes collection removes all attributes from an element's attribute collection, much like XmlAttributeCollection.RemoveAll. Listing 11.29 demonstrates how to use each method.

LISTING 11.29

```
C#
XmlDocument doc = new XmlDocument();
doc.LoadXml("<root><child att1=\"val\" " +
             "att2=\"val\" att3=\"val\"/> " +
                 "<child att=\"val\" "+
                 "att3=\"val\"/></root>");

XmlElement firstChild = (XmlElement)doc.DocumentElement.FirstChild;
```

```
XmlAttributeCollection atts = firstChild.Attributes;
firstChild.RemoveAttributeNode(atts[0]);
firstChild.RemoveAttribute("att2");
firstChild.RemoveAttributeAt(0);

XmlElement lastChild = (XmlElement)doc.DocumentElement.LastChild;
atts = lastChild.Attributes;
lastChild.RemoveAllAttributes();

doc.Save("out.xml");

VB
Dim doc As New XmlDocument()
doc.LoadXml("<root><child att1="&Chr(34)&"val"&Chr(34)& _
            "att2="&Chr(34)&"val"&Chr(34)& _
            "att3="&Chr(34)&"val"&Chr(34)&"/>" & _
             "<child att="&Chr(34)&"val"&Chr(34)& _
            "att3="&Chr(34)&"val"&Chr(34)&"/></root>")

Dim firstChild As XmlElement
firstChild = doc.DocumentElement.FirstChild
Dim atts As firstChild.Attributes
firstChild.RemoveAttributeNode(atts(0))
firstChild.RemoveAttribute("att2")
firstChild.RemoveAttributeAt(0)

Dim lastChild As XmlElement
lastChild = doc.DocumentElement.LastChild
atts = lastChild.Attributes
lastChild.RemoveAllAttributes()

doc.Save("out.xml")
```

Writing the XmlDocument

Once you have built and modified your XmlDocument, you can save it to a stream through the Save method on the XmlDocument class. The Save method allows you to specify the destination by using the name of a file, a Stream object, a TextWriter object, or an XmlWriter object. It is important to note that all operations that execute on an XmlDocument are made on the in-memory node tree, and changes will not be persisted until the Save method is called. Listing 11.30 shows how an XmlDocument can be saved using each overload of the Save method.

LISTING 11.30

```
C#
doc.Save("books.xml");
doc.Save(new FileStream("books.xml", FileMode.Open));
doc.Save(new StreamWriter("books.xml"));
doc.Save(new XmlTextWriter("books.xml"));

VB
doc.Save("books.xml")
doc.Save(New FileStream("books.xml", FileMode.Open))
doc.Save(New StreamWriter("books.xml"))
doc.Save(New XmlTextWriter("books.xml"))
```

In Brief

- The `XmlDocument` builds a tree data structure of the XML document it parses.
- The `XmlDocument` allows navigating the nodes in all directions: parent to child, child to parent, and sibling to sibling.
- The `XmlDocument` provides a rich API for accessing the data in each XML node.
- Searching the XML document is difficult since the .NET Compact Framework does not provide `XPath` support.
- Since `XPath` is not supported, care must taken when designing your XML data.
- The `XmlDocument` provides methods for creating every type of XML node.
- After manipulating your XML data, the contained XML data can be persisted using the `Save` method.

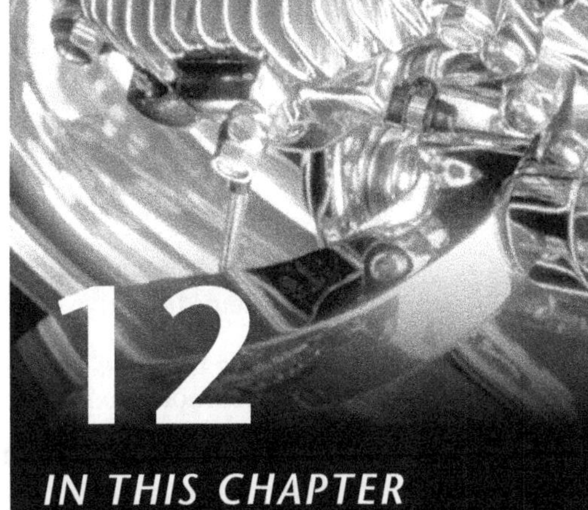

Interacting with Native Code

CHAPTER 12

IN THIS CHAPTER

Understanding When to Call Native Code

Chapter 2, "Introducing the .NET Compact Framework," discussed the architecture of the .NET Compact Framework and its implementation of the Common Language Runtime. Managed applications and managed DLLs that support them, such as `System.dll`, are composed of Intermediate Language (.IL) byte codes. As a program calls class methods, the runtime compiles the byte codes of the methods into machine code instructions that can be executed by the CPU on the device running the program.

Chapter 2 explains the rules of how the Just-In-Time (JIT) compilation works. Understanding these rules is crucial to writing complex managed applications that perform well.

By developing with managed code, developers enjoy a variety of helpful features: Automatic garbage collection, access to a rich class framework, and the use of the powerful Visual Studio IDE to develop applications for devices. These features cut development and debugging time drastically, but they do not change the fact that the CLR and .NET Compact Framework have limitations that can only be circumvented by calling native code. The following list depicts situations in which accessing native code from managed applications is helpful:

- Managed code does not execute as rapidly as native code. Applications that include computationally intense subsystems can benefit substantially by writing the computationally intense subsystems in native code.

- There are many legacy DLLs written in native code that offer functionality that would be difficult or impossible to rewrite in managed code.
- Calling into native code makes it possible to interact with COM components on the device.
- The Windows CE operating system's application programming interface (API) is exposed as a set of function calls whose code is housed in several DLLs. In order to directly call with Windows CE OS, developers must invoke the native code housed in the DLLs. This activity is called Platform Invoking, or P/Invoking for short.

Calling Native Code: Quick Start

This section describes the minimum steps needed to call a function in a native DLL binary. Before jumping directly into the tutorial, there are several fundamental assumptions we are making about the DLL into which we are calling.

The most fundamental requirement is that the DLL is compiled for the CPU and Windows CE OS version for the device on which you plan to use it. It is an easy mistake to make to accidentally use the DLL binary for the wrong platform and then wonder why you get exceptions when you try to call into from native code.

It is assumed that the DLL does not have C++-style mangled names. To understand what this means, consider what happens when a group of functions written in standard C are compiled into a DLL. The resulting DLL has a table indicating which functions are available to call and where in the DLL binary the function code begins. A DLL written with the C++ programming language has the table as well, but the names of the functions that are exported can be mangled. For example, the function `MyFunction1` might be mangled to look like `MyFunctioned87883ffe1`. The name mangling supports the polymorphic features of the C++ language, where a class can override a method name from a parent. Unfortunately, it can be problematic when trying to call the function from managed code.

HOW TO TELL IF A DLL HAS NAME MANGLING

You can tell if a DLL has name mangling by using the `DEPENDS` dependency walker utility that comes with Visual Studio. To use the dependency walker utility, start a Visual Studio command line, and change to the directory holding the DLL you want to examine. Then type `DEPENDS [dllname]`. The dependency walker application shows all of the native function names exported by the DLL. Figure 12.1 shows the dependency walker examining the `SquareAnInt.dll` file, which is used later in this chapter. As Figure 12.1 shows, it is easy to see whether the function names look "clean," or whether they have been mangled. The `SquareAnInt.dll` exports one function, `SquareAnInt`, whose name appears cleanly in the dependency walker utility.

All of the Windows CE OS functions are held in DLLs that do not mangle their names. Also, all of the binary DLLs that are included with the examples do not mangle their names. If you are building the DLL that you plan to call into from managed code, then consult your compiler's documentation to see how to disable name mangling.

It is assumed that the DLL has been placed in a location where the .NET CF runtime can find it. Two safe choices are in the `\Windows` directory or in the same directory as the managed application which will call into the DLL.

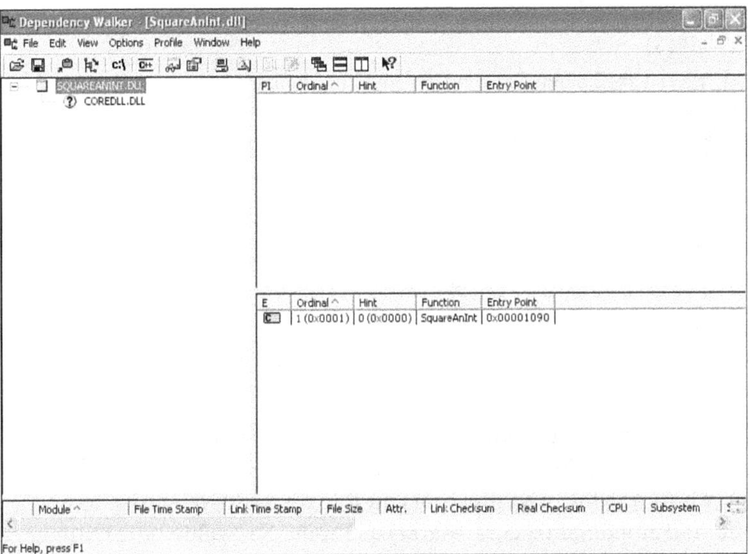

FIGURE 12.1 The dependency walker demonstrates that `SquareAnInt.dll` does not have mangled names.

Calling Native Code with Minimal Steps

If you have a DLL that adheres to all of the preceding restrictions, then you are ready to call functions in the DLL from managed code. This section examines the minimum steps you need to take in your managed application in order to call native code functions from managed code. The next section focuses on the specifics of passing various data types into native code in the next section after we understand the basics of calling into native code from managed code.

To call a native function from managed code, first declare the native function in the class where it will be called. In the declaration, specify what data types are passed into and out of the function, what the return type is, and what DLL the function is located in. After the native function is declared, it can be used in the class just as any other class method. From a programmer's perspective, the declaration creates the illusion that the native function is simply a class method.

When the class actually does call the native function, the runtime locates the DLL holding the native function. It loads the DLL and marshals the input arguments into the native function. The native code is then executed, and the output arguments and return values are marshalled back into the managed side. As long as everything goes well, this process is transparent to the user. Otherwise, an exception is thrown. Table 12.1 shows exceptions commonly thrown while trying to call native code.

TABLE 12.1

Exceptions Thrown When Calling into Native Code

EXCEPTION	CAUSE(S)
`MissingMethodException`	(1) The DLL file holding the function could not be found. (2) The DLL file was found, but it does not contain the function specified in the declaration of the native code function. Make sure that the DLL exports the function by using the `DEPENDS` utility.
`NotSupportedException`	The managed application attempted to pass by value an item greater than 32 bits long. For example, an attempt was made to pass a 64-bit `long` integer into a native function. See the section "Marshalling the Fundamental Data Types" for more information.

The C# keyword for declaring the function is the `DllImport`. For Visual Basic, use the `Declare Sub` construct.

The code in Listing 12.1 is derived from the ComputeSquare sample application. The code snippet shows the declaration and usage for the `SquareAnInt` function, which is held in `SquareAnInt.dll`. It does not return a value, and it accepts an integer and a reference to an integer (`int *`, in the C code that implements `SquareAnInt`) as input. The value of the integer passed in is squared and written to the referenced integer as an out parameter.

LISTING 12.1 Declaration and usage for the `SquareAnInt` function

```
C#
public class Form1 : System.Windows.Forms.Form
{
        private System.Windows.Forms.Button btnComputeSquare;
        private System.Windows.Forms.Label label1;
        private System.Windows.Forms.Label label2;
        private System.Windows.Forms.TextBox txtInput;
        private System.Windows.Forms.TextBox txtResult;
        private System.Windows.Forms.MainMenu mainMenu1;

        [DllImport("SquareAnInt.dll")]
        private static extern void SquareAnInt(int input,
                ref int output);
```

```
        // Rest of the class declarations deleted for brevity
        // ...
        // ...
        // Usage of the SquareAnInt function occurs as part of
        // btnComputeSquare_Click
        private void btnComputeSquare_Click(object sender,
                 System.EventArgs e)
        {
                int output = 0;
                int input = Convert.ToInt32(this.txtInput.Text);
                SquareAnInt(input, ref output);
                this.txtResult.Text = Convert.ToString(output);
        }
// Rest of class cut for brevity...
```

VB

```
Public Class Form1
    Inherits System.Windows.Forms.Form
    Friend WithEvents txtResult As System.Windows.Forms.TextBox
    Friend WithEvents txtInput As System.Windows.Forms.TextBox
    Friend WithEvents label2 As System.Windows.Forms.Label
    Friend WithEvents label1 As System.Windows.Forms.Label
    Friend WithEvents btnComputeSquare As System.Windows.Forms.Button
    Friend WithEvents MainMenu1 As System.Windows.Forms.MainMenu

    Declare Sub SquareAnInt Lib "SquareAnInt.dll"
            (ByVal input As Int32, ByRef output As Int32)
        ' Rest of the class declarations deleted for brevity
        ' ...
        ' ...
        ' Usage of the SquareAnInt function occurs as part of
        ' btnComputeSquare_Click
    Private Sub btnComputeSquare_Click(ByVal sender As
            System.Object, ByVal e As System.EventArgs)
            Handles btnComputeSquare.Click
        Dim output As Integer = 0
        Dim input As Integer = Convert.ToInt32(Me.txtInput.Text)
        SquareAnInt(input, output)
        Me.txtResult.Text = Convert.ToString(output)
    End Sub
```

Computing a Square with Native Code—Sample Application

The ComputeSquare sample application is a complete end-to-end demonstration of how to call native code from a managed application. The managed application is called SquareAnInt, and it is in the directory `\SampleApplications\Chapter12`. There are C# and Visual Basic versions.

The SquareAnInt application uses the `SquareAnInt.dll` binary to compute the square of a 32-bit integer passed into it. The source code for `SquareAnInt.dll` is held in the directory `\SampleApplications\Chapter12\NativeBinaries\SquareAnInt`.

Because `SquareAnInt.dll` is native code, a version for each supported CPU and Windows CE OS version is required. There are multiple subdirectories in the `\SampleApplications\Chapter12\NativeBinaries\SquareAnInt` directory. Each holds the native DLL file for a specific CPU type and OS version. For example, the 400_x86 directory holds the binaries for Windows CE 4.x running on 80x86 hardware. The popular Compaq iPaq running PocketPC 2002 would use the binaries in the 300_ARM directory.

To use the SquareAnInt application, open the managed project, build it, and deploy it to your device. Copy `SquareAnInt.dll` from the appropriate folder to your device. You can choose the `\Windows` directory or the directory where the `SquareAnInt.exe` file resides.

When you launch the application, it shows a simple interface into which you can enter an integer value. Click the button labeled Computer Square, and the integer value is squared by calling into the `SquareAnInt.dll` library and by using the native function, `SquareAnInt`.

ERROR HANDLING IN THE SAMPLE APPLICATIONS

Note this application and all of the others in this chapter are made to be as simple as possible while demonstrating the intended concepts. Thus, there is minimal input checking or error handling code, which can clog the project with extra code and detract from the central concept being demonstrated. For example, if you enter an alphabetic character into a field that is expected to be numeric, then you are likely to see an exception. Obviously, true production code should have much more error handling code built into it.

Marshalling the Fundamental Data Types

The previous section explained the fundamentals of calling into a native DLL. It touched on such topics as where to put the DLL, what the restrictions are for making a call, and how to declare a native function as part of a managed class. A simple example proved that everything worked as expected.

It is now time to dig deeper into the subject of *marshalling,* the act of moving function call arguments and return values back and forth between the managed runtime and the native side. This section discusses how to marshal fundamental data types. That is, those data types that are not part of a more complex structure. When marshalling fundamental data types, there are certain rules and definitions that must be adhered to. The two most important rules are discussed in the paragraphs that follow.

Fundamental data types can be passed by value or by reference. When a data type is passed by value, a copy of the data is made when it is copied across the native/managed code boundary. When a data type is passed by reference, only the address of the data object in the managed world is passed to the native side. The native code accesses the data object by referring to the address that was passed to it.

In the .NET Compact Framework, only data objects that are 32 bits or smaller in size may be passed by value. Anything larger must be passed by reference. Attempting to pass a data object larger than 32 bits long will result in a `NotSupportedException`. For example, long integers, which are 64 bits wide, must be passed by reference.

Data objects that are passed by reference to the native side can be altered by the native code. Thus, you can use pass by reference to create out or in/out parameters for calls into native code.

In order to be able to pass a fundamental data type into a native function, you must first declare the function by using `DllImport` (C#) or `Declare Sub` (VB), as described in the previous section. The next section shows how to set up declarations to pass a wide variety of fundamental data types by value and by reference. The easiest way to do this is by showing small snippets of code for the declarations and use of the native functions. As the text shows code snippets to demonstrate how to pass fundamental data types, we will call out special caveats for the specific data types.

Passing 32-bit Signed Integers

The 32-bit integer (`int` in C#, `Integer` in Visual Basic) can be passed by value or by reference. Following are sample declarations to pass by value into the `IntInFunc` and sample usage of the `IntInFunc`:

```
C#
[DllImport("Foo.dll")]
private static extern void IntInFunc(int in_Int);

// Usage, where l_Int is an integer
IntInFunc(l_Int)

VB
Declare Sub IntInFunc Lib "Foo.dll" (ByVal in_Int As Integer)
' Usage, where l_Int is an Integer
IntInFunc(l_Int)
```

Passing 32-bit Unsigned Integers

The C# keyword `uint` corresponds to the `UInt32` data type, which is available in both C# and Visual Basic. This is a 32-bit data type. Sample declarations for pass by value into the `UIntInFunc` and sample usage of the `UIntInFunc` include the following:

```
C#
[DllImport("Foo.dll")]
private static extern void UIntInFunc(uint in_UInt);

// Usage, where l_UInt is an unsigned integer
UIntInFunc(l_UInt);

VB
Declare Sub UIntInFunc Lib "Foo.dll" (ByVal in_UInt As UInt32)

' Usage, where l_UInt is an unsigned integer
UIntInFunc(l_UInt)
```

Passing Single Precision Floating-Point Types

The single precision floating-point data type (`float` in C#, `Single` in Visual Basic) uses 64 bits to provide single precision floating-point representation of a number. As such, it cannot be passed by value and must be passed by reference. The following sample declaration for `FloatInFunc` shows how to set up to call into a native function that expects a reference to a `float` as input. The sample code also shows how to pass a reference to a floating-point value into the function.

```
C#
[DllImport("Foo.dll")]
private static extern void FloatInFunc(ref float in_Float);

// Usage...
float in_Float = 4.00;

FloatInFunc(ref in_Float);

VB
Declare Sub FloatInFunc Lib "Foo.dll" (ByRef in_Float As Single)

' Usage...
Dim in_Float As Single = 4.00
FloatInFunc(in_Float)
```

Passing Signed Long Integer Types

The long integer data type (`long` in C#, `Long` in Visual Basic), also available as the `Int64` type, uses 64-bit-wide representation for a signed integer. Thus, it can only be passed to native code by reference. The following is a sample declaration and usage of a function that is passed a reference to a `long` data type:

```
C#
[DllImport("Foo.dll")]
private static extern void LongInFunc(ref long in_Long);

// Usage...
long in_Long = Convert.ToInt64(4);
LongToString(ref in_Long, out_StringBuilder);

VB
Declare Sub LongInFunc Lib "Foo.dll" (ByRef in_Long As Long)

' Usage...
Dim in_Long As Long = Convert.ToInt64(4)
LongToString(in_Long, out_StringBuilder)
```

Passing Unsigned Long Data Types

Unsigned long integer data types (`ulong` in C#, `UInt64` in C# and Visual Basic) is a 64-bit-wide data object that can only be passed by reference. The following is a sample declaration and usage of a function that is passed a reference to an unsigned long data type:

```
C#
[DllImport("FundamentalToString.dll")]
private static extern void ULongToString(ref ulong in_ULong);

// Usage...
ulong in_ULong = Convert.ToUInt64(4);
ULongToString(ref in_ULong, out_StringBuilder);

VB
Declare Sub ULongToString Lib "Foo.dll" (ByRef in_ULong As UInt64)

' Usage...
Dim in_ULong As UInt64 = Convert.ToUInt64(4)
ULongInFunc(in_ULong)
```

Passing Short Integers

The short integer data type (short in C#, Short in Visual Basic) uses 16 bits to represent a signed integer value. Thus, it can be passed by value or by reference. The following sample demonstrates how to declare a function that receives a short by value and includes a sample invocation:

```
C#
[DllImport("Foo.dll")]
private static extern void ShortInFunc(short in_Short);

// Usage...
short in_Short = Convert.ToInt16(this.txtShortIn.Text);
ShortInFunc(in_Short);

VB
Declare Sub ShortInFunc Lib "Foo.dll" (ByVal in_Short As Short)

' Usage...
Dim in_Short As Short = Convert.ToInt16(Me.txtShortIn.Text)
ShortInFunc(in_Short)
```

Passing Unsigned Short Integers

The unsigned short integer (ushort in C#, Uint16 in C# and Visual Basic) is like the short data type, except it uses its 16 bits to represent positive integers only, as seen in the following sample:

```
C#
[DllImport("Foo.dll")]
private static extern void UShortInFunc(ushort in_UShort);

// Usage...
ushort in_UShort = Convert.ToUInt16(this.txtUShortIn.Text);
UShortInFunc(in_UShort);

VB
Declare Sub UShortInFunc Lib "Foo.dll" (ByVal in_UShort As UInt16)

' Usage...
Dim in_UShort As UInt16 = Convert.ToUInt16(Me.txtUShortIn.Text)
UShortInFunc(in_UShort)
```

Passing Single Characters

Single characters (`char` in C#, `Char` in Visual Basic) use 8 bits to represent a character value. Thus, they can be passed by value or by reference. The following sample code shows a declaration and usage of a native function to which a character value is passed by value:

```
C#
[DllImport("Foo.dll")]
private static extern void CharInFunc(char in_Char);

// Usage...
char in_Char = Convert.ToChar("C");
CharInFunc(in_Char);

VB
Declare Sub CharInFunc Lib "Foo.dll" (ByVal in_Char As Char)

' Usage...
Dim in_Char As Char = Convert.ToChar(Me.txtCharIn.Text)
CharInFunc(in_Char)
```

Passing Bytes

Single byte values (`byte` in C#, `Byte` in Visual Basic) are 8-bit unsigned values. Thus, they can be passed by value or by reference. The code in Listing 12.2 demonstrates the declaration and usage of a native code function that receives a byte value by reference:

LISTING 12.2 Declaration and usage of a native code function that receives a `byte` value by reference

```
C#
[DllImport("Foo.dll")]
private static extern void ByteInFunc(byte in_Byte);

// Usage...
byte in_Byte = Convert.ToByte(this.txtByteIn.Text);
ByteInFunc(in_Byte);

VB
Declare Sub ByteInFunc Lib "Foo.dll" (ByVal in_Byte As Byte)

' Usage...
Dim in_Byte As Byte = Convert.ToByte(Me.txtByteIn.Text)
ByteInFunc(in_Byte)
```

Passing Boolean Values

Boolean values (bool in C#, Boolean in Visual Basic) consume 8 bits of storage, and so they can be passed by value or by reference. The following sample code demonstrates the declaration and usage of a native function that receives a Boolean value by value.

```
C#
[DllImport("Foo.dll")]
private static extern void BoolInFunc(bool in_Bool);

// Usage...
bool in_Bool = false;
BoolInFunc(in_Bool, out_StringBuilder);

VB
Declare Sub BoolToString Lib "Foo.dll" (ByVal in_Bool As Boolean)

' Usage...
Dim in_Bool As Boolean = False
BoolInFunc(in_Bool)
```

Passing a DWORD

The DWORD data type is commonly used in native code on Windows CE platforms. This data type is compatible with the UInt32 data type in the .NET Compact Framework.

Passing an IntPtr

The IntPtr data type corresponds to handles in the Windows CE operating system. For example, the Windows CE API PlaySound() function, discussed later in this section, receives an HMODULE as a parameter. Chapter 14, "Cryptography," uses Windows CE API functions that receive such handle types as HCRYPTKEY and HCRYPTHASH. These are all examples of handles in the Windows CE operating system. Handle types in Windows CE are designated with all capital letters and begin with an "H."

You can pass in an IntPtr type by value to a Windows CE API function that expects a handle of some sort. The IntPtr is internally a pointer to an integer. Thus, if native code changes the value of an IntPtr, then the changes are reflected on the managed side. If you want to pass a NULL value into native code, you can pass IntPtr.Zero. To extract the numeric value of an IntPtr, you can use IntPtr.ToInt32 or IntPtr.ToInt64, both of which are supported by the .NET Compact Framework.

Typically, an IntPtr gets a value from the Windows CE operating system. For example, you might receive a handle to an encryption key, and the handle is stored in an IntPtr on the managed side.

The PlaySound example later in this chapter demonstrates the use of an `IntPtr`, and Chapter 14 uses them extensively.

Passing Non-Mutable Strings

String values (`string` in C#, `String` in Visual Basic) can be passed into native code, where they appear as pointers to an array of wide (16-bit) characters. That is, a string passed in from managed code appears as `WCHAR *` in native C language.

When you pass a string type into native code, it is read only on the native side. Altering the string in native code will not change the value of the string on the managed side, and it is not recommended to try this because it can result in illegal memory accesses. If you want to see changes made to a string by native code in managed code, then use a `StringBuilder` in the manner described in following sections.

Passing Mutable Strings with `System.Text.StringBuilder`

The `StringBuilder` class can be passed by value to native code, where it appears as a pointer to an array of wide (16-bit) characters. That is, it appears as `WCHAR *` in native C language. The native code can alter the string, and the changes are reflected in the `StringBuilder` on the managed side. On the managed side, developers can extract a string altered by native code simply by calling `StringBuilder.ToString()`.

In order to use a `StringBuilder` correctly to pass string values to and from native code, instantiate the `StringBuilder` with a specified amount of storage before passing it to native code. Make sure the native side somehow knows how much space the `StringBuilder` has to avoid buffer overruns. This tactic is an example of an overall important strategy when dealing with native code from the managed world: Allocate memory on the managed side. Allocating memory in native code can get very messy because you are not using the managed heap allocation system with garbage collection. This means that memory allocated by native code must be freed by the native code later, or it will be leaked memory.

The code in Listing 12.3 is taken from the FundamentalToString sample application. The code demonstrates the declaration and usage of the native `IntToString` function. The function is passed an `Int32` and writes a string representation of the integer to the `StringBuilder` passed to it. The native C code that performs the conversion is also shown.

LISTING 12.3 Code taken from the FundmamentalToString sample application

```
C#
// Declaration
[DllImport("FundamentalToString.dll")]
private static extern void IntToString(int in_Int, StringBuilder
        out_IntAsString);
```

```
// Usage...
private void DoIntConversion()
{
   StringBuilder out_StringBuilder = new StringBuilder(MAX_CAPACITY);
   int in_Int = Convert.ToInt32(this.txtIntegerIn.Text);

   IntToString(in_Int, out_StringBuilder);

   this.txtIntegerOut.Text = out_StringBuilder.ToString();
}

VB
' Declaration
Declare Sub IntToString Lib "FundamentalToString.dll"
        (ByVal in_Int As Integer,
        ByVal out_IntAsString As System.Text.StringBuilder)

' Usage...
Private Sub DoIntConversion()
    Dim out_StringBuilder As System.Text.StringBuilder = New
            System.Text.StringBuilder(MAX_CAPACITY)
    Dim in_Int As Integer = Convert.ToInt32(Me.txtIntegerIn.Text)
    IntToString(in_Int, out_StringBuilder)
    Me.txtIntegerOut.Text = out_StringBuilder.ToString()
End Sub

Native C code
void __cdecl IntToString(int in_Int, WCHAR * out_IntAsString)
{
        swprintf(out_IntAsString, L"%d", in_Int);
}
```

Converting Data to Strings with the FundamentalToString Sample Application

The FundamentalToString sample application is in the folder `\SampleApplications\Chapter12`. There is a C# and Visual Basic version of the managed sample application. FundamentalToString uses a native DLL, `FundamentalToString.dll`. The source code for this DLL is in the folder `\SampleApplications\Chapter12\NativeBinaries\FundamentalToString`. Just like the SquareAnInt sample application, there are also pre-compiled DLL binaries in the

`\SampleApplications\Chapter12\NativeBinaries\FundamentalToString` directory. Each binary is in its own subdirectory according to the hardware and operating system for which the DLL was built.

The FundamentalToString sample application demonstrates how to pass each of the fundamental types discussed in this section into native code. The native code DLL converts each of the fundamental types into a string representation. The string is marshalled back to the managed side through a `StringBuilder`, as described in the previous chunk of sample code.

To use the FundamentalToString application, build and deploy the managed project to your device. Copy the appropriate `FundamentalToString.dll` library to your device, either to the `\Windows` directory or the directory where the managed executable, `FundamentalToString.exe`, resides. You can now launch the application.

When the FundamentalToString application launches, it shows a form with labels demarking these data types: `Int`, `Float`, `Short`, `UShort`, `Byte`, `DWORD`, `Unsigned Int`, `Long`, `Unsigned Long`, `Char`, and `Bool`. Next to each data type is a white text box into which default values are placed. Next to each text box is an empty gray read-only text box.

When the button labeled Convert is clicked, the application makes a series of calls into native code. There is a unique native function for each input data type.

Each call passes two things to the native side: a data value from one of the white text boxes and a `StringBuilder`. The native function receives the data, converts it into a string representation, and writes the string representation to the array of wide characters it sees passed in to it. When the native function call returns, the managed code writes the value of the `StringBuilder` into the gray read-only text box.

Marshalling Simple Structures

The .NET Compact Framework can marshal simple structures between the managed and native worlds. Unlike the full desktop .NET Framework, there are strong restrictions governing what kinds of structures can be marshalled automatically.

First, the .NET Compact Framework marshals all simple structures by reference. Thus, the native side sees parameters that are structures as pointers to the structure. As such, the native code can alter the contents of structures, and the alterations are visible as side effects on the managed side.

Another important rule is that only references to "shallow" structures can be marshalled automatically. This means that the structure can contain an arbitrary number of fundamental data types, as described in previous sections, but the structure cannot contain strings or nested structures. For example, the first structure shown below can be marshalled automatically, but the second cannot.

The following is a structure that can be automatically marshalled:

```
C#
public struct MarshallableStruct
{
   public int m_Int;
   public char m_Char;
   public Int32 m_DWORD;
   public bool m_Bool;
}

VB
Public Structure MarshallableStruct
   Public m_Int As Integer
   Public m_Char As Char
   Public m_DWORD As Int32
   Public m_Bool As Boolean
End Structure
```

The following is a structure that cannot be automatically marshalled:

```
C#
public struct NestedStruct
{
   public int m_Int;
   public char m_Char;
}

public struct CannotAutoMarshal
{
   public int m_Int;
   public NestedStruct m_Nested;
   public Int32 m_DWORD;
   public bool m_Bool;
}

VB
Public Structure NestedStruct
   Public m_Int As Integer
   Public m_Char As Char
End Structure

Public Structure CannotAutoMarshal
```

```
    Public m_Int As Integer
    Public m_Nested As NestedStruct
    Public m_DWORD As Int32
    Public m_Bool As Boolean
End Structure
```

Complex structures that cannot be marshalled automatically and structures with strings in them can be passed to native code, but doing so requires manual marshalling. The next section describes how to do manual marshalling.

Once you understand the rules, passing shallow structures into native code is not significantly different from passing fundamental data types into native code. You need to declare the native function that is passed the structure. You also need to create a structure in managed code that has a compatible layout to the structure you create in native code. For example, if your managed structure is composed of two `Int32` values, then you also need a structure that houses two 32-bit integers on the native side.

The code in Listing 12.4 is derived from the MarshalShallowStruct sample application. It shows the structure and method declarations in C# and Visual Basic to set up for a call to a native C function called `ManipulateStruct`. It also shows the corresponding C code for the `ManipulateStruct`.

LISTING 12.4 Structure and method declarations in C# and Visual Basic to set up for a call to a native C function called `ManipulateStruct`

```
C#
// Structure declaration
public struct ShallowStruct
{
    public int m_Int;
    public char m_Char;
    public Int32 m_DWORD;
    public bool m_Bool;
}

// Method declaration
// Method invocation on native function
// Set up input...
ShallowStruct in_Struct = new ShallowStruct();
in_Struct.m_Int = Convert.ToInt32(4);
in_Struct.m_Char = Convert.ToChar('C');
in_Struct.m_DWORD = Convert.ToInt32(4434);
in_Struct.m_Bool = true;
ShallowStruct out_Struct = new ShallowStruct();
```

```
// Call native function
ManipulateStruct(ref in_Struct, ref out_Struct);
// out_Struct has been altered by the native code implementation
// of ManipulateStruct

VB
' Structure declaration
Public Structure ShallowStruct
   Public m_Int As Integer
   Public m_Char As Char
   Public m_DWORD As Int32
   Public m_Bool As Boolean
   End Structure

' Method declaration
Declare Sub ManipulateStruct Lib "ManipulateStruct.dll" (ByRef
        in_ShallowStruct As ShallowStruct,
        ByRef out_ShallowStruct As ShallowStruct)

' Method invocation on native function
' Set up input...
Dim in_Struct As ShallowStruct = New ShallowStruct
in_Struct.m_Int = Convert.ToInt32(4)
in_Struct.m_Char = Convert.ToChar("C")
in_Struct.m_DWORD = Convert.ToInt32(4434)
in_Struct.m_Bool = True
Dim out_Struct As ShallowStruct = New ShallowStruct

' Call native function
ManipulateStruct(in_Struct, out_Struct)
' out_Struct has been altered by the native code implementation
' of ManipulateStruct

Native C implementation for ManipulateStruct
/* Structure declaration */
struct ShallowStruct
{
   int    m_Int;
   char   m_Char;
   DWORD  m_DWORD;
   bool   m_Bool;
};
```

```
/* ManipulateStruct function implementation */
void __cdecl ManipulateStruct(ShallowStruct * in_ShallowStruct,
        ShallowStruct * out_ShallowStruct)
{
   out_ShallowStruct->m_Int   = in_ShallowStruct->m_Int + 1;
   out_ShallowStruct->m_Char  = in_ShallowStruct->m_Char + 1;
   out_ShallowStruct->m_DWORD = in_ShallowStruct->m_DWORD + 1;
   out_ShallowStruct->m_Bool  = (in_ShallowStruct->m_Bool !=
           true);
}
```

Marshalling Shallow Structures with the MarshalShallowStruct Sample Application

The MarshalShallowStruct sample application demonstrates an end-to-end scenario for calling a native function and passing automatically marshalled shallow structures as parameters. The sample application is located in the folder `\SampleApplications\Chapter12`. There are C# and Visual Basic versions.

MarshalShallowStruct uses a native DLL, `ManipulateStruct.dll`. The source code for this DLL is in the folder `\SampleApplications\Chapter12\NativeBinaries\ManipulateStruct`. Just like the SquareAnInt sample application, there are also pre-compiled DLL binaries in the `\SampleApplications\Chapter12\NativeBinaries\ManipulateStruct` directory. Each binary is in its own subdirectory according to the hardware and operating system for which the DLL was built.

To use the MarshalShallowStruct application, build and deploy the managed project to your device. Then copy the appropriate `ManipulateStruct.dll` library to your device, either to the `\Windows` directory or the directory where the managed executable, `ManipulateStruct.exe`, resides. You can now launch the application.

When MarshalShallowStruct launches, it shows a form with four fields: one for an integer, one for a character, one for a `DWORD` value, and one for a Boolean value. Each holds a default value. When you click the button labeled Update Struct, the application passes a structure holding each field's value to the native function `ManipulateStruct()` that resides in `ManipulateStruct.dll`.

The `ManipulateStruct()` function increments each field in the shallow structure passed to it. When the function returns, the managed application updates the main form to display the new values that were set by the native code.

Passing Nested Structures by Using Custom Marshalling Code

As previously noted, the .NET Compact Framework cannot automatically marshal deep structures between the native and managed worlds. If you are in a position where you absolutely must marshal a deep structure between native and managed code, then you will have to write custom marshalling code. This sounds daunting, and it is tricky.

The fundamental idea to writing custom marshalling code is that the nested structure has a specific layout in memory. On the native side, creating a nested structure means that you have deliberately created this layout and populated it with values. On the managed side, you must duplicate the exact same memory layout with managed structures and then pass a reference to the main managed structure to the native side. When the native code accesses the structure, it makes assumptions about where in memory the data members reside. Since you have duplicated the memory layout with your managed structures, everything works fine.

THE unsafe KEYWORD

In order to do custom marshalling, the managed code must explicitly deal with memory layouts and pointers to specific memory locations. This is allowed in C# by using the `unsafe` keyword. Blocks of code in an unsafe block allow you to use real memory pointers to access memory directly. Doing this is not a recommended practice, and it cannot be done at all with Visual Basic, because there is no notion at all of a memory pointer in Visual Basic.

This is easy to say but harder to do. The idea will be illustrated by working through a tutorial.

The tutorial starts with a look at the native code that will be called into:

```
struct AllIntsInside
{
   int m_one;
   int m_two;
   int m_three;
};

struct DeepStruct
{
   char * m_message;
   AllIntsInside * m_someInts;
};

// Caution! Very unsafe code here used for demonstration purposes only.
// inout_DeepStruct is assumed to exist as a valid pointer, and m_message is a string
// with at least 1 character
```

```
void __cdecl ManipulateDeepStruct(DeepStruct * inout_DeepStruct)
{
   inout_DeepStruct->m_message[0] = 'H';
   inout_DeepStruct->m_someInts->m_one++;
   inout_DeepStruct->m_someInts->m_two++;
   inout_DeepStruct->m_someInts->m_three++;
}
```

The function to be called is named `ManipulateDeepStruct`. It is passed a `DeepStruct`. It changes the first character of the string to an "H" and increments each integer in the AllIntsInside structure that `DeepStruct.m_someInts` points to.

Instances of `DeepStruct` cannot be automatically marshalled by the .NET Compact Framework because it holds a pointer to another struct and a pointer to a null-terminated string. In order to access the `ManipulateDeepStruct` native function, the same memory layout used on the native side must be simulated in managed code.

The first thing needed in the managed world is a way to turn strings into null-terminated byte arrays and vice versa. The following two methods perform this task. Notice the use of the `unsafe` keyword because the manage code is actually accessing pointers to memory.

```
C#
// Puts a null terminator on a string, which makes it usable to
// native c-style code
private char[] NullTerminateString(string in_managedString)
{
   in_managedString += "\0";
   return (in_managedString.ToCharArray());
}

private unsafe string NativeStringToManaged(char* in_NativeNullTerminatedString)
{
   System.Text.StringBuilder l_SB = new System.Text.StringBuilder(100);
   int i = 0;
   while (in_NativeNullTerminatedString[i] != '\0')
   {
      char[] l_toInsert = new char[1];
      l_toInsert[0] = in_NativeNullTerminatedString[i];
      l_SB.Insert(i, l_toInsert);
      i++;
   }
   return l_SB.ToString();
}
```

NullTerminateString simply adds a null at the end of the string and then converts it to a character array. NativeStringToManaged uses a StringBuilder to insert the characters of the in_NativeNullTerminatedString one-by-one and then returns the string held inside the StringBuilder.

Here is the managed code definition for the ManipulateDeepStruct function:

```
C#
[DllImport("ManipulateStruct.dll")]
private static extern void ManipulateDeepStruct(ref DeepStruct
        inout_ShallowStruct);
```

The next thing needed is a managed version of the native AllIntsInside structure. This turns out to be easy:

```
C#
public struct AllIntsInside
{
   public int m_one;
   public int m_two;
   public int m_three;
}
```

There also needs to be a managed code version of DeepStruct. This is also easy, but the unsafe keyword must be used in order to be allowed to use pointers:

```
public unsafe struct DeepStruct
{
   public char * m_message;
   public AllIntsInside * m_someInts;
}
```

The managed DeepStruct is just like the native version. There is a pointer to char and a pointer to an AllIntsInside structure. To call the native ManipulateDeepStruct function, instantiate an instance of DeepStruct and pass a reference to it into ManipulateDeepStruct. But how does one instantiate a valid DeepStruct whose internal pointers are also valid? A block of code from the MarshalDeepStruct sample application holds the answer:

```
DeepStruct l_DeepStruct;
l_DeepStruct = new DeepStruct();
char [] l_messageBuffer = NullTerminateString(" ello World!");
AllIntsInside l_Ints = new AllIntsInside();
l_Ints.m_one = 1;
l_Ints.m_two = 2;
l_Ints.m_three = 3;
```

```
unsafe
{
   fixed (char * l_ptrMessageBuf = &l_messageBuffer[0])
   {
      l_DeepStruct.m_message = l_ptrMessageBuf;
      l_DeepStruct.m_someInts = &l_Ints;

      MessageBox.Show(NativeStringToManaged(l_DeepStruct.m_message), "BEFORE P/INVOKE");

      String l_IntValuesString = "Int Values: "
              + Convert.ToString(l_DeepStruct.m_someInts->m_one)
         + "," + Convert.ToString(l_DeepStruct.m_someInts->m_two) + ","
         + Convert.ToString(l_DeepStruct.m_someInts->m_three);
      MessageBox.Show(l_IntValuesString, "BEFORE P/INVOKE");

      ManipulateDeepStruct(ref l_DeepStruct);

      MessageBox.Show(NativeStringToManaged(l_DeepStruct.m_message), "AFTER P/INVOKE");
      l_IntValuesString = "Int Values: "
              + Convert.ToString(l_DeepStruct.m_someInts->m_one) + ","
         + Convert.ToString(l_DeepStruct.m_someInts->m_two) + ","
         + Convert.ToString(l_DeepStruct.m_someInts->m_three);

   MessageBox.Show(l_IntValuesString, "AFTER P/INVOKE");
   }
}
```

The code is divided into two chunks: the part outside of the `unsafe` block and the part inside. The outside part creates a new instance of a DeepStruct, `l_DeepStruct`. Right now, `l_DeepStruct` does not have valid values in its member pointers. The block also creates a null-terminated character array, `l_MessageBuffer`, which holds the value `Hello World!` and an instance of `AllIntsInside` named `l_Ints`. The member variables of `l_Ints` are set to `1`, `2`, and `3`.

The first thing that happens in the unsafe block is to acquire a pointer to `l_messageBuffer` and store it into `l_DeepStruct.m_message`. The code stores the address of the `l_Ints` structure into `l_DeepStruct.m_someInts`. Now the `l_DeepStruct` structure is correctly instantiated.

Note that this code is also inside a `fixed` block that prevents `l_ptrMessageBuf` from being moved by the managed memory allocation system while the block is executing. This is necessary because if the managed memory is moved around at this time, then `l_DeepStruct` would no longer hold valid pointers.

The next block of code uses a MessageBox to display the character array value and the integer values pointed to by `l_DeepStruct`. The native function `ManipulateDeepStruct` is then called. Finally, the contents of `l_DeepStruct` are again displayed with a MessageBox. The end user will notice the side effect caused by the native code.

Marshalling Deep Structures with the MarshalDeepStruct Sample Application

The MarshalDeepStruct sample application is available in the folder `\SampleApplications\Chapter12\MarshalDeepStruct_CSharp`. There is only a C# version of this sample because of the language limitations of Visual Basic.NET. MarshalDeepStruct relies on the `ManipulateStruct.dll` native library, which is available in the folder `\SampleApplications\Chapter12\NativeBinaries\ManipulateStruct`. There are binaries for a variety of CPU types and Windows CE versions.

MarshalDeepStruct is an end-to-end application that implements the code in the preceding tutorial. To use it, copy the appropriate version of `ManipulateStruct.dll` to the device, either in `\Windows` or the same directory as the managed executable. When you launch the application, it shows only a form with a single button labeled Manipulate Deep Struct. When you click this button, the code discussed in the tutorial of the previous section executes. You will see a message box displaying the contents of a `DeepStruct` before and after calling the native `ManipulateDeepStruct()` function.

Calling the Windows CE Operating System

The desktop version of the .NET Framework is very rich. It includes classes capable of performing advanced windowing, graphics manipulation, network connectivity, and much more. The .NET Framework is nearly a platform of its own, insulating developers from needing to interact with the operating system below. The price of this power and programming convenience is a footprint too large for most devices running Windows CE.

Users of the .NET Compact Framework can invoke the Windows CE operating system to perform actions that are out of reach by using the .NET Compact Framework alone. Situations where doing so would be useful include the following:

- Performing cryptographic manipulations, such as encrypting data and computing hashes (this subject is treated in great detail in Chapter 14)
- Accessing the device registry
- Playing sounds
- Performing advanced graphics manipulations

Calling the Windows CE operating system is no different from calling native code, as we have described in detail already. In order to call the Windows CE operating system, you need to follow these steps:

1. Determine what native Windows CE function you want to call. You may need to consult documentation, such as MSDN, to find out what function(s) you must call to get what you want. For example, the `PlaySound()` function can be used to play a sound file in Windows CE.
2. Determine what DLL the Windows CE function you want to call resides in.
3. Create a method declaration in your managed code for the function you want to call. This means understanding what parameters are passed into the function call and choosing the correct managed data types to pass in.
4. Call the function from managed code, and handle the return value and output parameters as appropriate.

USING MSDN DOCUMENTATION

MSDN Documentation and the documentation included with Embedded Visual C++ 3.0 and 4.0 are invaluable for working through steps 1 and 2. The online documentation is easily searchable and usually states what DLL a Windows CE API function resides in.

LOCATING THE CORRECT DLL FOR WINDOWS CE API FUNCTIONS

Many API functions in Windows CE reside in `coredll.dll`. It is a good first guess if you can't determine the DLL for a given API function call. Many Windows CE API functions reside in `coredll.dll`, even if the corresponding function for the desktop versions of Windows is held in a different DLL.

If you understand all of the previous sections of this chapter, then the real challenge is in steps 1, 2, and 4. To solidify the experience of calling the Windows CE operating system, a tutorial approach is used. The goal is to use the Windows CE API to play a `.wav` sound file.

Calling the Windows CE API to Play a Sound

This section presents a tutorial that implements all four steps needed to call into the Windows CE operating system, as described in the preceding discussion. The goal of the tutorial is to play a `.wav` sound file.

Determining Which Windows CE Function to Call

The Windows CE function for playing a `.wav` file is `PlaySound()`. You can learn all about `PlaySound()` by consulting the documentation included with Embedded Visual C++ 3.0.

The `PlaySound()` function resides in `coredll.dll`. The function prototype, including descriptions of the input parameters, are described as follows:

```
BOOL WINAPI PlaySound(LPCSTR pszSound, HMODULE hmod, DWORD fdwSound)
```

`pszSound` This parameter is a pointer to a null-terminated string that holds the filename of the sound to play. If the parameter is `NULL`, then any sound currently playing is stopped.

`hmod` This parameter provides the ability to specify a sound file embedded in the resource portion of an `.exe` file for playing. The documentation states that it can be safely set to `NULL` if the goal is not to play a sound stored in the resource portion of an `.exe` file. The managed code value `IntPtr.Zero` represents the `NULL` value to pass in.

`fdwSound` This `DWORD` is the sum of flags we want to pass in to control how the sound is played. Flags of interest include

`SND_LOOP` This flag tells `PlaySound()` to play the sound continuously in a loop.

`SND_ASYNC` This flag causes the function to execute asynchronously. The function call returns immediately, even if the sound is still playing. The default behavior without this flag for the function to block until the sound is finished playing.

One question you may be asking is, How do you know the numeric values of the flags in managed code? In native code, for example, C++, you just need to include the appropriate header file and the flags as defined constants. That doesn't help if you need to know the values for use from managed code.

You may find these values in the documentation. For the constants `SND_LOOP` and `SND_ASYNC`, there were no such values in the documentation. Thus, for the tutorial, it was necessary to create a simple application called ConstFinder by using Embedded Visual C++ 3.0. In this C++ application, the constant values `SND_LOOP` and `SND_ASYNC` were assigned into DWORD variables. Then, it is easy to use the debugger to step through the code and examine the values and find that `SND_LOOP = 8` and `SND_ASYNC = 1`. Now one can use these numeric values within managed code and pass them into the `PlaySound()` function from managed code.

You can find the PlaySound ConstFinder project in the directory `\SampleApplications\Chapter12\PlaySound_ConstFinder`.

Creating the Managed Code Declaration for `PlaySound`

The next step is to create a managed code declaration for the `PlaySound` function. The following declaration displays this code:

```
C#
[DllImport "coredll.dll"]
private static extern bool PlaySound(string pszSound,
        IntPtr hmod, uint fdwSound);
```

```
VB
Declare Sub PlaySound Lib "coredll.dll" (
        ByVal pszSound As String, ByVal hmod As IntPtr,
        ByVal fdwSound As UInt32)
```

Calling the Native `PlaySound` Function

Now call the function as appropriate. The following code sample plays the `"\ASound.wav"` sound file once. The function blocks until the sound has been played.

```
C#
IntPtr l_NULL = IntPtr.Zero;
PlaySound("\\ASound.wav", l_NULL, 0);
```

```
VB
Dim l_NULL As IntPtr = IntPtr.Zero
PlaySound("\Asound.wav", l_NULL, Convert.ToUInt32(0))
```

The following code sample plays the `"\ASound.wav"` file continuously. The function returns immediately, even as the sound plays in the background. The calling thread sleeps for five seconds, and then another call to `PlaySound()` stops the sound from playing.

```
C#
const int SND_LOOP = 8;
const int SND_ASYNC = 1;
IntPtr l_NULL = IntPtr.Zero;
PlaySound("\\ASound.wav", l_NULL, SND_LOOP + SND_ASYNC);
System.Threading.Thread.Sleep(5000);
PlaySound(null, l_NULL, 0);
```

```
VB
Const SND_LOOP As Integer = 8
Const SND_ASYNC As Integer = 1
Dim l_NULL As IntPtr = IntPtr.Zero
PlaySound("\ASound.wav", l_NULL,
        Convert.ToUInt32(SND_LOOP + SND_ASYNC))
System.Threading.Thread.Sleep(5000)
PlaySound(Nothing, l_NULL, Convert.ToUInt32(0))
```

Putting It All Together with a Sample Application: PlaySound

The PlaySound sample application is available in the folder `\SampleApplications\Chapter12`, where there are C# and Visual Basic versions.

The PlaySound application is a complete implementation of the issues tackled in the previous tutorial. To use it, build, deploy, and launch it on a device.

The application shows a form with a text box in which you can enter the path to a WAV file to play. The default value is a WAV file included with the Pocket PC version of Windows CE.

To play the sound just once, click the button labeled Play Sound Once. The application calls the native `PlaySound` function in synchronous mode. The main thread of the managed application blocks until the sound is finished being played.

To play a sound continuously, click the button labeled Play Sound Continuously. The application plays the sound by calling the Windows CE API function, `PlaySound()`, passing arguments to play the sound in a loop asynchronously. The Play Sound Continuously button's label changes to Stop Playing Sound!!!, and the managed application's main thread remains responsive, even as the sound plays. Click the button relabeled Stop Playing Sound!!! to stop playing the sound. The managed application responds by calling `PlaySound()` again with a `NULL` value for the sound file name argument. This causes the sound to stop playing.

This sample application helps illustrate an important point when calling Windows CE functions: Pay attention to whether the function returns immediately and how long the function will take to execute. It is easy to accidentally call a function that performs some sort of I/O in blocking mode and ties up the managed thread that makes the native call for a long time. If you call the function from the main managed thread, then your application will stop responding to user input for a potentially long time. In such cases, consider spinning off a new thread to call the Windows CE operating system. Chapter 4, "Using Threads and Timers in the .NET Compact Framework," describes threading in great detail.

Accessing COM Components from the .NET Compact Framework

Unlike the desktop .NET Framework, the .NET Compact Framework does not have a built-in capability for consuming COM components. However, Odyssey Software (`http://www.odysseysoftware.com`) has developed a product called CFCom that makes it possible for the .NET Compact Framework to access COM components. Odyssey Software offers a variety of licensing schemes for developers. For example, a shareware developer can purchase a license to use the Odyssey product for one project. Alternately, a user can purchase CFCom for access only to specific COM components, such as the popular Pocket Outlook Object Model (POOM). Of course, developers can also buy a full-fledged license by which they can access any COM component with the .NET Compact Framework, and they can ship this capability with their product. See the Odyssey Software Web site for more details.

In Brief

- There are a variety of situations where it is necessary for managed code to call into native code, such as for calling the Windows CE API.
- While managed binaries are portable to all of the CPU types supported by the .NET Compact Framework, developers must be careful to use native binaries compatible with the hardware they are running on.
- DLLs written in C++ sometimes use name mangling on the function names they expose, making it harder to call into such DLLs from managed code.
- To call a native function from managed code, you first declare the native function in the class from which it will be called. Then the native function is treated as one of the class' methods.
- Managed code throws exceptions if it cannot find the native DLL to call into, or if it cannot marshal the parameters passed between the managed and native code.
- Fundamental data types 32 bits wide or less can be passed by value or reference. Larger fundamental types, such as long integers, must be passed by reference.
- The `DWORD` type, commonly used in native code, is compatible with the UInt32 type in the .NET Compact Framework.
- The managed `IntPtr` data type is compatible with handle types in the Windows CE operating system.
- To pass a non-mutable string into native code, use the C# `string` (`String` in Visual Basic) type. To pass a mutable string into native code, use a `StringBuilder`.
- The .NET Compact Framework can automatically marshal simple structures, which are structures with no nested sub structures. All simple structures must be passed by reference.
- It is possible to pass complex structures that contain nested structures into native code by reference, but custom marshalling code is required. Doing this is only possible with C# because Visual Basic does not allow access to pointers under any circumstances.
- Calling the Windows CE operating system is no different from calling native code in any other DLL. Many user-callable functions in the Windows CE OS are in the `core.dll` library.
- Odyssey Software offers a product called CFCom with which managed code can access COM components.

Exploring .NET Reflection

IN THIS CHAPTER

Understanding the .NET Compact Framework Reflection API

The .NET Compact Framework Reflection API provides access to the metadata in a .NET assembly. .NET assemblies contain modules, modules contain types, and types contain members, such as methods, fields, properties, and events. The Reflection API provides objects that encapsulate and provide access to the metadata located within assemblies, modules, types, and members. The Reflection API is commonly used to dynamically create an instance of a type or get the type from an existing object. After you have that type, you can invoke methods and constructors or access fields and properties.

The desktop .NET Framework provides the ability to dynamically generate new code by using the `System.Reflection.Emit` namespace. This is not supported by the .NET Compact Framework.

Loading .NET Assemblies

The `System.Reflection.Assembly` class represents a physical .NET executable, either an `.exe` or a `.dll` file. The `System.Reflection.Assembly` class can be used to load assemblies, load modules, or discover a type from the assembly and create an instance of it. The `Assembly` class does not have a public constructor, so in order to get a reference to an Assembly instance, you must use a static method on the Assembly class itself. The static method LoadFrom loads an assembly given its filename. This method loads the assembly into the AppDomain of the caller and returns an Assembly instance that represents the loaded assembly. Listing 13.1 demonstrates how to load an assembly by using the LoadFrom method.

LISTING 13.1

```
C#
Assembly anAssembly = Assembly.LoadFrom("\\sample.dll");
MessageBox.Show("Loaded " + anAssembly.FullName);

VB
Dim anAssembly as Assembly.LoadFrom("\\sample.dll")
MessageBox.Show("Loaded " & anAssembly.FullName)
```

Discovering Type Information

The System.Type class is the primary means by which type metadata is accessed. It provides an API that will allow you to manipulate newly discovered types. A System.Type object instance can represent all .NET languages types, including classes, interfaces, arrays, values, and enumerations.

First, you will need to get a reference to a System.Type object instance. There are several different ways to do this. In this section we will discuss two of the most popular.

Retrieving Type Information with the GetType Method

The easiest way to get a reference to a System.Type object instance is to call the GetType method on an already existing object instance. The GetType method is a public virtual method declared on the Object class. This means that every .NET class inherits this method, and in turn, this gives you access to its underlying type. Listing 13.2 shows how to use the GetType method.

LISTING 13.2

```
C#
string str = "Hello, Reflection!";
Type strType = str.GetType();
MessageBox.Show(strType.ToString(),"Name of String Type");

VB
Dim str as string
Dim strType as Type
```

```
str = "Hello, world!"
strType = str.GetType()
MessageBox.Show(strType.ToString(),"Name of String Type")
```

RETRIEVING TYPE INFORMATION WITH THE C# typeof OPERATOR

If you are using C#, then you can use the `typeof` operator to retrieve the `System.Type` object. A `typeof` expression takes the form `typeof(`*`type`*`)`. The *type* parameter is the name of the type. This is not the name of the type contained in a .NET string. Rather, it is the name of the type without quotation marks. Listing 13.3 exemplifies how to use the `typeof` operator.

LISTING 13.3

```
C#
public class TypeOf_Operator {
  public int integer;
  public TypeOfOperator () {
  }
}

public class Test {
  public static void Main() {
    Type t = typeof(TypeOf_Operator);
    MessageBox.Show(t.ToString());
  }
}
```

It is important to realize that the `typeof` operator is *not* resolved at compile time. There is work performed at runtime to return the correct `System.Type` object instance. In some cases this work can be quite expensive, so do not choose the `typeof` operator over the `GetType` method under the assumption that the `typeof` operator will perform more efficiently. They both require a substantial amount of work to ensure that the correct `System.Type` object instance is returned.

Retrieving Type Information from a Loaded Assembly

You can also load a type from a loaded assembly. This is very valuable if you do not know the name of the type you are loading or if you do not have an instance of the type you are working with. The `GetTypes` method will return an array of `System.Type` objects that contains all of the types found in the assembly. Listing 13.4 demonstrates how to use this method.

LISTING 13.4

```
C#
Assembly anAssembly = Assembly.LoadFrom("\\sample.dll");
Type[] types = anAssembly.GetTypes();

foreach(Type type in types)
  MessageBox.Show("Found " + type.ToString());

VB
Dim anAssembly as Assembly.LoadFrom("\\sample.dll")
Dim Types[] as anAssembly.GetTypes()

for (i = 0 to types.Length)
  MessageBox.Show("Found " & types[i].ToString())
Next i
```

The `GetType` method on the `Assembly` class allows you to retrieve a specific `System.Type` from the assembly if you know the type's name. This method is an overload of the `GetType` method inherited from the `object` class. The `GetType` method on the `Assembly` class takes the name of the type as a parameter and searches the `Assembly` for a type with the matching name. If one is found, a `System.Type` object instance representing that type is returned. Otherwise, `null` is returned. Listing 13.5 demonstrates how to use the `GetType` method.

LISTING 13.5

```
C#
Assembly anAssembly = Assembly.LoadFrom("\\sample.dll");
Type type = anAssembly.GetType("TypeClass");
MessageBox.Show("Found " + type.ToString());

VB
Dim anAssembly as Assembly.LoadFrom("\\sample.dll")
Dim Type as anAssembly.GetType("TypeClass")
MessageBox.Show("Found " & type.ToString())
```

There is a similar static `GetType` defined on the `Type` class. This `GetType` method also takes the type name as a parameter. The name of the type should be assembly name qualified because the method needs to know in which assembly to search. An assembly qualified name has the following format:

```
TopNamespace.SubNamespace.Class+NestedClass,AssembleName
```

Listing 13.6 shows how to use the `Type.GetType` with an assembly-qualified type name.

LISTING 13.6

```
C#
Type type = Type.GetType("TypeClass,Sample");
MessageBox.Show("Found " + type.ToString());

VB
Dim Type as anAssembly.GetType("TypeClass, Sample")
MessageBox.Show("Found " & type.ToString())
```

Creating Type Instances by Using ConstructorInfo

Once you have a `System.Type` object instance, you can use it to create an instance of the type it represents. This can be done by querying for a `ConstructorInfo` object. The `ConstructorInfo` class is used to discover the attributes of a constructor, and it provides access to constructor metadata. Most importantly, a `ConstructorInfo` instance can be used to invoke a type's constructor.

Before we start invoking constructors, let's discuss how you can get a hold of one of these powerful objects. The `Type` class provides two methods for retrieving `ConstructorInfos`: `GetConstructors` and `GetConstructor`.

Using the GetConstructors Method

The `GetConstructors` method comes in two flavors. The easiest flavor to use accepts no parameters and returns all of the public constructors defined for the type. You can iterate through the array of `ConstructorInfo` objects and then invoke the constructor you deem appropriate to use. We will discuss invoking the constructor shortly. Listing 13.7 demonstrates how to use `GetConstructors` to retrieve a list of all public constructors.

LISTING 13.7

```
C#
public class Type_GetConstructors {
  public int integer;
  public Type_GetConstructors() : this(0) {
  }

  public Type_GetConstructors(int anInt) {
    integer = anInt;
  }
}
```

```
public class Test {
  public static void Main() {
    Type t = typeof(Type_GetConstructors);
    ConstructorInfo[] cstors = t.GetConstructors();
    foreach(ConstructorInfo cstor in cstors) {
      MessageBox.Show(
        String.Format(
        "Found a {0} constructor that takes {1} parameters",
        (cstor.IsPublic ? "Public" : "Non-Public"),
        cstor.GetParameters().Length));
    }
  }
}
```

VB

```
Module Module1
    Sub Main()
        Dim t As Type
        Dim cstors As ConstructorInfo()
        Dim ndx As Int32
        Dim tgc As New Type_GetConstructors()

        t = tgc.GetType()
        cstors = t.GetConstructors()

        For ndx = 0 To cstors.Length - 1
            Dim visibility As String
            If cstors(ndx).IsPublic Then
                visibility = "Public"
            Else
                visibility = "Non-Public"
            End If

            MessageBox.Show( _
              String.Format( _
              "Found a {0} constructor that takes {1} parameters", _
               visibility, _
               cstors(ndx).GetParameters().Length))
        Next ndx
    End Sub
```

```
    Public Class Type_GetConstructors
        Public m_Int As Int32
        Public Sub New()
            m_Int = 0
        End Sub

        Public Sub New(ByVal int As Int32)
            m_Int = int
        End Sub
    End Class
End Module
```

The Type class also provides an overload of the GetConstructors method that takes one parameter. This parameter is of the type BindingFlags, an enumeration that controls the way in which the search for members and types is conducted by reflection. As we investigate reflection more in this chapter, you will see the BindingFlags enumeration appear again and again. This enumeration is defined with a FlagsAtribute attribute that allows a bitwise combination of its member values. Table 13.1 contains the list of BindingFlags values that are appropriate to pass to the GetConstructors method and how they affect the search.

TABLE 13.1

BindingFlags Members Relevant to the GetConstructors Method

MEMBER	NAME DESCRIPTION
Default	Specifies no binding flag
Instance	Specifies that instance constructors are to be included in the search
NonPublic	Specifies that nonpublic constructors are to be included in the search
Public	Specifies that public constructors are to be included in the search
Static	Specifies that static constructors are to be included in the search

Listing 13.8 demonstrates how to use GetConstructors to search for all constructors of a given type. This will include both static and instance constructors as well as both public and nonpublic constructors.

LISTING 13.8

```
C#
public class ConstructorInfo_GetConstructors {
  public static int staticInteger;
  public int integer;
  public ConstructorInfo_GetConstructors () : this(0) {
  }
```

```
    public ConstructorInfo_GetConsructors (int anInt) {
      integer = anInt;
    }

    private ConstructorInfo_GetConsructors (int x, int y) {
      integer = x;
      staticInteger = y;
    }

    static ConstructorInfo_GetConsructors () {
      staticInteger = 13;
    }

    public static void Main() {
      Type t = typeof(ConstructorInfo_GetConstructors);
      ConstructorInfo[] cstors =
        t.GetConstructors(BindingFlags.Instance | BindingFlags.Static |
                          BindingFlags.Public | BindingFlags.NonPublic);

      foreach (ConstructorInfo cstor in cstors) {
          MessageBox.Show(
          String.Format
          ("Found a {0} {1} constructor that takes {2} parameters",
            (cstor.IsStatic ? "Static" : "Instance"),
            (cstor.IsPublic ? "Public" : "Non-Public"),
            cstor.GetParameters().Length));
      }
    }
}
```

VB

```
Module Module1
    Public Class ConstructorInfo_GetConsructors
        Public Shared sharedInteger As Int32
        Public m_Int As Int32
        Public Sub New()
            m_Int = 0
        End Sub

        Public Sub New(ByVal anInt As Int32)
            m_Int = anInt
        End Sub
```

```
    Public Sub New(ByVal x As Int32, ByVal y As Int32)
        m_Int = x
        sharedInteger = y
    End Sub

    Shared Sub New()
        sharedInteger = 13
    End Sub
End Class

Sub Main()
    Dim t As Type
    Dim cstors As ConstructorInfo()
    Dim ndx As Int32
    Dim cigc = New ConstructorInfo_GetConsructors()
    t = cigc.GetType()
    cstors = t.GetConstructors(BindingFlags.Instance Or _
                               BindingFlags.Static Or _
                               BindingFlags.Public Or _
                               BindingFlags.NonPublic)

    For ndx = 0 To cstors.Length - 1
        Dim visibility As String
        Dim isStatic As String

        If cstors(ndx).IsPublic Then
            visibility = "Public"
        Else
            visibility = "Non-Public"
        End If

        If cstors(ndx).IsStatic Then
            isStatic = "Static"
        Else
            isStatic = "Instance"
        End If

        MessageBox.Show( _
         String.Format( _
         "Found a {0} {1} constructor that takes {2} parameters", _
         visibility, _
```

```
            isStatic, _
            cstors(ndx).GetParameters().Length))
        Next ndx
    End Sub
End Module
```

Retrieving Individual ConstructorInfo Objects with GetConstructor

So far we have been searching for constructors based on visibility and scope. There will be times when you need to query for a constructor based on the number and type of parameters the constructor accepts. For instance, if you are building a custom serializer, you may need to search for a default constructor with which to construct an object being deserialized. The `Type` class provides the `GetConstructor` method to provide this functionality. The `GetConstructor` method returns a single `ConstructorInfo` object, or else `null` if the constructor could not be found. The method takes one parameter, a `Type` array that represents the number, order, and type of the parameters for the constructor to get. You can pass in an empty array of the type `Type` to get a constructor that takes no parameters. Do not pass in `null` for this; it would result in an `ArgumentNullException`. It is important to note that `GetConstructor` will look for public instance constructors and cannot be used to obtain a class initializer (static constructor). Listing 13.9 demonstrates how to use `GetConstructor` to query for a public instance constructor that takes zero parameters.

LISTING 13.9

```
C#
public class ConstructorInfo_GetConstructor {
  public static int staticInteger;
  public int integer;
  public ConstructorInfo_GetConstructor () : this(0) {
  }

  private ConstructorInfo_GetConstructor (int anInt) {
    integer = anInt;
  }

  static ConstructorInfo_GetConstructor () {
    staticInteger = 13;
  }

  public static void Main() {
    Type t = typeof(ConstructorInfo_GetConstructor);
```

```
    ConstructorInfo cstor = t.GetConstructor(new Type[0]);

    if(cstor == null)
      MessageBox.Show("No constructor found.");
    else
      MessageBox.Show
        ("Found a public instance constructor with no params");
  }
}

VB
Module Module1
    Public Class ConstructorInfo_GetConstructor
        Public Shared sharedInteger As Int32
        Public m_int As Int32
        Public Sub New()
            m_int = 0
        End Sub

        Public Sub New(ByVal anInt As Int32)
            m_int = anInt
        End Sub

        Shared Sub New()
            sharedInteger = 13
        End Sub
    End Class

    Sub Main()
        Dim t As Type
        Dim cstor As ConstructorInfo
        Dim empty() = New Type() {}
        Dim cigc As New ConstructorInfo_GetConstructor()

        t = cigc.GetType()
        cstor = t.GetConstructor(empty)

        If cstor Is Nothing Then
            MessageBox.Show("No constructor found.")
        Else
            MessageBox.Show _
              ("Found a public instance constructor with no params")
```

```
        End If
    End Sub
End Module
```

Creating Object Instances with the ConstructorInfo Class

Now that we know how to find a ConstuctorInfo, we are ready to use it to create an object instance. The ConstructorInfo class provides the Invoke method to invoke the constructor it represents. The Invoke method takes one parameter, an array of objects that represent values of the parameters to be passed to the constructor. The values should match the number, type, and order of the parameters for the constructor reflected by the ConstructorInfo object. Before calling the constructor, Invoke verifies that the parameters are valid. If the constructor takes no parameters, then you should pass in an empty array of objects. The ConstructorInfo method returns a reference to an object, so you must cast it to the correct type. Listing 13.10 code shows how to find a specific constructor, invoke the constructor, and inspect the new object instance.

LISTING 13.10

```
C#
public class ConstructorInfo_Invoke {
  public int integer;
  public string str;

  public ConstructorInfo_Invoke () : this(0, string.Empty) {
  }

  public ConstructorInfo_Invoke (int int1, string str1) {
    integer = int1;
    str = str1;
  }

  public static void Main() {
    Type[] ts = {typeof(int), typeof(string)};
    Type t = typeof(ConstructorInfo_Invoke);
    ConstructorInfo cstor = t.GetConstructor(ts);

    if (cstor == null) {
      MessageBox.Show("Could not create the object instance");
      return;
    }
```

```
    object[] os = {13, "This object was created with reflection."};
    object objTest = cstor.Invoke(os);

    ConstructorInfo_Invoke test = objTest as ConstructorInfo_Invoke;
    if( test == null) {
      MessageBox.Show("Could not cast the object to it correct type");
      return;
    }

    MessageBox.Show("New ConstructorInfo_Invoke object created:\n" +
                    "\tinteger: " + test.integer +
                    "\n\tstr: " + test.str);
  }
}
```

VB

```
Module Module1
    Public Class ConstructorInfo_Invoke
        Public m_int As Int32
        Public str As String

        Public Sub New()
            m_int = 0
            str = String.Empty
        End Sub

        Public Sub New(ByVal anInt As Int32, ByVal aStr As String)
            m_int = anInt
            str = aStr
        End Sub
    End Class

    Sub Main()
        Dim t As Type
        Dim ts() = New Type() {0.GetType(), String.Empty.GetType()}
        Dim os() = New Object() { _
                    13, _
                    "This object was created with reflection"}
        Dim ret As Object
        Dim cstor As ConstructorInfo
        Dim cii As New ConstructorInfo_Invoke()
```

```
        t = cii.GetType()
        cstor = t.GetConstructor(ts)

    If cstor Is Nothing Then
            MessageBox.Show("Could not create the object instance")
            Return
        End If

        os(0) = 13
        os(1) = "This object was created with reflection"
        ret = cstor.Invoke(os)

        Dim test = CType(ret, ConstructorInfo_Invoke)
        If test Is Nothing Then
            MessageBox.Show _
              ("Could not cast the object to its " & _
               "correct type instance")
            Return
        End If

        MessageBox.Show("New ConstructorInfo_Invoke object created:"& _
                        Chr(13) & "integer: " & test.m_int & _
                        Chr(13) & "str: " & test.str)
    End Sub
End Module
```

Invoking Methods by Using the MethodInfo Class

The Reflection API also provides the ability to search for and invoke methods that you may not have knowledge of at compile time. Before you can invoke a method, you must successfully request a `MethodInfo` object from the `Type` object. In this section we will investigate three ways to retrieve a `MethodInfo` from a `Type` object.

Working with the GetMethod Method

The easiest way to retrieve a `MethodInfo` object is to search for the method by name. The `Type` object provides the `GetMethod(string methodName)` to do just this. The string parameter contains the case-sensitive name of the public method to get. If you request a nonpublic method or a method that does not exist, then a `null` reference is returned. Beware that for overloaded methods an `AmbiguousMatchException` will be thrown. Listing 13.11 demonstrates how to find a specific method by name.

LISTING 13.11

C#

```
public class Type_GetMethod {
  public int m_Integer;
  public string m_String;

  public Type_GetMethod(int i, string s) {
    m_Integer = i;
    m_String = s;
  }

  public int GetSecretCode() {
    char[] chars = m_String.ToCharArray();
    int charsLength = chars.Length;
    int ret = m_Integer;

    for (int i = 0; i < charsLength; ++i)
      ret += (int)chars[i];

    return ret;
  }

  public static void Main() {
    Type t = typeof(Type_GetMethod);
    MethodInfo methodInfo = t.GetMethod("GetSecretCode");

    if (methodInfo == null) {
      MessageBox.Show("Could not find the GetSecretCode method.");
      return;
    }

    MessageBox.Show(
      String.Format("Found {0} method name {1}",
      (methodInfo.IsPublic ? "Public" : "Non-Public"),
      methodInfo.Name));
  }
}
```

```
VB
Module Module1
    Public Class Type_GetMethod
        Public m_Integer As Int32
        Public m_String As String

        Public Sub New(ByVal i As Int32, ByVal s As String)
            m_Integer = i
            m_String = s
        End Sub

        Public Function GetSecretCode() As Int32
            Dim chars = m_String.ToCharArray()
            Dim charsLength = chars.Length
            Dim ret = m_Integer
            Dim i As Int32

            For i = 0 To charsLength
                ret = ret + CInt(chars(i))
            Next i
        End Function
    End Class

    Public Sub Main()
        Dim tgm = New Type_GetMethod(0, 0)
        Dim t = tgm.GetType()
        Dim methodInfo = t.GetMethod("GetSecretCode")
        Dim visibility As String

        If MethodInfo Is Nothing Then
            MessageBox.Show("Could not find the " & _
                            "GetSecretCode method.")
            Return
        End If

        If MethodInfo.IsPublic Then
            visibility = "Public"
        Else
            visibility = "Non-Public"
        End If

        MessageBox.Show( _
```

```
            String.Format("Found {0} method name {1}", _
            visibility, _
            methodInfo.Name))
    End Sub
End Module
```

FIND MethodInfo OBJECTS FOR OVERLOADED METHODS

An overload of the GetMethod method exists that can also handle searching for overloaded methods. This overload allows you to look up a public method by name and by the type of parameters it accepts. Listing 13.12 demonstrates how to search for an overloaded method.

LISTING 13.12

```
C#
public class Type_GetMethod {
  public void OverloadedMethod(int param1) {
  }

  public void OverloadedMethod(int param1, bool param2) {
  }
}

public class Test {
  public static void Main() {
    Type t = typeof(Type_GetMethod);
    Type[] args = {typeof(int), typeof(bool)};
    MethodInfo methodInfo = t.GetMethod("OverloadedMethod", args);

    if (methodInfo == null) {
      MessageBox.Show("Could not find the OverloadedMethod method.");
        return;
    }

    MessageBox.Show(
      String.Format("Found {0} method name {1}",
        (methodInfo.IsPublic ? "Public" : "Non-Public"),
         methodInfo.Name));
  }
}

VB
Module Module1
    Public Class Type_GetMethod
        Public Sub OverloadedMethod(ByVal param1 As Int32)
        End Sub
```

```
        Public Sub OverloadedMethod(ByVal param1 As Int32, _
ByVal param2 As Int32)
        End Sub
    End Class

    Public Sub Main()
        Dim tgm As New Type_GetMethod()
        Dim t = tgm.GetType()
        Dim args() = New Type() {0.GetType(), 0.GetType()}
        Dim methodInfo = t.GetMethod("OverloadedMethod", args)
        Dim visibility As String

        If methodInfo Is Nothing Then
            MessageBox.Show("Could not find the " & _
                            "OverloadedMethod method.")
            Return
        End If

        If methodInfo.IsPublic Then
            visibility = "Public"
        Else
            visibility = "Non-Public"
        End If

        MessageBox.Show( _
            String.Format("Found {0} method name {1}", _
            visibility, _
            methodInfo.Name))
    End Sub
End Module
```

Another overload of the `GetMethod` method allows you to search for a method by name and by a set of binding constraints. The binding constraints are specified by passing a combination of `BindingFlags` values. Just like with `GetType`, there is a list of `BindingFlags` values that affects the execution of the search for the method. The `BindingFlags` values that affect the `GetMethod` method fall into two categories: those that define which methods to include in the search and those that change how the search works. Table 13.2 shows the list of `BindingFlags` values that define which methods to include in the search. Table 13.3 lists those that change how the search works.

TABLE 13.2

BindingFlags Members that Define Which Methods Are Searched via GetMethod

MEMBER	DESCRIPTION
Instance	Includes instance methods in the search
Static	Includes static methods in the search
Public	Includes public methods in the search
NonPublic	Includes private and protected methods in the search
FlattenHierarchy	Includes static methods up the class hierarchy

TABLE 13.3

BindingFlags Members That Change How GetMethod Searches

MEMBER	MEANING
IgnoreCase	Ignores the case of the specified name parameter
DeclaredOnly	Searches only the methods declared on the current Type and not methods that were inherited

The BindingFlags enumeration values can be bitwise combined to allow a very flexible way to customize the search for a desired method. If the request type is nonpublic or does not exist, then null will be returned. Listing 13.13 demonstrates how to find a public static method by using a case-insensitive search.

LISTING 13.13

```
C#
public class Type_GetMethod {
  public static void StaticMethod() {
  }

  public void InstanceMethod(){
  }
}

public class Test {
  public static void Main() {
    Type t = typeof(Type_GetMethod);
    BindingFlags flags =
      BindingFlags.Static|BindingFlags.Public|BindingFlags.IgnoreCase;

    MethodInfo methodInfo = t.GetMethod("sTaTiCmEtHoD", flags);
```

```
      if (methodInfo == null) {
        MessageBox.Show("Could not find the method named StaticMethod.");
          return;
      }

      MessageBox.Show(
        String.Format("Found {0} {1} method name {2}",
          (methodInfo.IsPublic ? "public" : "non-public"),
          (methodInfo.IsStatic ? "static" : "Instance"),
           methodInfo.Name));
    }
  }
```

VB

```
Module Module1
    Public Class Type_GetMethod
        Public Shared Sub SharedMethod()
        End Sub

        Public Sub InstanceMethod()
        End Sub
    End Class

    Public Sub Main()
        Dim tgm = New Type_GetMethod()
        Dim t = tgm.GetType()
        Dim flags = BindingFlags.Static Or _
                    BindingFlags.Public Or _
                    BindingFlags.IgnoreCase
        Dim methodInfo = t.GetMethod("sHaReDmEtHoD", flags)
        Dim visibility As String
        Dim isShared As String

        If methodInfo Is Nothing Then
            MessageBox.Show("Could not find the SharedMethod method.")
            Return
        End If

        If methodInfo.IsPublic Then
            visibility = "Public"
        Else
            visibility = "Non-Public"
```

```
        End If

        If methodInfo.IsStatic Then
            isShared = "Shared"
        Else
            isShared = "Instance"
        End If

        MessageBox.Show( _
            String.Format("Found {0} {1} method name {2}", _
            visibility, isShared, methodInfo.Name))
    End Sub
End Module
```

Invoking a Method with the MethodInfo Class

Now that you have learned how to retrieve a MethodInfo object by searching over Type object, we can now investigate how to invoke that method. Much like the ConstructorInfo class, the MethodInfo class provides the Invoke method to allow users to invoke the method it reflects. The Invoke method accepts two parameters. The first parameter represents an instance of the type on which the method exists. The second parameter is an array of objects that represent the methods argument list. For methods that do not accept parameters, you can pass an empty array of objects. The Invoke method returns an object that represents the return value of the function. If the method does not have a return value, then null is returned. Listing 13.14 demonstrates how to invoke a method through a MethodInfo object.

LISTING 13.14

```
C#
public class MethodInfo_Invoke {
  public int AddParameterValues(int bar, int zoo) {
       return bar + zoo;
  }
}

public class Test {
  public static void Main() {
    Type t = typeof(MethodInfo_Invoke);
    Type[] argTypes = {typeof(int), typeof(int)};
    MethodInfo methodInfo =
      t.GetMethod("AddParameterValues", argTypes);
```

```
      if (methodInfo == null) {
        MessageBox.Show("Could not find the AddParameterValues method.");
          return;
      }

      MethodInfo_Invoke c = new MethodInfo_Invoke();
      Object[] argValues = {1, 1};
      object ret = methodInfo.Invoke(c, argValues);

      if(ret == null)
        MessageBox.Show("The invoked method returned null");
      else
        MessageBox.Show("The invoked method returned " + ret);
    }
  }

VB
Module Module1
    Public Class MethodInfo_Invoke
        Public Function AddParameterValues(ByVal bar As Int32, _
                                           ByVal zoo As Int32)
            Return (bar + zoo)
        End Function
    End Class

    Public Sub Main()
        Dim mii = New MethodInfo_Invoke()
        Dim t = mii.GetType()
        Dim argTypes() = New Type() {0.GetType(), 0.GetType}
        Dim MethodInfo = t.GetMethod("AddParameterValues", argTypes)

        If MethodInfo Is Nothing Then
            MessageBox.Show("Could not find the " & _
                            "AddParameterValues method.")
            Return
        End If

        Dim c As New MethodInfo_Invoke()
        Dim argValues() = New Object() {1, 1}
        Dim ret = MethodInfo.Invoke(c, argValues)

        If ret Is Nothing Then
```

```
            MessageBox.Show("The invoke method returned nothing")
        Else
            MessageBox.Show("The invoke method returned " & _
                            ret.ToString())
        End If
    End Sub
End Module
```

Using Reflection to Manipulate Object State

The Reflection API provides the ability to manipulate an object's state, meaning its properties and fields. Using the Reflection API, you can call property accessors as well as directly change the value of an object's fields.

The two classes that represent fields and properties are the `FieldInfo` class and the `PropertyInfo` class, respectively. As with most of the reflection classes, you first have to search and discover a `FieldInfo` or `PropertyInfo` object before it can be invoked. Again, the `Type` class provides methods for discovering both fields and properties.

Discovering a Type's Field Information by Using the `GetFields` method

The `Type.GetFields` method has two overloads. One overload takes no parameters and returns an array of `FieldInfo` objects that represents all of the public fields defined for the `Type`. If there are no public fields defined, then an empty array of type `FieldInfo` is returned. Listing 13.15 code demonstrates how to use the `GetFields` method to discover all of the public fields on a class:

LISTING 13.15

```
C#
public class Type_GetFields {
  public int intType;
  public bool boolType;
  public string stringType;
}

public class Test {
  public static void Main(string[] args) {
    Type t = typeof(Type_GetFields);
    FieldInfo[] fields = t.GetFields();
```

```
      foreach(FieldInfo field in fields) {
          MessageBox.Show(String.Format(
            "Found {0} field {1} of type {2} ",
            (field.IsPublic ? "Public" : "Non-Public"),
            field.Name,
            field.FieldType));
      }
    }
}
```

VB

```
Module Module1
    Public Class Type_GetFields
        Public intType As Int32
        Public boolType As Boolean
        Public stringType As String
    End Class

    Public Sub Main()
        Dim tgf = New Type_GetFields()
        Dim t = tgf.GetType()
        Dim fields() = t.GetFields()
        Dim visibility As String
        Dim i As Int32

        For i = 0 To fields.Length - 1
            If fields(i).IsPublic Then
                visibility = "Public"
            Else
                visibility = "Non-Public"
            End If

            MessageBox.Show(String.Format( _
                "Found {0} field {1} of type {2} ", _
                visibility, _
                fields(i).Name, _
                fields(i).FieldType))
        Next i
    End Sub
End Module
```

The GetFields method has an overload that allows you to customize how the search for the desired fields is conducted. This customization is controlled by passing a BindingFlags value to the GetFields method. The BindingFlags values can either define which fields to include in the search, values described in Table 13.4, or change how the search works, values described in Table 13.5.

TABLE 13.4

BindingFlags Members That Define Which Methods Are Searched by GetFields

MEMBER	MEANING
Instance	Includes instance fields in the search
Static	Includes static fields in the search
Public	Includes public fields in the search
NonPublic	Includes private and protected fields in the search
FlattenHierarchy	Includes static fields up the class hierarchy

TABLE 13.5

BindingFlags Members That Change How GetFields Searches

MEMBER	MEANING
DeclaredOnly	Searches only the fields declared on the current Type and not fields that were inherited

It should be noted that if neither BindingFlags.Instance nor BindingFlags.Static is specified, then no FieldInfo objects will be returned. If no fields are found that adhere to the specified binding constraints, or if no fields are defined for the Type, then a zero length array of type FieldInfo is returned. Listing 13.16 demonstrates how to search for a field by using a binding constraint.

LISTING 13.16

```
C#
public class Type_GetFields {
  public static int StaticIntType;
  public static bool StaticBoolType;
  public static string StaticStringType;
  protected static short StaticShortType;

  public int IntType;
  public bool boolType;
  public string stringType;
}
```

```
public class Test {
  public static void Main() {
    Type t = typeof(Type_GetFields);
    FieldInfo[] fields =
      t.GetFields(BindingFlags.Static | BindingFlags.Public);

    foreach(FieldInfo field in fields) {
      MessageBox.Show(String.Format(
        "Found {0} {1} field {2} of type {3} ",
        (field.IsPublic ? "public" : "non-public"),
        (field.IsStatic ? "static" : "instance"),
        field.Name,
        field.FieldType));
    }
  }
}
```

VB

```
Module Module1
    Public Class Type_GetFields
        Public Shared SharedIntType As Int32
        Public Shared SharedBoolType As Boolean
        Public Shared SharedStringType As String
        Protected Shared SharedShortType As String

        Public intType As Int32
        Public boolType As Boolean
        Public stringType As String
    End Class

    Public Sub Main()
        Dim tgf = New Type_GetFields()
        Dim t = tgf.GetType()
        Dim fields = t.GetFields(BindingFlags.Static Or _
                                 BindingFlags.Public)
        Dim visibility As String
        Dim i As Int32

        For i = 0 To fields.Length - 1
            If fields(i).IsPublic Then
                visibility = "Public"
```

```
            Else
                visibility = "Non-Public"
            End If

            MessageBox.Show(String.Format( _
                "Found {0} field {1} of type {2} ", _
                visibility, _
                fields(i).Name, _
                fields(i).FieldType))
        Next i
    End Sub
End Module
```

Discovering a Type's Field Information by Using `GetField` Method

The `Type` class also provides the `GetField` method for discovering a `Type`'s fields by name and an optional binding constraint. The `GetField` type comes in two overloaded versions. The simplest overload accepts the string name of the field for which to search. The search for the name is case-sensitive, and if the field is not found or the requested field is nonpublic, then `null` is returned. Listing 13.17 demonstrates how to search for a field by its name only.

LISTING 13.17

```
C#
public class Type_GetField {
  public int IntType;
  public bool boolType;
  public string stringType;
}

public class Test {
  public static void Main() {
    Type t = typeof(Type_GetField);
    FieldInfo field = t.GetField("stringType");

    MessageBox.Show(String.Format(
      "Found {0} {1} field {2} of type {3} ",
        (field.IsPublic ? "public" : "non-public"),
        (field.IsStatic ? "static" : "instance"),
        field.Name,
```

```
            field.FieldType));
    }
}

VB
Module Module1
    Public Class Type_GetFields
        Public intType As Int32
        Public boolType As Boolean
        Public stringType As String
    End Class

    Public Sub Main()
        Dim tgf = New Type_GetFields()
        Dim t = tgf.GetType()
        Dim field = t.GetField("stringType")
        Dim visibility As String

        If field.IsPublic Then
            visibility = "Public"
        Else
            visibility = "Non-Public"
        End If

        MessageBox.Show(String.Format( _
            "Found {0} field {1} of type {2} ", _
            visibility, _
            field.Name, _
            field.FieldType))
    End Sub
End Module
```

The second overload of GetField allows you to search for fields given a binding constraint. As with all reflection methods, the binding constraints are specified by the BindFlags enumeration. The BindingFlags values that apply to the GetField method can either define which fields to include in the search or change how the search works. Tables 13.6 and 13.7 describe both types of BindingFlags enumeration.

TABLE 13.6

BindingFlags Members that Define Which Methods Are Searched by GetField

MEMBER	MEANING
`Instance`	Includes instance fields in the search
`Static`	Includes static fields in the search
`Public`	Includes public fields in the search
`NonPublic`	Includes private and protected fields in the search
`FlattenHierarchy`	Includes static fields up the class hierarchy

TABLE 13.7

BindingFlags Members That Change How GetField Searches

MEMBER	MEANING
`IgnoreCase`	Ignores the case of the specified name
`DeclaredOnly`	Searches only the fields declared on the current Type and not fields that were inherited

As with `GetFields`, either `BindingFlags.Instance` or `BindingFlags.Static` must be supplied, or `null` will be returned. If a field that matches the specified binding constraints cannot be found, `null` is returned. Listing 13.18 demonstrates how to find a static field through a case-insensitive search.

LISTING 13.18

```
C#
public class Type_GetField {
  public static int StaticIntType;
  public static bool StaticBoolType;
  public static string StaticStringType;
  public static short StaticShortType;
}

public class Test {
  public static void Main() {
    Type t = typeof(Type_GetField);
    FieldInfo field =
      t.GetField("sTATICsHORTtYPE",
      BindingFlags.Static |
      BindingFlags.IgnoreCase |
      BindingFlags.Public);
```

```
      if(field == null) {
        MessageBox.Show("Could not find the field StaticShortType");
        return;
      }

      MessageBox.Show(String.Format(
        "Found {0} {1} field {2} of type {3} ",
        (field.IsPublic ? "public" : "non-public"),
          (field.IsStatic ? "static" : "instance"),
          field.Name,
          field.FieldType));
    }
  }

VB
Module Module1
    Public Class Type_GetField
        Public Shared SharedIntType As Int32
        Public Shared SharedBoolType As Boolean
        Public Shared SharedStringType As String
        Public Shared SharedShortType As Short
    End Class

    Public Sub Main()
        Dim tgf = New Type_GetField()
        Dim t = tgf.GetType()
        Dim field = t.GetField("sHaReDsHORTtYPE", _
                               BindingFlags.Static Or _
                               BindingFlags.IgnoreCase Or _
                               BindingFlags.Public)
        Dim visibility As String

        If field Is Nothing Then
            MessageBox.Show("Could not find the field StaticShortType")
            Return
        End If

        If field.IsPublic Then
            visibility = "Public"
        Else
            visibility = "Non-Public"
        End If
```

```
        MessageBox.Show(String.Format( _
            "Found {0} field {1} of type {2} ", _
            visibility, _
            field.Name, _
            field.FieldType))
    End Sub
End Module
```

Using the FieldInfo Class to Retrieve a Field's Value

Now that we have a FieldInfo object, we can use it to get or set a field's value. The FieldInfo object provides the GetValue method and the SetValue methods to expose this functionality.

The GetValue method takes a single parameter of type object. This object is the type instance whose field will be retrieved. The object should be an instance of a class that inherits or declares the field. If the field is static, then the object parameter is ignored. The GetValue method returns an object instance representing the field value. Listing 13.19 demonstrates retrieving a Type's public field.

LISTING 13.19

```
C#
public class FieldInfo_GetValue {
  public int intType = 10;
  public bool boolType = true;
  public string stringType = "Initial Value";
}

public class Test {
  public static void Main()
  {
    Type t = typeof(FieldInfo_GetValue);
    FieldInfo field = t.GetField("stringType");

    if(field == null) {
        MessageBox.Show("Could not find the field named stringType");
        return;
    }

    FieldInfo_GetValue obj = new FieldInfo_GetValue();
    object value = field.GetValue(obj);
```

```
        MessageBox.Show(
          string.Format("Found the value, '{0}', in the {1} field",
            value.ToString(),
            field.Name));
    }
}

VB
Module Module1
    Public Class Field_GetValue
        Public intType = 10
        Public boolType = True
        Public stringType = "Initial Value"
    End Class

    Public Sub Main()
        Dim tgv = New Field_GetValue()
        Dim t = tgv.GetType()
        Dim field = t.GetField("stringType")

        If field Is Nothing Then
            MessageBox.Show("Could not find the field stringType")
            Return
        End If

        Dim obj = New Field_GetValue()
        Dim value = field.GetValue(obj)

        MessageBox.Show( _
            String.Format("Found the value, '{0}', in the {1} field", _
            value.ToString(), _
            field.Name))
    End Sub
End Module
```

Using the `FieldInfo` Class to Change a Field's Value

Using the `SetValue` method is just as simple. Unlike `GetValue`, which takes one parameter, `SetValue` takes two. The first parameter is of type object, and it is the type instance whose field will be changed. If the field is static, then this parameter is ignored. Again, the object should be an instance of a class that inherits or declares the field. The second parameter is also of type object, but this object contains the new value to assign to the field. This object should contain an instance that is the same type as the field. Listing 13.20 demonstrates how to use the `SetValue` method.

LISTING 13.20

```
C#
public class FieldInfo_SetValue {
  public int IntType = 10;
  public bool boolType = true;
  public string stringType = "Initial Value";
}

public class Test {
  public static void Main() {
    Type t = typeof(FieldInfo_SetValue);
    FieldInfo field = t.GetField("stringType");

    if(field == null) {
      MessageBox.Show("Could not find the field named stringType");
      return;
    }

    FieldInfo_SetValue obj = new FieldInfo_SetValue();
    object value = field.GetValue(obj);
    MessageBox.Show(string.Format(
      "Found the value, '{0}', in the {1} " +
        "field before changing the field",
        value.ToString(),
        field.Name));

    string newValue = "New Value";
    field.SetValue(obj, newValue);
    value = field.GetValue(obj);
    MessageBox.Show(string.Format(
      "Found the value, '{0}', in the {1} " +
       "field after changing the field",
       value.ToString(),
       field.Name));
  }
}
```

```
VB
Module Module1
    Public Class FieldInfo_SetValue
        Public intType = 10
        Public boolType = True
        Public stringType = "Initial Value"
    End Class

    Public Sub Main()
        Dim tsv = New FieldInfo_SetValue()
        Dim t = tsv.GetType()
        Dim field = t.GetField("stringType")

        Dim obj = New FieldInfo_SetValue()
        Dim value = field.GetValue(obj)

        MessageBox.Show(String.Format( _
            "Found the value, '{0}', in the {1} field" & _
            " before changing the field", _
            value.ToString(), _
            field.Name))

        Dim newValue = "New Value"
        field.SetValue(obj, newValue)
        value = field.GetValue(obj)
        MessageBox.Show(String.Format( _
            "Found the value, '{0}', in the {1} " & _
            "field after changing the field", _
            value.ToString(), _
            field.Name))
    End Sub
End Module
```

Discovering a Type's Property Information by Using the `GetProperties` Method

Let us now discuss manipulating an object's properties. The `Type.GetProperties` method has two overloads. One overload takes no parameters and returns an array of `PropertyInfo` objects that represents all of the public properties defined for the `Type`. If there are no public properties defined, then an empty array of type `PropertyInfo` is returned. Listing 13.21 demonstrates how to use the `GetProperties` method to discover all of the public fields in a class.

LISTING 13.21

C#

```
public class Type_GetProperties {
  int intType = 0;
  bool boolType = false;
  string stringType = "Initial Value";

  public int Int {
    get{ return intType; }
    set{ intType = value; }
  }

  public bool Bool {
    set{ boolType = value; }
  }

  public string String {
    get{ return stringType; }
  }
}

public class Test {
  public static void Main() {
    Type t = typeof(Type_GetProperties);
    PropertyInfo[] properties = t.GetProperties();

    foreach(PropertyInfo property in properties)
    {
      MessageBox.Show(String.Format(
        "Found property {2} of type {3} with {0}{1} access",
        (property.CanRead ? "Read" : ""),
        (property.CanWrite ? "Write" : ""),
        property.Name,
        property.PropertyType));
    }
  }
}
```

VB

```
Module Module1
    Public Class Type_GetProperties
        Dim intType = 0
```

```
        Dim boolType = False
        Dim strType = "Initial Value"

        Public Property Int32Type() As Int32
            Get
                Return intType
            End Get
            Set(ByVal Value As Int32)
                intType = Value
            End Set
        End Property

        Public WriteOnly Property BooleanType() As Boolean
            Set(ByVal Value As Boolean)
                boolType = Value
            End Set
        End Property

        Public ReadOnly Property StringType() As String
            Get
                Return strType
            End Get
        End Property

    End Class

    Sub Main()
        Dim tgp = New Type_GetProperties()
        Dim t = tgp.GetType()
        Dim properties() = t.GetProperties()
        Dim i As Int32
        Dim canRead As String
        Dim canWrite As String

        For i = 0 To properties.Length - 1
            If properties(i).CanRead Then
                canRead = "Read"
            Else
                canRead = String.Empty
            End If

            If properties(i).CanWrite Then
                canWrite = "Write"
```

```
            Else
                canWrite = String.Empty
            End If
            MessageBox.Show(String.Format( _
             "Found property {2} of type {3} with {0}{1} access", _
             canRead, _
             canWrite, _
             properties(i).Name, _
             properties(i).PropertyType))
        Next i
    End Sub
End Module
```

The `GetProperties` method has an overload that allows you to specify the binding constraints by passing a `BindingFlags` value to the `GetProperties` method. The `BindingFlags` values can either define which fields to include in the search, values described in Table 13.8, or change how the search works, values described in Table 13.9.

TABLE 13.8
`BindingFlags` Members That Define Which Methods Are Searched by `GetProperties`

MEMBER	MEANING
`Instance`	Includes instance properties in the search
`Static`	Includes static properties in the search
`Public`	Includes public properties in the search
`NonPublic`	Includes private and protected properties in the search
`FlattenHierarchy`	Includes static properties up the class hierarchy

TABLE 13.9
`BindingFlags` Members That Change How `GetProperties` Searches

MEMBER	MEANING
`DeclaredOnly`	Searches only the properties declared on the current `Type` and not properties that were inherited

The meaning of `BindingFlags.Public` and `BindingFlags.NonPublic` is a little different for properties. A property is considered public to reflection if it has at least one accessor that is public. Otherwise, the property is considered private.

If a property that matches the specified binding constraints cannot be found, then an empty array is returned. Listing 13.22 demonstrates how to retrieve a list of static properties from a given type.

LISTING 13.22

```
C#
public class Type_GetProperties {
  int intType = 0;
  static bool boolType = false;
  static string stringType = "Initial Value";

  public int Int {
    get{ return intType; }
    set{ intType = value; }
  }

  public static bool Bool {
    set{ boolType = value; }
  }

  public static string String {
    get{ return stringType; }
  }
}

public class Test {
  public static void Main() {
    Type t = typeof(Type_GetProperties);
    PropertyInfo[] properties =
      t.GetProperties(BindingFlags.Static | BindingFlags.Public);

    foreach(PropertyInfo property in properties) {
        MessageBox.Show(String.Format(
          "Found property {0} of type {1} with {2}{3} access",
          property.Name,
          property.PropertyType,
          (property.CanRead ? "Read" : ""),
          (property.CanWrite ? "Write" : "")));
    }
  }
}

VB
Module Module1
    Public Class Type_GetProperties
        Dim intType = 0
```

```
    Shared boolType = False
    Shared strType = "Initial Value"

    Public Property Int32Type() As Int32
        Get
            Return intType
        End Get
        Set(ByVal Value As Int32)
            intType = Value
        End Set
    End Property

    Public Shared WriteOnly Property BooleanType() As Boolean
        Set(ByVal Value As Boolean)
            boolType = Value
        End Set
    End Property

    Public Shared ReadOnly Property StringType() As String
        Get
            Return strType
        End Get
    End Property

End Class

Sub Main()
    Dim tgp = New Type_GetProperties()
    Dim t = tgp.GetType()
    Dim properties() = _
      t.GetProperties(BindingFlags.Static Or BindingFlags.Public)
    Dim i As Int32
    Dim canRead As String
    Dim canWrite As String

    For i = 0 To properties.Length - 1
        If properties(i).CanRead Then
            canRead = "Read"
        Else
            canRead = String.Empty
        End If
```

```
            If properties(i).CanWrite Then
                canWrite = "Write"
            Else
                canWrite = String.Empty
            End If
            MessageBox.Show(String.Format( _
             "Found property {2} of type {3} with {0}{1} access", _
             canRead, _
             canWrite, _
             properties(i).Name, _
             properties(i).PropertyType))
        Next i
    End Sub
End Module
```

Discovering a Type's Field Information by Using the `GetProperty` Method

The `Type` class also provides the `GetProperty` method. The first overload we will discuss takes one parameter, the name of the property. This method searches all public properties attempting to match the name of the property to the specified name parameter. The search is case-sensitive, and if the property cannot be found, a `null` reference is returned. Listing 13.23 demonstrates how to use this method.

LISTING 13.23

```
C#
public class Type_GetProperty {
  int intType = 0;
  static bool boolType = false;
  static string stringType = "Initial Value";

  public int Int {
    get{ return intType; }
    set{ intType = value; }
  }

  public bool Bool {
    set{ boolType = value; }
  }
```

```
  public string String {
    get{ return stringType; }
  }
}

public class Test {
  public static void Main(string[] args) {
    Type t = typeof(Type_GetProperty);
    PropertyInfo property = t.GetProperty("Int");

    MessageBox.Show(String.Format(
      "Found property {0} of type {1} with {2}{3} access",
      property.Name,
      property.PropertyType,
      (property.CanRead ? "Read" : ""),
      (property.CanWrite ? "Write" : "")));
  }
}
```

VB

```
Module Module1
    Public Class Type_GetProperties
        Dim intType = 0
        Shared boolType = False
        Shared strType = "Initial Value"

        Public Property Int32Type() As Int32
            Get
                Return intType
            End Get
            Set(ByVal Value As Int32)
                intType = Value
            End Set
        End Property

        Public WriteOnly Property BooleanType() As Boolean
            Set(ByVal Value As Boolean)
                boolType = Value
            End Set
        End Property
```

```
        Public ReadOnly Property StringType() As String
            Get
                Return strType
            End Get
        End Property
    End Class

    Sub Main()
        Dim tgp = New Type_GetProperties()
        Dim t = tgp.GetType()
        Dim prop = t.GetProperty("Int32Type")
        Dim canRead As String
        Dim canWrite As String

        If prop.CanRead Then
            canRead = "Read"
        Else
            canRead = String.Empty
        End If

        If prop.CanWrite Then
            canWrite = "Write"
        Else
            canWrite = String.Empty
        End If
        MessageBox.Show(String.Format( _
            "Found property {2} of type {3} with {0}{1} access", _
            canRead, _
            canWrite, _
            prop.Name, _
            prop.PropertyType))
    End Sub
End Module
```

The second override allows for searching with given binding context specified by passing a `BindingFlags` enumeration value. The `BindingFlags` values that apply to the `GetProperty` method are specified in Tables 13.10 and 13.11.

TABLE 13.10

`BindingFlags` Members That Define Which Methods Are Searched by `GetProperty`

MEMBER	MEANING
`Instance`	Includes instance properties in the search
`Static`	Includes static properties in the search
`Public`	Includes public fields in the search
`NonPublic`	Includes private and protected properties in the search
`FlattenHierarchy`	Includes static properties up the class hierarchy

TABLE 13.11

`BindingFlags` Members That Change How `GetProperty` Searches

MEMBER	MEANING
`IgnoreCase`	Ignores the case of the specified name
`DeclaredOnly`	Searches only the properties declared on the current Type and not properties that were inherited

This method is demonstrated by the code in Listing 13.24.

LISTING 13.24

```
C#
public class Type_GetProperty {
  int intType = 0;
  static bool boolType = false;
  static string stringType = "Initial Value";

  public int Int {
    get{ return intType; }
    set{ intType = value; }
  }

  public static bool Bool {
    set{ boolType = value; }
  }

  public static string String {
    get{ return stringType; }
  }
}
```

```
public class Test {
  public static void Main() {
    Type t = typeof(Type_GetProperty);
    PropertyInfo property =
    t.GetProperty("bOOl",
    BindingFlags.Static ¦
    BindingFlags.IgnoreCase ¦
    BindingFlags.Public);

    MessageBox.Show(String.Format(
      "Found property {0} of type {1} with {2}{3} access",
      property.Name,
      property.PropertyType,
      (property.CanRead ? "Read" : ""),
      (property.CanWrite ? "Write" : "")));
  }
}
```

VB

```
Module Module1
    Public Class Type_GetProperty
        Dim intType = 0
        Shared boolType = False
        Shared strType = "Initial Value"

        Public Property Int32Type() As Int32
            Get
                Return intType
            End Get
            Set(ByVal Value As Int32)
                intType = Value
            End Set
        End Property

        Public Shared WriteOnly Property BooleanType() As Boolean
            Set(ByVal Value As Boolean)
                boolType = Value
            End Set
        End Property

        Public ReadOnly Property StringType() As String
            Get
```

```
                Return strType
            End Get
        End Property

    End Class

    Sub Main()
        Dim tgp = New Type_GetProperty()
        Dim t = tgp.GetType()
        Dim prop = t.GetProperty("BoOlEaNType", _
                                 BindingFlags.Static Or _
                                 BindingFlags.IgnoreCase Or _
                                 BindingFlags.Public)

        Dim canRead As String
        Dim canWrite As String

        If prop.CanRead Then
            canRead = "Read"
        Else
            canRead = String.Empty
        End If

        If prop.CanWrite Then
            canWrite = "Write"
        Else
            canWrite = String.Empty
        End If
        MessageBox.Show(String.Format( _
            "Found property {2} of type {3} with {0}{1} access", _
            canRead, _
            canWrite, _
            prop.Name, _
            prop.PropertyType))
    End Sub
End Module
```

Using the `PropertyInfo` Class to Retrieve and Change the Value of the Property

Now that we have a `PropertyInfo` object in hand, we can get and set its values by using the `PropertyInfo`'s `SetValue` and `GetValue`. These methods are very similar to their `FieldInfo` counterparts except that they each take an extra parameter of type `Object[]`. This object array represents index values for indexed properties. If the property is not indexed, then this parameter should be null. Listing 13.25 demonstrates how to use both the `GetValue` and `SetValue` methods.

LISTING 13.25

```
C#
public class PropertyInfo_SetValue {
  int intType = 0;
  bool boolType = false;
  string stringType = "Initial Value";

  public int Int {
    get{ return intType; }
    set{ intType = value; }
  }

  public bool Bool {
    set{ boolType = value; }
  }

  public string String {
    get{ return stringType; }
    set{ stringType = value; }
  }
}

public class Test {
  public static void Main(string[] args) {
    Type t = typeof(PropertyInfo_SetValue);
    PropertyInfo property = t.GetProperty("String");

    if(property == null) {
        MessageBox.Show("Could not find the property named String");
        return;
    }

    PropertyInfo_SetValue obj = new PropertyInfo_SetValue();
    object value = property.GetValue(obj, null);
    MessageBox.Show(string.Format(
        "Found the value, '{0}', in the {1} " +
        "property before changing the property",
```

```
            value.ToString(),
            property.Name));

        string newValue = "New Value";
        property.SetValue(obj, newValue, null);
        value = property.GetValue(obj, null);
        MessageBox.Show(string.Format(
            "Found the value, '{0}', in the {1} " +
            "property after changing the field",
            value.ToString(),
            property.Name));
    }
}
```

VB

```
Module Module1
    Public Class PropertyInfo_SetValue
        Dim intType = 0
        Shared boolType = False
        Shared strType = "Initial Value"

        Public Property Int32Type() As Int32
            Get
                Return intType
            End Get
            Set(ByVal Value As Int32)
                intType = Value
            End Set
        End Property

        Public WriteOnly Property BooleanType() As Boolean
            Set(ByVal Value As Boolean)
                boolType = Value
            End Set
        End Property

        Public Property StringType() As String
            Get
                Return strType
            End Get
            Set(ByVal Value As String)
                strType = Value
            End Set
```

```
        End Property
    End Class

    Sub Main()
        Dim tgp = New PropertyInfo_SetValue()
        Dim t = tgp.GetType()
        Dim prop = t.GetProperty("StringType")

        If prop Is Nothing Then
            MessageBox.Show("Could not find the property named String")
            Return
        End If

        Dim obj = New PropertyInfo_SetValue()
        Dim value = prop.GetValue(obj, Nothing)
        MessageBox.Show(String.Format( _
            "Found the value, '{0}', in the {1} " & _
            "property before changing the property", _
            value.ToString(), _
            prop.Name))

        Dim newValue = "New Value"
        prop.SetValue(obj, newValue, Nothing)
        value = prop.GetValue(obj, Nothing)
        MessageBox.Show(String.Format( _
            "Found the value, '{0}', in the {1} " & _
            "property after changing the field", _
            value.ToString(), _
            prop.Name))
    End Sub
End Module
```

Adding Custom Metadata to .NET Elements

The Reflection APIs work by querying the metadata stored in a .NET assembly. Custom attributes are a simple way to extend the metadata of any given managed element. Using custom attributes, you can add extra information to an assembly's metadata and then query for this extra information at runtime.

Defining a Custom Attribute

A custom attribute is a declarative programming construct that allows you to extend a language element's metadata. This information is stored in an assembly's metadata and can be retrieved at runtime. A corresponding attribute class must exist before an attribute can be used to decorate a language element. All attribute classes inherit from `System.Attribute`. The attribute class contains properties that store and retrieve the extra declared metadata. Listing 13.26 demonstrates how to define a custom attribute.

LISTING 13.26

```
C#
[AttributeUsage(AttributeTargets.Class¦AttributeTargets.Struct)]
class BusinessObjectAttribute : System.Attribute {
  private string m_DBName;
  private string m_TableName;
  private string m_QueryString;

  public BusinessObjectAttribute(string dbName,
                                 string tableName) {
    m_DBName = dbName;
    m_TableName = tableName;
  }

  public string Database{
    get{ return m_DBName; }
    set{ m_DBName = value; }
  }

  public string Table {
    get{ return m_TableName; }
    set{ m_TableName = value; }
  }

  public string QueryString {
    get{ return m_QueryString; }
    set{ m_QueryString = value; }
  }
}

VB
<AttributeUsage(AttributeTargets.Class Or AttributeTargets.Struct)> _
Class BusinessObjectAttribute
```

```
    Inherits System.Attribute

    Private m_DBName As String
    Private m_TableName As String
    Private m_QueryString As String

    Public Sub New(ByVal dbName As String, _
                   ByVal tableName As String)
        m_DBName = dbName
        m_TableName = tableName
    End Sub

    Public Property Database() As String
        Get
            Return m_DBName
        End Get
        Set(ByVal Value As String)
            m_DBName = Value
        End Set
    End Property

    Public Property Table() As String
        Get
            Return m_TableName
        End Get
        Set(ByVal Value As String)
            m_TableName = Value
        End Set
    End Property

    Public Property QueryString() As String
        Get
            Return m_QueryString
        End Get
        Set(ByVal Value As String)
            m_QueryString = Value
        End Set
    End Property
End Class
```

The code declares a new `Attribute` named `BusinessObjectAttribute`. This attribute is intended to be applied to classes or structs. Astute readers will notice that the attribute class itself is in turn decorated with an attribute, `AttributeUsage`. The `AttributeUsage` attribute describes how a custom attribute can be used. The `AttributeUsage` has three properties: the required `AttributeTarget` property, the optional `AllowMultiple` property, and the optional `Inherited` property.

The `AttributeTargets` property specifies the language elements on which the attribute can be applied. The values of the `AttributeTargets` enumeration can be combined to specify multiple targets. In the previous example, the `BusinessObjectAttribute` can be applied to classes and structs. Table 13.12 shows all the possible `AttributeTargets` values.

TABLE 13.12

`AttributeTargets` Enumeration Value

MEMBER
All
Assembly
Class Module
Constructor
Delegate Struct
Enum
Field
Interface
Method
Module
Parameter
Event
Property
ReturnValue
Struct

CLARIFY WHAT LANGUAGE ELEMENT AN `Attribute` APPLIES TO

Usually, the attribute will directly precede the language element to which it applies. However, position of the attribute is not always enough to determine to which element the attribute applies. For instance, consider this snippet:

```
C#
[Attribute()]
public int Function(int) {…}
```

In this instance, there is no way to tell whether the attribute is intended for the method element or for the method element's return value. To clarify which element the attribute applies to, you prefix the attribute name with the AttributeTargets enumeration value that describes which language element to which it applies.

```
C#
[returnvalue:Attribute]
public int Function(int) {…}
```

The AllowMultiple property specifies whether the attribute can be used more than once on the same language element. This value is option and false by default. In the preceding example, the BusinessObjectAttribute can appear only once on the same class or struct.

The inherited property specifies whether the attribute is inherited by derived classes. This value is optional and is false by default. In the example the BusinessObjectAttribute will not be inherited by classes that derive from a class that is decorated with this attribute.

All custom attributes must inherit from the System.Attribute class. Also, attributes should use the Attribute suffix in their names. When the attribute is used, the Attribute suffix does not need to be included. Instead, the class name minus the Attribute suffix becomes an alias for the class. Listing 13.27 demonstrates how the BusinessObjectAttribute class can be applied to a class.

LISTING 13.27

```
C#
[BusinessObject("CustomerRecordsDB", "Customers")]
class Customer {
}

VB
<BusinessObject("CustomerRecordsDB", "Customers")> _
Public Class Customer
End Class
```

Here the BusinessObject name is actually an alias for the BusinessObjectAttribute class. It is no mistake that the snippet's attribute declaration appears similar to a construction call. The two parameters correspond to the two parameters to the attribute class's constructor. The parameters must appear in the same order as the parameters in the constructor declaration. It is also possible to specify properties that do not appear in the constructor's parameter list. Listing 13.28 demonstrates how to initialize the QueryString property of the BusinessObjectAttribute when it is applied to a class.

LISTING 13.28

```
C#
[BusinessObject("CustomerRecordsDB",
                "Customers",
                QueryString="SELECT * FROM CUSTOMERS")]
class Customers {
}

VB
<BusinessObject("CustomerRecordsDB", _
    "Customers", _
    QueryString="SELECT * FROM CUSTOMERS")> _
Public Class Customer
End Class
```

This code demonstrates how to use a named parameter to initialize a property that does not appear in the attributes constructor. The name of the named parameter corresponds to the property's accessor name, not the field name.

Retrieving Custom Attributes

Now that you have defined a custom attribute and applied the attribute to a language element, it is time to retrieve that attribute at runtime. You can retrieve custom attributes by using the `GetCustomAttributes` methods of the `System.MemberInfo` class. This method comes in these two flavors:

```
object [] GetCustomAttributes(bool)
object [] GetCustomAttributes(Type, bool)
```

To retrieve all of the custom attributes of a class, you can use the `GetCustomAttributes` method that takes one bool parameter. This `bool` parameter specifies whether to search the member's inheritance chain to find the attributes. This method returns either an array of all the custom attributes or an array of zero elements if no attributes are defined. Listing 13.29 demonstrates how to use the `GetCustomAttributes` method.

LISTING 13.29

```
C#
[BusinessObject("CustomerRecordsDB", "Customers")]
class Customer {
}

class CustomAttributeTest {
  public static void Main() {
```

```
        Type customerType = typeof(Customer);
        Object[] atts = customerType.GetCustomAttributes(false);
        foreach(Attribute att in atts) {
          if(att is BusinessObjectAttribute) {
            BusinessObjectAttribute b = (BusinessObjectAttribute)att;
              MessageBox.Show("Database: " + b.Database +
                              "\nTable: " + b.Table);
            }
        }
    }
}

VB
Module Module1
    <BusinessObject("CustomerRecordsDB", "Customers")> _
    Public Class Customer
    End Class

    Sub Main()
        Dim cust As New Customer()
        Dim customerType = cust.GetType()
        Dim atts() = customerType.GetCustomAttributes(False)
        Dim i As Int32

        For i = 0 To atts.Length - 1
            If TypeOf atts(i) Is BusinessObjectAttribute Then
                Dim b = CType(atts(i), BusinessObjectAttribute)
                MessageBox.Show("Database: " & b.Database & _
                       Chr(13) & "Table: " + b.Table)
            End If
        Next i
    End Sub
End Module
```

To retrieve all of the custom attributes of a class that can be assigned to a given type, use the `GetCustomAttributes` method that takes two parameters. The first parameter is the type of the attribute for which to search. Only attributes that are assignable to this type will be returned. The second parameter specifies whether to search the member's inheritance chain to find the attribute. This parameter is identical to the sole parameter of the other `GetCustomAttributes` method overload. This method returns either an array of all the custom attributes or, if no attributes that are assignable to the specified type are defined, an array of zero elements. Listing 13.30 demonstrates how to use `GetCustomAttributes` to retrieve only `BusinessObjectAttribute` attributes.

LISTING 13.30

```
C#
[BusinessObject("CustomerRecordsDB", "Customers")]
class Customer {
}

class CustomAttributeTest {
  public static void Main() {
    Type customerType = typeof(Customer);
    Object[] atts = customerType.GetCustomAttributes(
        typeof(BusinessObjectAttribute), false);

    foreach(BusinessObjectAttribute att in atts) {
      MessageBox.Show("Database: " + att.Database +
                      "\nTable: " + att.Table);
    }
  }
}

VB
Module Module1
  <BusinessObject("CustomerRecordsDB", "Customers")> _
  Public Class Customer
  End Class

  Sub Main()
    Dim boa = New BusinessObjectAttribute(String.Empty, String.Empty)
    Dim cust As New Customer()
    Dim customerType = cust.GetType()
    Dim atts() = customerType.GetCustomAttributes(boa.GetType(), False)
    Dim i As Int32

    For i = 0 To atts.Length - 1
      If TypeOf atts(i) Is BusinessObjectAttribute Then
        Dim b = CType(atts(i), BusinessObjectAttribute)
        MessageBox.Show("Database: " & b.Database & _
                        Chr(13) & "Table: " + b.Table)
      End If
    Next i
  End Sub
End Module
```

In Brief

- An assembly can be used to discover information about the type it contains.
- A `System.Type` object can be used to discover information about its constructors, methods, properties, and fields.
- A `ConstructorInfo` object can be used to create object instances.
- A `MethodInfo` object can be used to invoke methods on a given type.
- A `PropertyInfo` object and a `FieldInfo` object can be used to manipulate an object's state.
- Custom attributes can be defined to extend the metadata of a language element.
- Custom attributes can also be used to retrieve extended metadata at runtime.

Cryptography

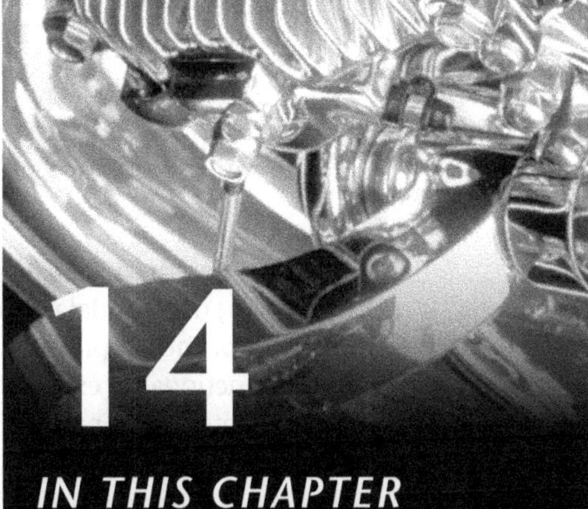

IN THIS CHAPTER

Cryptography on the .NET Compact Framework

The full .NET Framework includes rich support for cryptographic functions, such as computing hashes and encrypting data using a variety of algorithms. For example, developers can use the RC2 or RC4 algorithms for data encrtption and the MD5 algorithm for computing hashes. In contrast, the .NET Compact Framework contains only enough cryptography support to allow the `HttpWebRequest` to access secure Web pages (see Chapter 5, "Network Connectivity with the .NET Compact Framework").

Most Windows CE devices and all Pocket PC devices include an implementation of the CryptoAPI, which is a native code library that performs a wide variety of cryptographic functions. Its architecture allows developers to add new algorithm packages, thus extending the cryptography capabilities while retaining a familiar API for accessing new algorithms.

Because the .NET Compact Framework itself does not support most cryptography functions, performing cryptography requires platform invoking into the CryptoAPI. Readers will need a good understanding of how platform invoking works on the .NET Compact Framework. This subject is covered in detail in Chapter 12, "Interacting with Native Code."

Performing Cryptography with the `ManagedCryptoAPI` Class

The CryptoAPI is a very rich programming interface, but it is rife with pitfalls and has a very steep learning curve.

This chapter presents a deep enough explanation of the CryptoAPI for you to understand how to encrypt and decrypt data, share session keys, and compute hashes. To isolate developers from complexities of the CryptoAPI, this chapter uses a wrapper class called `ManagedCryptoAPI` to access the functionality exposed by the CryptoAPI. The `ManagedCryptoAPI` wrapper class provided with this chapter is not part of the .NET Compact Framework. All of the examples in this chapter use the `ManagedCryptoAPI` class to perform cryptography-related work.

The `ManagedCryptoAPI` wrapper class demonstrates the central concepts needed to understand the CryptoAPI. CryptoAPI is complex enough to deserve a book in its own right, and there are some optimizations that `ManagedCryptoAPI` does not employ. Specifically, `ManagedCryptoAPI` always allocates enough memory before making calls into CryptoAPI to ensure that the calls always succeed. CryptoAPI supports a more advanced approach in which programs inquire how much memory would be required for a given call into CryptoAPI to succeed, allocate the needed memory, and then make the call. Because this level of complexity would obfuscate the central concepts that `ManagedCryptoAPI` is meant to demonstrate, such optimizations are omitted.

Invoking the CryptoAPI

The CryptoAPI is a Windows CE API for performing common cryptographic functions such as computing hashes and encrypting or decrypting data. The .NET Compact Framework has no support for cryptographic functions that are accessible to the outside user. Thus, .NET Compact Framework developers must invoke the native CryptoAPI from their managed applications in order to perform cryptographic functions.

The CryptoAPI is a brittle API that will cause confusion for developers who don't understand some basic principles behind its use. The following sections review those basics before the rest of the chapter jumps into using the CryptoAPI.

Using Handles to Interact with CryptoAPI

Like many other functional areas in the Windows operating system, the CryptoAPI is driven by the concept of a handle. For example, encrypting data requires a handle to the cryptographic provider and a handle to the specific encryption key to be used. Programming against the CryptoAPI is really just the process of acquiring the correct handles from some functions and passing them in as arguments to other functions to achieve the desired results. However, the rules for using the handles correctly are very specific, and function calls fail if the handles are used incorrectly.

Understanding the CryptoAPI Context

The CryptoAPI architecture provides a standard interface for calling into cryptography functions, regardless of what the underlying algorithms are. Groups of algorithms are bundled together into packages called cryptographic service providers, or CSPs. The CSP used in the `ManagedCryptoAPI` wrapper class is the `PROV_RSA_FULL` provider. `PROV_RSA_FULL` provides encryption and hashing algorithms that are patented by the RSA and provided by Microsoft under license. It is a good general-purpose provider, and it is ubiquitous on Windows CE devices.

Other commonly used CSPs are shown in Table 14.1. Additionally, developers can write their own CSPs, and then anyone can access their algorithms through CryptoAPI.

Not all of the CSPs shown in Table 14.1 are included by default with Windows CE and Pocket PC devices. It is because the `PROV_RSA_FULL` provider is so common that we use it exclusively in the `ManagedCryptoAPI` wrapper.

TABLE 14.1

Cryptographic Service Providers Commonly Used with CryptoAPI

PROV_RSA_FULL	PROV_RSA_SIG
PROV_DSS	PROV_DSS_DH
PROV_SSL	PROV_EC_ECDSA_SIG
PROV_EC_ECNRA_SIG	PROV_EC_ECDSA_FULL
PROV_SPYRUS_LYNKS	PROV_FORTEZZA
PROV_MSEXCHANGE	PROV_RSA_CHANNEL

To perform any cryptographic function with the CryptoAPI, you must first acquire a context by calling `CryptAcquireContext`. Acquiring a context serves two purposes. First, it communicates to the CryptoAPI which CSP you want to use. Second, it tells the CryptoAPI which key container to use. A *key container* is a location where encryption keys are internally stored by CryptoAPI. As a developer, you usually interact with a key through its handle, and you let the CryptoAPI store the bytes of the key internally.

The DllImport definition for CryptAcquireContext in `ManagedCryptoAPI` looks like this:

```
C#
[DllImport("coredll.dll")]
private static extern bool CryptAcquireContext(ref IntPtr phProv,
        string pszContainer, string pszProvider, Int32 dwProvType,
        uint dwFlags);


VB
Declare Function CryptAcquireContext Lib "coredll.dll" (ByRef phProv As IntPtr,
        ByVal pszContainer As String, ByVal pszProvider As String, ByVal
        dwProvType As UInt32, ByVal dwFlags As UInt32) As Boolean
```

When acquiring a context, you may choose to use the default key container, or you may pass in a string name for a key container. It is strongly recommended that you pass in a name for the key container because other software on your device could already be using the default key container and you could inadvertently clobber the keys used by the other software. Specifically, the Pocket PC driver software for several brands of wireless network cards uses the default key container. If you also use the default container, you risk clobbering the keys that the driver software uses to encrypt wireless traffic, which would cause your wireless card to stop working suddenly.

`ManagedCryptoAPI` has two methods for acquiring a context. `AcquireDefaultContext` returns an `IntPtr`, which holds the handle to the default key container for the `PROV_RSA_FULL` provider. `AcquireNamedContext` returns an `IntPtr`, which holds the handle to a named key container that you pass in as a string. Programs retain the returned handle and pass it in to other methods that are discussed later in this chapter. The code for `AcquireDefaultContext` is shown in Listing 14.1.

LISTING 14.1 `AcquireDefaultContext`

```
C#
public IntPtr AcquireDefaultContext()
{
   IntPtr hProvider = IntPtr.Zero;

   if (!CryptAcquireContext(ref hProvider, null, "Microsoft Base Cryptographic
            Provider v1.0", PROV_RSA_FULL, 0))
   {
      // We might have to create a new keyset...
      if (!CryptAcquireContext(ref hProvider, null, "Microsoft Base
              Cryptographic Provider v1.0", PROV_RSA_FULL, CRYPT_NEWKEYSET))
      {
         // Big trouble, could not create a new
         // keyset and could not acquire default keyset
         throw new Exception("ManagedCryptoAPI cannot access default
                  keyset or create new default keyset!");
      }
   }
   return hProvider;
}

VB
   Public Function AcquireDefaultContext() As IntPtr
      Dim hProvider As IntPtr = IntPtr.Zero

      If (CryptAcquireContext(hProvider, Nothing, "Microsoft Base Cryptographic
               Provider v1.0", PROV_RSA_FULL, Convert.ToUInt32(0)) = False) Then
```

```
      Dim l_Failure As String = TranslateErrorCode(Convert.ToInt64
            (GetLastError()))
      ' We might have to create a new keyset...
      If (CryptAcquireContext(hProvider, Nothing, "Microsoft Base Cryptographic
            Provider v1.0", PROV_RSA_FULL, CRYPT_NEWKEYSET) = False) Then
         ' Big trouble, could not create a new keyset and could not acquire
         ' default keyset
         Throw New Exception("ManagedCryptoAPI cannot access default keyset or
               create new default keyset!" + TranslateErrorCode
               (Convert.ToInt64(GetLastError())))
      End If

   End If
   Return hProvider
End Function
```

The code for `AcquireNamedContext` is in Listing 14.2.

LISTING 14.2 `AcquireNamedContext`

```
C#
public IntPtr AcquireNamedContext(string in_ContainerName)
{
   IntPtr hProvider = IntPtr.Zero;
   if (!CryptAcquireContext(ref hProvider, in_ContainerName,
         "Microsoft Base Cryptographic Provider v1.0", PROV_RSA_FULL,
         0))
   {
      if (!CryptAcquireContext(ref hProvider, in_ContainerName,
            "Microsoft Base Cryptographic Provider v1.0", PROV_RSA_FULL,
            CRYPT_NEWKEYSET))
      {
         throw new Exception("ManagedCryptoAPI cannot access default
         keyset or create new default keyset!");
      }
   }
   return hProvider;
}

VB
Public Function AcquireNamedContext(ByVal in_ContainerName As String) As IntPtr
   Dim hProvider As IntPtr = IntPtr.Zero
```

```
    If (CryptAcquireContext(hProvider, in_ContainerName, "Microsoft Base
            Cryptographic Provider v1.0", PROV_RSA_FULL,
            Convert.ToUInt32(0)) = False) Then
        Dim l_Failure As String = TranslateErrorCode(Convert.ToInt64
                (GetLastError()))
        ' We might have to create a new keyset...
        If (CryptAcquireContext(hProvider, in_ContainerName, "Microsoft Base
                Cryptographic Provider v1.0", PROV_RSA_FULL, CRYPT_NEWKEYSET) = False)
                Then
            ' Big trouble, could not create a new keyset and could not acquire
            ' default keyset
            Throw New Exception("ManagedCryptoAPI cannot access default keyset or
                    create new default keyset!" + TranslateErrorCode(Convert.ToInt64
                    (GetLastError())))
        End If
    End If

    Return hProvider
End Function
```

Avoiding Pitfalls Associated with `CryptAcquireContext`

`CryptAcquireContext` returns a handle to a key container within a CSP. If the context for the key container that is asked for has never been acquired in the past, then the key container does not exist yet. In this case the key container must be created. `AcquireDefaultContext` and `AcquireNamedContext` both check for failure when calling `CryptAcquireContext` and try to create the key container if the first call fails. If the attempt to create the key container fails, then the methods throw an exception.

It is not a good practice to always attempt to create a new key container when calling `CryptAcquireContext`, because if the key container already exists, then the call to `CryptAcquireContext` will fail. Thus, it is very important to try to access the key container, assuming it does exist, and to try to create it only if necessary.

Computing a Hash

The `ManagedCryptoAPI` class includes a method called `ManagedComputeHash`, which computes a hash for an array of bytes. To use `ManagedComputeHash`, follow these steps:

1. Acquire an instance of the `ManagedCryptoAPI` class.
2. Use the `ManagedCryptoAPI` instance to acquire a handle to a key container within a CSP.
3. Pass the context handle, the array of bytes for which to compute a hash, and a reference to an `Int32` to the `ManagedComputeHash` method.

4. `ManagedComputeHash` returns an array of bytes holding the hash bytes and sets the reference to the `Int32` to the number of bytes in the returned array that hold valid hash data.

Listing 14.3, taken from the sample application ComputeHash, demonstrates computing a hash with `ManagedCryptoAPI`. The code first acquires the hash bytes and then paints them into a textbox.

LISTING 14.3 Computing a hash with `ManagedCryptoAPI`

```
C#
ManagedCryptoAPI l_Crypto = new ManagedCryptoAPI();
IntPtr l_hProvider = l_Crypto.AcquireNamedContext("KICKSTART");
Int32 l_TotalHashBytes = 0;

byte[] l_TextBytes =
        System.Text.Encoding.ASCII.GetBytes(this.txtTextToHash.Text);

byte[] l_HashBytes = l_Crypto.ManagedComputeHash(l_hProvider,
        l_TextBytes, ref l_TotalHashBytes);

this.txtHashedBytes.Text = "";
for (int i = 0; i < l_TotalHashBytes; i++)
{
   this.txtHashedBytes.Text += "[" + l_HashBytes[i] + "] ";
}

VB
Dim l_Crypto As ManagedCryptoAPI = New ManagedCryptoAPI
Dim l_hProvider As IntPtr = l_Crypto.AcquireNamedContext("KICKSTART")
Dim l_TotalHashBytes As Int32 = 0

Dim l_TextBytes() As Byte = System.Text.Encoding.ASCII.GetBytes
        (Me.txtTextToHash.Text)

Dim l_HashBytes() As Byte = l_Crypto.ManagedComputeHash(l_hProvider,
        l_TextBytes, l_TotalHashBytes)

Me.txtHashedBytes.Text = ""
Dim i As Integer
For i = 0 To l_TotalHashBytes - 1
   Me.txtHashedBytes.Text += "[" + Convert.ToString(l_HashBytes(i)) + "] "
Next i
```

Looking Inside ManagedComputeHash

Examining the code inside ManagedComputeHash teaches specifically how to compute a hash through CryptoAPI. ManagedComputeHash will also serve as a case study to observe how ManagedCryptoAPI manipulates the CryptoAPI. The discussion will not drill down into every method in ManagedCryptoAPI, because it would fill the chapter with redundant information. Instead, there is only a drill for computing hashes and encrypting data by using a password. Other parts of the chapter will demonstrate how to perform cryptographic actions with the help of ManagedCryptoAPI without examining the internal actions performed by ManagedCryptoAPI.

To compute the hash for an array of bytes, ManagedComputeHash follows these steps:

1. ManagedComputeHash creates a buffer large enough to hold the hash data that it will compute.
2. A new hash object is created by calling CryptCreateHash. CryptCreateHash accepts a handle to a key container in a CSP, returns true if it has succeeded, and returns the handle to a hash object through a reference to an IntPtr. The handle to the key container is passed into ManagedComputeHash. ManagedCryptoAPI uses the CALG_MD5 hash algorithm. The DllImport statement for CryptCreateHash is as follows:

```
C#
[DllImport("coredll.dll")]
private static extern bool CryptCreateHash(IntPtr hProv, uint Algid,
        IntPtr hKey, uint dwFlags, ref IntPtr phHash);

VB
Declare Function CryptCreateHash Lib "coredll.dll" (ByVal hProv As
        IntPtr, ByVal Algid As Int32, ByVal hKey As IntPtr, ByVal
        dwFlags As UInt32, ByRef phHash As IntPtr) As Boolean
```

3. ManagedComputHash calls CryptHashData to hash the input bytes. CryptHashData accepts the handle to the hash algorithm that was derived in step 2. The CryptoAPI computes the hash and retains the resulting hash bytes internally. The DllImport statement for CryptHashData follows:

```
C#
private static extern bool CryptHashData(IntPtr hHash, byte[] pbData,
        Int32 dwDataLen, uint dwFlags);

VB
Declare Function CryptHashData Lib "coredll.dll" (ByVal hHash As
        IntPtr, ByVal pbData() As Byte, ByVal dwDataLen As Int32,
        ByVal dwFlags As UInt32) As Boolean
```

4. ManagedComputeHash acquires the resulting hash bytes by calling CryptGetHashParam. This is the DllImportStatement for CryptGetHashParam:

```
C#
private static extern bool CryptGetHashParam(IntPtr hHash,
        uint dwParam, byte[] pbData, ref Int32 out_NumHashBytes,
        uint dwFlags);

VB
Declare Function CryptGetHashParam Lib "coredll.dll" (ByVal hHash As
        IntPtr, ByVal dwParam As UInt32, ByVal pbData() As Byte,
        ByRef out_NumHashBytes As Int32, ByVal dwFlags
        As UInt32) As Boolean
```

5. The resulting hash bytes are returned to the caller.

Programs can call the CryptHashData function repeatedly before acquiring the resulting hash bytes. Each time CryptHashData is called, the internal hash result is updated to reflect the act of hashing additional data. For example, a program that computes the hash for a large text file could call CryptHashData on each line of text and acquire the resulting hash bytes only after the last line of data has been hashed.

The code for ManagedComputeHash is shown in Listing 14.4.

LISTING 14.4 ManagedComputeHash

```
C#
// Computes a hash of the bytes passed in
public byte[] ManagedComputeHash(IntPtr in_hProvider, byte [] in_DataToHash,
        ref Int32 out_NumHashBytes)
{
   byte [] l_HashBuffer = new byte[MAX_HASH];
   out_NumHashBytes = (int)MAX_HASH;

   try
   {
      // Step 1: Get a handle to a new hash object that will
      // hash the password bytes
      IntPtr l_hHash = IntPtr.Zero;

      if (!CryptCreateHash(in_hProvider, CALG_MD5, IntPtr.Zero, 0, ref l_hHash))
      {
         throw new Exception("Could not create a hash object!");
      }
```

```
        // Step 2: hash the password data....
        // l_hHash - reference to hash object   in_passwordBytes -
        // bytes to add to hash
        // in_passwordBytes.Length - length of data to compute hash on
        // 0 - extra flags
        if (!CryptHashData(l_hHash, in_DataToHash, in_DataToHash.Length, 0))
        {
            throw new Exception("Failure while hashing password bytes!");
        }

        // Step 3: Retrieve the hash bytes
        if (!CryptGetHashParam(l_hHash,  HP_HASHVAL, l_HashBuffer,
                ref out_NumHashBytes, 0))
        {
            throw new Exception("Failure when retrieving hash bytes with
                CryptGetHashParam!");
        }
    }
    finally
    {
        // Release hash object.
        CryptDestroyHash(hHash);
    }

    return l_HashBuffer;
}

VB
' Computes a hash of the bytes passed in
Public Function ManagedComputeHash(ByVal in_hProvider As IntPtr, ByVal in_
        DataToHash() As Byte, ByRef out_NumHashBytes As Int32) As Byte()

    Dim l_HashBuffer(Convert.ToInt32(MAX_HASH)) As Byte
    out_NumHashBytes = Convert.ToInt32(MAX_HASH)

    Try
        ' Step 1: Get a handle to a new hash object that will hash the
        ' password bytes
        Dim l_hHash As IntPtr = IntPtr.Zero
```

```
        If (CryptCreateHash(in_hProvider, Convert.ToInt32(CALG_MD5),
                IntPtr.Zero, Convert.ToUInt32(0), l_hHash) = False) Then
            Throw New Exception("Could not create a hash object!")
        End If

        ' Step 2: hash the password data....
        ' l_hHash - reference to hash object   in_passwordBytes - bytes to add to hash
        ' in_passwordBytes.Length - length of data to compute hash on
        ' 0 - extra flags
        If (CryptHashData(l_hHash, in_DataToHash, in_DataToHash.Length,
                Convert.ToUInt32(0)) = False) Then
            Throw New Exception("Failure while hashing password bytes!")
        End If

        ' Step 3: Retrieve the hash bytes
        If (CryptGetHashParam(l_hHash, HP_HASHVAL, l_HashBuffer, out_
                NumHashBytes, Convert.ToUInt32(0)) = False) Then
            Throw New Exception("Failure when retrieving hash bytes with
                    CryptGetHashParam!")
        End If
    Finally
        ' Release hash object.
        CryptDestroyHash(hHash);
    End Try
    Return l_HashBuffer
End Function
```

Computing a Hash with a Sample Application

The ComputeHash sample application is located in the folder `SampleApplications\Chapter14`. There is a C# and a Visual Basic version. ComputeHash is a stand-alone application that demonstrates how to compute a hash by using the techniques described previously. To use this program, enter text into the textbox labeled Text to Hash and then press the Compute Hash button. A hash for the ASCII encoding of the input bytes is computed and displayed in the Hash Bytes textbox.

Encrypting and Decrypting Data by Using a Password

Encrypting and decrypting data is achieved by calling `CryptEncrypt` and `CryptDecrypt`. These two functions require as an input a handle to a key that is used for encrypting or decrypting the data. The easiest way to get a handle to a key is to derive one using a password. This section discusses how to perform encryption and decryption by deriving the keys with passwords.

BE CAREFUL WITH PASSWORD BASED ENCRYPTION

Although password-based encryption and decryption is convenient, the resulting encrypted text is not necessarily portable to other devices. If your application calls for encrypting and decrypting data that can be securely exchanged between two devices, see the section titled "Encrypting and Decrypting by Using a Handle to a Session Key."

The simplest way to perform encryption and decryption with a password is to use the `ManagedCryptoAPI` class to do all of the work for you. To encrypt data using a password as the encryption key, follow these steps:

1. Acquire an instance of the `ManagedCryptoAPI` class.
2. Acquire byte arrays for text to encrypt and acquire the text password.
3. Acquire a handle to a key container in a CSP by calling `ManagedCryptoAPI.AcquireNamedContext` or `ManagedCryptoAPI.AcquireDefaultContext`.
4. Acquire the encrypted bytes by calling `ManagedCryptoAPI.PasswordEncrypt`.
5. `PasswordEncrypt` returns an array of bytes holding the encrypted data. The `Int32` that is passed by reference into `PasswordEncrypt` holds the number of bytes in the returned array that are actual encrypted data. The size of the returned array can be greater than the number of bytes that actually hold encrypted data.

Listing 14.5 demonstrates how to encrypt data by using a password. It is taken from the sample application EncryptionDemo. The sample code acquires the data to be encrypted and the password from the user interface and then paints the encrypted byte values into the user interface.

LISTING 14.5 How to encrypt data by using a password

```
C#
ManagedCryptoAPI l_Crypto = new ManagedCryptoAPI();

byte[] l_PlainTextBytes = System.Text.Encoding.ASCII.GetBytes
        (this.txtToEncrypt.Text);
byte[] l_PasswordBytes = System.Text.Encoding.ASCII.GetBytes
        (this.txtPassword.Text);

IntPtr l_hProvider = l_Crypto.AcquireNamedContext("KICKSTART");
m_TotalEncryptedBytes = 0;
m_EncryptedBytes = l_Crypto.PasswordEncrypt(l_PlainTextBytes, l_hProvider,
        l_PasswordBytes, ref m_TotalEncryptedBytes);
```

```
this.txtEncryptedBytes.Text = "";
for (int i = 0; i < m_TotalEncryptedBytes; i++)
{
    this.txtEncryptedBytes.Text += "[" + m_EncryptedBytes[i] + "] ";
}

VB
Dim l_Crypto As ManagedCryptoAPI = New ManagedCryptoAPI

Dim l_PlainTextBytes() As Byte = System.Text.Encoding.ASCII.GetBytes
         (Me.txtToEncrypt.Text)
Dim l_PasswordBytes() As Byte = System.Text.Encoding.ASCII.GetBytes
         (Me.txtPassword.Text)

Dim l_hProvider As IntPtr = l_Crypto.AcquireNamedContext("KICKSTART")
m_TotalEncryptedBytes = 0
m_EncryptedBytes = l_Crypto.PasswordEncrypt(l_PlainTextBytes, l_hProvider,
        l_PasswordBytes, m_TotalEncryptedBytes)

Me.txtEncryptedBytes.Text = ""

Dim i As Integer
For i = 0 To m_TotalEncryptedBytes - 1
    Me.txtEncryptedBytes.Text += "[" + Convert.ToString(m_EncryptedBytes(i))
         + "] "
Next i
```

To decrypt data by using a password as the encryption key, follow these steps:

1. Acquire an instance of the `ManagedCryptoAPI` class.
2. Acquire byte arrays for text to decrypt and the password.
3. Acquire a handle to a key container in a CSP by calling `ManagedCryptoAPI.AcquireNamedContext` or `ManagedCryptoAPI.AcquireDefaultContext`.
4. Decrypt the data by calling `ManagedCryptoAPI.PasswordDecrypt`.
5. `PasswordDecrypt` returns an array of bytes holding the decrypted data. The `Int32` that is passed by reference into `PasswordDecrypt` holds the number of bytes in the returned array that are actual decrypted data. The size of the returned array can be greater than the number of bytes that actually hold decrypted data.

Listing 14.6 demonstrates how to decrypt data by using a password as the decryption key. It is taken from the sample application EncryptionDemo.

LISTING 14.6 How to decrypt data using a password as the decryption key

```
C#
ManagedCryptoAPI l_Crypto = new ManagedCryptoAPI();

byte[] l_PasswordBytes = System.Text.Encoding.ASCII.GetBytes
         (this.txtPassword.Text);

IntPtr l_hProvider = l_Crypto.AcquireNamedContext("KICKSTART");

Int32 l_TotalDecryptedBytes = 0;
byte[] l_DecryptedBytes = l_Crypto.PasswordDecrypt(m_EncryptedBytes,
        m_TotalEncryptedBytes, l_hProvider, l_PasswordBytes,
        ref l_TotalDecryptedBytes);

this.txtDecryptedText.Text = System.Text.Encoding.ASCII.GetString
        (l_DecryptedBytes, 0, l_DecryptedBytes.Length);

VB
Dim l_Crypto As ManagedCryptoAPI = New ManagedCryptoAPI
Dim l_PasswordBytes() As Byte = System.Text.Encoding.ASCII.GetBytes
         (Me.txtPassword.Text)

Dim l_hProvider As IntPtr = l_Crypto.AcquireNamedContext("KICKSTART")

Dim l_TotalDecryptedBytes As Int32 = 0
Dim l_DecryptedBytes() As Byte = l_Crypto.PasswordDecrypt(m_EncryptedBytes,
        m_TotalEncryptedBytes, l_hProvider, l_PasswordBytes,
        l_TotalDecryptedBytes)

Me.txtDecryptedText.Text = System.Text.Encoding.ASCII.GetString
         (l_DecryptedBytes, 0, l_DecryptedBytes.Length)
```

Looking Inside PasswordEncrypt

Examining the code inside PasswordEncrypt teaches how to encrypt data by using a password as the basis for the encryption key. PasswordEncrypt encrypts the data passed into it, using the following steps:

1. A handle to a new hash object is obtained by calling the CryptoAPI function `CreateHash`. `CreateHash` is passed a handle to the key container of the CSP, which was passed in to the `PasswordEncrypt` method. The handle to the hash object is stored in the `l_hHash` variable.
2. A hash is computed for the password bytes by calling the CryptoAPI function `CryptHashData`. The `l_hHash` handle is passed into `CryptHashData`.
3. An encryption key for the `CALG_RC4` algorithm is derived based on the hash by calling the CryptoAPI function `CryptDeriveKey`. The `l_hHash` handle is passed into `CryptDeriveKey` and a handle to an encryption key named `l_hKey` is passed out of `CryptDeriveKey`.
4. A byte array named `l_encryptedBytes` is set up. It is large enough to hold the encrypted data even if there is some overflow from the encryption process.
5. The data is encrypted by calling the CryptoAPI function `CryptEncrypt`. The handle to the encryption key, `l_hKey` is passed into `CryptEncrypt`.
6. `CryptEncrypt` fills `l_encryptedBytes` with encrypted data and sets the value `out_NumEncryptedBytes` with the number of bytes inside `l_encryptedBytes` with encrypted data.
7. `l_encryptedBytes` is returned to the caller.

The source code for `PasswordEncrypt` is shown in Listing 14.7.

LISTING 14.7 `PasswordEncrypt`

```
C#
// Encrypts data by using the bytes passed in as the key for the encryption
// This method allocates enough memory to encrypt the data even if CryptoAPI
// needs more bytes than in_BytesToEncrypt holds. The allocation is done
// dynamically
public byte[] PasswordEncrypt(byte [] in_BytesToEncrypt, IntPtr in_hProvider,
        byte[] in_passwordBytes, ref Int32 out_NumEncryptedBytes)
{
   byte[] l_encryptedBytes = new byte[2 * in_BytesToEncrypt.Length];
   // We are not going to catch exceptions. If things go wrong,
   // any system-generated exceptions might help the user of this code
   // understand more about what is going on. We need a finally clause because
   // we want to release the CryptoAPI resources if something goes wrong
   try
   {
      // Step 1: Get a handle to a new hash object that will hash
      // the password bytes
```

```
IntPtr l_hHash = IntPtr.Zero;
if (!CryptCreateHash(in_hProvider, CALG_MD5, IntPtr.Zero, 0,
        ref l_hHash))
{
   throw new Exception("Could not create a hash object!");
}

// Step 2: hash the password data....
// l_hHash - reference to hash object
// in_passwordBytes - bytes to add to hash
// in_passwordBytes.Length - length of data to compute hash on
// 0 - extra flags

// Note: We don't actually get the hash bytes back, we just have a
// handle to the hash object that did the computation.  It is holding
// those bytes internally, so we don't want or need them
if (!CryptHashData(l_hHash, in_passwordBytes, in_passwordBytes.Length,
        0))
{
   throw new Exception("Failure while hashing password bytes!");
}

// Step 3: Derive an encryption key based on hashed data
// in_hProvider - Handle to provider we previously acquired
// CALG_RC4 - Popular encryption algorithm
// l_hHash - Handle to hash object which will hand over the
//  hash as part of the key derivation
// CRYPT_EXPORTABLE - Means the key's bytes could be exported into a
//  byte array, using CryptExportKey
// l_hKey - Handle to the key we are deriving
IntPtr l_hKey = IntPtr.Zero;
if (!CryptDeriveKey(in_hProvider, CALG_RC4, l_hHash, CRYPT_EXPORTABLE,
        ref l_hKey))
{
   throw new Exception("Failure when trying to derive the key!");
}

// Step 4: Acquire enough memory to assure that encryption succeeds
// even allowing for some overflow
for (int i = 0; i < in_BytesToEncrypt.Length; i++)
{
   l_encryptedBytes[i] = in_BytesToEncrypt[i];
```

```
        }
        out_NumEncryptedBytes = in_BytesToEncrypt.Length;

        // Step 5: Do the encryption
        // l_hKey - Previously acquired key for encryption
        // IntPtr.Zero - Indicates that we don't want any additional hashing
        // true - Passed in because this is the only and last data to be
        // encrypted in this session
        // 0 - Additional flags (none)
        // l_encryptedBytes - in/out - bytes to be encrypted in place in this
        //    buffer
        // l_datalength - Length of data (number of bytes) to be encrypted
        // l_encryptedBytes.Length - Lets CryptoAPI know how big the buffer it.
        //  2X data size is plenty
        if (!CryptEncrypt(l_hKey, IntPtr.Zero, true ,0, l_encryptedBytes,
                ref out_NumEncryptedBytes,l encryptedBytes.Length))
        {
            throw new Exception("Failure when calling CryptEncrypt!");
        }
    }
    finally
    {
        // Release resources
    }
    return l_encryptedBytes;
}
```

VB

```
' Encrypts data by using the bytes passed in as the key for the encryption
' This method allocates enough memory to encrypt the data even if CryptoAPI
' needs more bytes than in_BytesToEncrypt holds. The allocation is
' done dynamically
Public Function PasswordEncrypt(ByVal in_BytesToEncrypt() As Byte, ByVal
        in_hProvider As IntPtr, ByVal in_passwordBytes() As Byte, ByRef
        out_NumEncryptedBytes As Int32) As Byte()
    Dim l_encryptedBytes(2 * in_BytesToEncrypt.Length) As Byte ' = New Byte(2 *
        in_BytesToEncrypt.Length)

    ' We are not going to catch exceptions. If things go wrong, any
    ' system-generated exceptions might help the user of this code understand
    ' more about what is going on. We need a finally clause because we want
    ' to release the CryptoAPI resources if something goes wrong
```

```
Try
   ' Step 1: Get a handle to a new hash object that will hash the
   ' password bytes
   Dim l_hHash As IntPtr = IntPtr.Zero

   If (CryptCreateHash(in_hProvider, Convert.ToInt32(CALG_MD5), IntPtr.Zero,
            Convert.ToUInt32(0), l_hHash) = False) Then
      Throw New Exception("Could not create a hash object!")
   End If

   ' Step 2: hash the password data....
   ' l_hHash - reference to hash object   in_passwordBytes -
   ' bytes to add to hash
   ' in_passwordBytes.Length - length of data to compute hash on
   ' 0 - extra flags

   ' Note: We don't actually get the hash bytes back, we just have a
   ' handle to the hash object that did the computation.  It is holding
   ' those bytes internally, so we don't want or need them
   If (CryptHashData(l_hHash, in_passwordBytes, in_passwordBytes.Length,
           Convert.ToUInt32(0)) = False) Then
      Throw New Exception("Failure while hashing password bytes!")
   End If

   ' Step 3: Derive an encryption key based on hashed data
   ' in_hProvider - Handle to provider we previously acquired
   ' CALG_RC4 - Popular encryption algorithm
   ' l_hHash - Handle to hash object which will hand over the hash as
   ' part of the key derivation CRYPT_EXPORTABLE - Means the key's bytes
   ' could be exported into a byte array, using CryptExportKey
   ' l_hKey - Handle to the key we are deriving
   Dim l_hKey As IntPtr = IntPtr.Zero
   If (CryptDeriveKey(in_hProvider, CALG_RC4, l_hHash, CRYPT_EXPORTABLE,
           l_hKey) = False) Then
      Throw New Exception("Failure when trying to derive the key!")
   End If

   ' Step 4: Acquire enough memory to assure that encryption succeeds even
   ' allowing for some overflow
   Dim i As Integer
   For i = 0 To in_BytesToEncrypt.Length - 1
      l_encryptedBytes(i) = in_BytesToEncrypt(i)
   Next i
```

```
        out_NumEncryptedBytes = in_BytesToEncrypt.Length

        ' Step 5: Do the encryption
        ' l_hKey - Previously acquired key for encryption
        ' IntPtr.Zero - Indicates that we don't want any additional hashing
        ' true - Passed in because this is the only and last data to be encrypted
        ' in this session
        ' 0 - Additional flags (none)
        ' l_encryptedBytes - in/out - bytes to be encrypted in place in
        '     this buffer
        ' l_datalength - Length of data (number of bytes) to be encrypted
        ' l_encryptedBytes.Length - Lets CryptoAPI know how big the buffer it.
        '   2X data size is plenty
        If (CryptEncrypt(l_hKey, IntPtr.Zero, True, Convert.ToUInt32(0),
                l_encryptedBytes, out_NumEncryptedBytes, l_encryptedBytes.Length)
                 = False) Then
            Throw New Exception("Failure when calling CryptEncrypt!")
        End If
    Finally
        ' Release resources
    End Try
    Return l_encryptedBytes
End Function
```

Looking Inside PasswordDecrypt

Examining the code inside `PasswordDecrypt` teaches how to decrypt data by using a password as a basis for the decryption key. `PasswordEncrypt` decrypts the data passed into it through the following steps:

1. A byte array named `l_decryptedBytes` is set up that is as large as the number of encrypted bytes passed in.
2. A hash is created and used on the password bytes and, an encryption key is created as in steps 1–3 for `PasswordEncrypt`.
3. The data is decrypted by calling the CryptoAPI function `CryptDecrypt`. The handle to the encryption key, `l_hKey`, is passed into `CryptDecrypt`.
4. `CryptDecrypt` fills the `l_decryptedBytes` array with decrypted data and sets `out_NumDecryptedBytes` to the number of bytes in the array to the number of bytes actually holding decrypted data.
5. `PasswordDecrypt` returns `l_decryptedBytes`.

The source code for `PasswordDecrypt` is shown in Listing 14.8.

LISTING 14.8 Source code for `PasswordDecrypt`

```
C#
// Decrypts data by using the bytes passed in as the key for the decryption.
// This method allocates enough memory to decrypt the data even if CryptoAPI
// needs more bytes than in_BytesToDecrypt holds. The allocation is done
// dynamically.
public byte[] PasswordDecrypt(byte [] in_BytesToDecrypt, Int32
        in_NumBytesToDecrypt, IntPtr in_hProvider, byte[] in_passwordBytes,
        ref Int32 out_NumDecryptedBytes)
{
   byte[] l_decryptedBytes = new byte[in_BytesToDecrypt.Length];

   for (int i = 0; i < in_BytesToDecrypt.Length; i++)
   {
      l_decryptedBytes[i] = in_BytesToDecrypt[i];
   }

   // We are not going to catch exceptions. If things go wrong, any
   // system-generated exceptions might help the user of this code
   // understand more about what is going on. We need a finally
   // clause because we want to release the CryptoAPI resources if
   // something goes wrong.
   try
   {
      // Step 1: Get a handle to a new hash object that will hash the
      // password bytes.
      IntPtr l_hHash = IntPtr.Zero;

      if (!CryptCreateHash(in_hProvider, CALG_MD5, IntPtr.Zero, 0,
              ref l_hHash))
      {
         throw new Exception("Could not create a hash object!");
      }

      // Step 2: Hash the password data....
      // l_hHash - reference to hash object
      // in_passwordBytes - bytes to add to hash
      // in_passwordBytes.Length - length of data to compute hash on
      // 0 - extra flags
```

```
      // Note: We don't actually get the hash bytes back, we just have a
      // handle to the hash object that did the computation. It is holding
      // those bytes internally, so we don't want or need them.
      if (!CryptHashData(l_hHash, in_passwordBytes,
              in_passwordBytes.Length, 0))
      {
         throw new Exception("Failure while hashing password bytes!");
      }

      // Step 3: Derive an encryption key based on hashed data
      // in_hProvider - Handle to provider we previously acquired
      // CALG_RC4 - Popular encryption algorithm
      // l_hHash - Handle to hash object which will hand over the hash as part
      //   of the key derivation
      // CRYPT_EXPORTABLE - Means the key's bytes could be exported into a
      //   byte array, using CryptExportKey
      // l_hKey - Handle to the key we are deriving
      IntPtr l_hKey = IntPtr.Zero;
      if (!CryptDeriveKey(in_hProvider, CALG_RC4, l_hHash, CRYPT_EXPORTABLE,
              ref l_hKey))
      {
         throw new Exception("Failure when trying to derive the key!");
      }

      // And now decrypt the data
      out_NumDecryptedBytes = in_NumBytesToDecrypt;

      if (!CryptDecrypt(l_hKey, IntPtr.Zero, true , 0, l_decryptedBytes,
              ref out_NumDecryptedBytes))
      {
         throw new Exception("Failure when trying to decrypt the data");
      }
   }
   finally
   {
      // Release resources
   }

   return l_decryptedBytes;
}
```

```
VB
' Decrypts data by using the bytes passed in as the key for the decryption
' This method allocates enough memory to decrypt the data even if CryptoAPI
' needs more bytes than in_BytesToDecrypt holds. The allocation is done
' dynamically.
Public Function PasswordDecrypt(ByVal in_BytesToDecrypt() As Byte, ByVal
        in_NumBytesToDecrypt As Int32, ByVal in_hProvider As IntPtr,
        ByVal in_passwordBytes() As Byte, ByRef out_NumDecryptedBytes As Int32)
         As Byte()
   Dim l_decryptedBytes(in_BytesToDecrypt.Length) As Byte
   Dim i As Integer
   For i = 0 To in_BytesToDecrypt.Length - 1
      l_decryptedBytes(i) = in_BytesToDecrypt(i)
   Next i

   ' We are not going to catch exceptions. If things go wrong, any
   ' system-generated exceptions might help the user of this code understand
   ' more about what is going on. We need a finally clause because we want
   ' to release the CryptoAPI resources if something goes wrong.
   Try
      ' Step 1: Get a handle to a new hash object that will hash the
      ' password bytes
      Dim l_hHash As IntPtr = IntPtr.Zero

      If (CryptCreateHash(in_hProvider, Convert.ToInt32(CALG_MD5), IntPtr.Zero,
              Convert.ToUInt32(0), l_hHash) = False) Then
         Throw New Exception("Could not create a hash object!")
      End If

      ' Step 2: Hash the password data....
      ' l_hHash - reference to hash object   in_passwordBytes - bytes to
      '   add to hash
      ' in_passwordBytes.Length - length of data to compute hash on
      ' 0 - extra flags

      ' Note: We don't actually get the hash bytes back, we just have a
      ' handle to the hash object that did the computation.  It is holding
      ' those bytes internally, so we don't want or need them
      If (CryptHashData(l_hHash, in_passwordBytes, in_passwordBytes.Length,
              Convert.ToUInt32(0)) = False) Then
         Throw New Exception("Failure while hashing password bytes!")
      End If
```

```
        ' Step 3: Derive an encryption key based on hashed data
        ' in_hProvider - Handle to provider we previously acquired
        ' CALG_RC4 - Popular encryption algorithm
        ' l_hHash - Handle to hash object which will hand over the hash as part
        '   of the key derivation
        ' CRYPT_EXPORTABLE - Means the key's bytes could be exported into a byte
        '   array, using CryptExportKey
        ' l_hKey - Handle to the key we are deriving
        Dim l_hKey As IntPtr = IntPtr.Zero
        If (CryptDeriveKey(in_hProvider, CALG_RC4, l_hHash, CRYPT_EXPORTABLE,
                l_hKey) = False) Then
           Throw New Exception("Failure when trying to derive the key!")
        End If

        ' And now decrypt the data
        out_NumDecryptedBytes = in_NumBytesToDecrypt

        If (CryptDecrypt(l_hKey, IntPtr.Zero, True, Convert.ToUInt32(0),
                l_decryptedBytes, out_NumDecryptedBytes) = False) Then
           Throw New Exception("Failure when trying to decrypt the data")
        End If

     Finally
        ' Release resources
     End Try

     Return l_decryptedBytes
  End Function
```

Using Session Keys to Encrypt and Decrypt Data

To understand how to perform encryption and decryption with a session key, readers must first have a deep understanding of what session keys are, how to create them, and how to persist and share them. The session keys aren't actually used for encryption and decryption until the end of this section.

A session key is an encryption key that is shared between two parties to perform encryption and decryption or used by only a single machine for encryption and decryption. One way to derive a session key is to generate a unique password on the fly as needed. This technique was discussed in the previous section. The problem with this technique is that it doesn't provide a way to share a session key securely between two devices unless each side already knows the password. Also, if two different devices are using slightly different operating system versions, the same password can yield different session keys, so the two devices would be unable to share encrypted data.

In the previous section, which demonstrated how to perform encryption by using a password, only the handle to the session key was retained. The bytes of the actual key were never directly manipulated. Thus, it was impossible to move the session key to other machines or even to save it to the disk of the local device.

A session key is analogous to a shared private key used for encryption and decryption between two parties. Because each party has a copy of the shared session key, data encrypted by one side can be decrypted by the other. One problem that arises is how to get a shared session key on two devices in a secure fashion. If the bytes of a session key are transmitted to the remote party, an attacker who can listen to the transmission also has a copy of the session key and thus can decrypt all of the traffic that is encrypted with the session key.

CryptoAPI solves this problem by imposing strict rules on how session keys can be distributed. As seen in the previous section, getting a handle to a session key is very easy, but handles are not portable. To get the underlying bytes of a session key, developers must follow the strict rules imposed by CryptoAPI. This section examines two scenarios for acquiring a session key. The first scenario is when users want to store the bytes of the session key for use by only the same machine that created the session key in the first place. The second scenario is for when developers want to store the session key in such a way that another device can also use the session key.

Storing a Session Key for Use Only by the Device That Created It

`ManagedCryptoAPI` provides a simple way to store a session key in such a way that it is usable only by the same device that created it. If the bytes of the session key were moved to another device, the other device would be unable to load it.

To save a session key in such a way that it is usable only by the device that created the key using `ManagedCryptoAPI`, follow these steps:

1. Acquire an instance of the `ManagedCryptoAPI` class.
2. Acquire and retain a handle to a key store in a CSP by calling `ManagedCryptoAPI.AcquireNamedContext` or `ManagedCryptoAPI.AcquireDefaultContext`.
3. Acquire and retain a handle to a new session key by calling `ManagedCryptoAPI.GenerateSessionKey`. Pass the handle to the key store acquired in step 2 into `GenerateSessionKey`.
4. Call `ManagedCryptoAPI.LocalSaveSessionKey`, which returns a byte array holding the actual bytes of the session key. The third argument to `LocalSaveSessionKey` is a reference to an `Int32`. It is set to the number of bytes in the returned array that actually contain bytes of the session key.

5. The bytes of the session key are encrypted by the CryptoAPI before they are exported. CryptoAPI stores the information needed to decrypt and use the session key in a way not specified to outside users. Thus, *only the same device* can import and use the bytes of the session key that `LocalSaveSessionKey` returns.
6. Store the bytes of the session key using any convenient means.

Listing 14.9 is taken from the ExportSessionKey sample application. This code follows all of the steps just outlined and drops the bytes of a session key on the file system under the name `LocalSessionKey.blob`.

LISTING 14.9 Code from ExportSessionKey

```
C#
private void DropPrivateSessionKey(IntPtr in_hProvider, IntPtr in_hSessionKey,
        ManagedCryptoAPI in_Crypto)
{
   Int32 l_SessionKeySize = 0;
   // Get the bytes of the session key
   byte[] l_SessionKeyBytes =
   in_Crypto.LocalSaveSessionKey(in_hProvider, in_hSessionKey, ref
   l_SessionKeySize);

   // Store the bytes of the session key on disk
   System.IO.BinaryWriter l_Writer = new BinaryWriter(new
           System.IO.FileStream("\\LocalSessionKey.blob",
           System.IO.FileMode.CreateNew));

   l_Writer.Write(l_SessionKeyBytes, 0, (int)l_SessionKeySize);
   l_Writer.Close();
}

// This code is used to drop the session key.
ManagedCryptoAPI l_Crypto = new ManagedCryptoAPI();
IntPtr l_hProvider = IntPtr.Zero;
IntPtr l_hSessionKey = IntPtr.Zero;

l_hProvider = l_Crypto.AcquireNamedContext("KICKSTART");
l_hSessionKey = l_Crypto.GenerateSessionKey(l_hProvider);
DropPrivateSessionKey(l_hProvider, l_hSessionKey, l_Crypto);

VB
Private Sub DropPrivateSessionKey(ByVal in_hProvider As IntPtr, ByVal
        in_hSessionKey As IntPtr, ByVal in_Crypto As ManagedCryptoAPI)
```

```
    Dim l_SessionKeySize As Int32 = 0
    ' Get the bytes of the session key
    Dim l_SessionKeyBytes() As Byte = in_Crypto.LocalSaveSessionKey
            (in_hProvider, in_hSessionKey, l_SessionKeySize)

    If (System.IO.File.Exists("\LocalSessionKey.blob")) Then
        System.IO.File.Delete("\LocalSessionKey.blob")
    End If

    ' Store the bytes of the session key on disk
    Dim l_Writer As System.IO.BinaryWriter = New System.IO.BinaryWriter(New
            System.IO.FileStream("\LocalSessionKey.blob",
            System.IO.FileMode.CreateNew))
    l_Writer.Write(l_SessionKeyBytes, 0, Convert.ToInt32(l_SessionKeySize))
    l_Writer.Close()
End Sub

' This code is used to drop the session key.
Dim l_Crypto As ManagedCryptoAPI = New ManagedCryptoAPI
Dim l_hProvider As IntPtr = IntPtr.Zero
Dim l_hSessionKey As IntPtr = IntPtr.Zero

l_hProvider = l_Crypto.AcquireNamedContext("KICKSTART")
l_hSessionKey = l_Crypto.GenerateSessionKey(l_hProvider)
DropPrivateSessionKey(l_hProvider, l_hSessionKey, l_Crypto)
```

Once a session key has been saved, it can be reloaded by the same device in the future and used to perform encryption and decryption. This process is discussed in the section named "Importing a Previously Saved Session into the Same Device That Created It."

The ExportSessionKey sample application located in `SampleApplications\Chapter14`. There is a C# and a Visual Basic version. The application demonstrates exporting sessions keys for use by the same machine that created it and for use by other machines. To export a session key that can be imported only by the same device, click the button labeled Drop Private Session Key. The program writes the bytes of the session key into a file named `LocalSessionKey.blob`. The key can be imported by following the steps in the next section, "Importing a Previously Saved Session into the Same Device That Created It."

Clicking the button labeled Drop Exported Session Key writes a session key in a format that can be used by another device. The section titled "Sharing a Session Key with a Remote Device" outlines all of the steps needed to share a session key with another device. This sample application demonstrates step 4 of the "Overview of Steps for Sharing a Session Key" section.

Clicking the button labeled Drop Pair of Session Keys has the same effect as first clicking Drop Private Session Key and then clicking Drop Exported Session Key.

Importing a Previously Saved Session into the Same Device That Created It

The previous section described how to save the bytes of a session key as an array of bytes. The `ManagedCryptoAPI` class can easily import the session key easily from the bytes that were saved previously. Importing a session key yields a handle to an encryption key that can be used to encrypt and decrypt data. Data that was previously encrypted with the same key can be decrypted by loading the session key and using its handle to decrypt data. To import a private session key with ManagedCryptoAPI, follow these steps:

1. Acquire an instance of the `ManagedCryptoAPI` class.
2. Acquire and retain a handle to a key store in a CSP by calling `ManagedCryptoAPI.AcquireNamedContext` or `ManagedCryptoAPI.AcquireDefaultContext`.
3. Acquire a byte array that holds the bytes of the session key to import.
4. Call `ManagedCryptoAPI.LoadPrivateSessionKey`, passing in the byte array and the handle to the key store. `LoadPrivateSessionKey` returns a handle to an encryption key that you can use to encrypt and decrypt data.

Listing 14.10, derived from the ImportSessionKey sample application, demonstrates these steps.

LISTING 14.10 `ImportSessionKey`

```
C#
ManagedCryptoAPI l_Crypto = new ManagedCryptoAPI();
IntPtr l_hProvider = IntPtr.Zero;
IntPtr l_hPrivateSessionKey = IntPtr.Zero;

l_hProvider = l_Crypto.AcquireNamedContext("KICKSTART");

l_Reader = new BinaryReader(new System.IO.FileStream("\\LocalSessionKey.blob",
        System.IO.FileMode.Open));

byte[] l_PrivateKeyBytes = l_Reader.ReadBytes(1000);

l_Reader.Close();

l_hPrivateSessionKey = l_Crypto.LoadPrivateSessionKey(l_hProvider,
        l_PrivateKeyBytes);

// Use l_hPrivateSessionKey for encryption and decryption
```

```
VB
Dim l_Crypto As ManagedCryptoAPI = New ManagedCryptoAPI
Dim l_hProvider As IntPtr = IntPtr.Zero
Dim l_hPrivateSessionKey As IntPtr = IntPtr.Zero

l_hProvider = l_Crypto.AcquireNamedContext("KICKSTART")

' Get the bytes of the public exchange key which we already have
' gotten from another device.
Dim l_Reader As System.IO.BinaryReader = Nothing
l_Reader = New System.IO.BinaryReader(New System.IO.FileStream
        ("\LocalSessionKey.blob", System.IO.FileMode.Open))

Dim l_PrivateKeyBytes() As Byte = l_Reader.ReadBytes(1000)
l_Reader.Close()
l_hPrivateSessionKey = l_Crypto.LoadPrivateSessionKey(l_hProvider,
        l_PrivateKeyBytes)
' Use l_hPrivateSessionKey for encryption and decryption
```

The sample application ImportSessionKey demonstrates the end-to-end scenario of importing a private session key and using it for encryption and decryption. Because it can also import a session key exported from another device, it is discussed in detail at the end of the chapter.

Sharing a Session Key with a Remote Device

Sharing a session key with a remote device is complicated. Before jumping into the details of how to do it, this section outlines the overall steps that two devices must take in order to share a session key between them. Each step in the overall outline is covered in greater detail in this chapter, including code samples and sample applications.

Overview of Steps for Sharing a Session Key

1. Device A generates a session key and saves it locally, as depicted in the section "Importing a Previously Saved Session into the Same Device That Created It."
2. Device B generates a public key and saves it, as depicted in the section titled "Generating the Public Key."
3. The saved public key is transferred from Device B to Device A.
4. Device A "imports" the public key exchange key and uses it to save the same session key that was created in step 1. Only Device B can use the session key that is saved in

this way. Saving the key this way is depicted in the section titled "Importing a Public Key Exchange Key."

5. The saved session key from step 4 is transferred from Device A to Device B.
6. Device B imports the saved session key that was transferred to it in step 5 and uses it for encryption and decryption. This step is depicted in the section titled "Importing a Session Key from Another Device."
7. Device A imports the session key it generated in step 1 and uses it for encryption and decryption. This step is depicted in the section titled "Importing a Previously Saved Session into the Same Device That Created It."
8. Device A and Device B can decrypt each other's encrypted data by using the session keys they imported in steps 6 and 7. Encrypting and decrypting data using a handle to a session key is covered in the section titled "Encrypting and Decrypting by Using a Handle to a Session Key."

Generating the Public Key

This discussion assumes there are two devices called Device A and Device B. Device A has generated a session key that it can save for its own future use as described in the section titled "Storing a Session Key for Use Only by the Device That Created It." Device A also wants to give the session key to Device B in a secure way. How does this happen?

The answer is that Device B must generate a public-private key pair. Any data encrypted with the public key exchange key can be decrypted with the corresponding private key. Device B can send the bytes of the public key to Device A, and Device A can encrypt the bytes of the session key using the public key. The encrypted session key can then go to Device B, and only Device B holds the private key capable of decrypting the session key. Thus, it would not matter if an attacker saw the encrypted session key as it was transferred—it is useless without the private key.

Encryption using public-private key pairs is roughly a thousand times slower than using a shared session key. For this reason, the only support in CryptoAPI for public-private key pairs is to use them to exchange session keys safely, which are then used for encryption. There is no support in CryptoAPI for using the exported public keys to do anything but encrypt other session keys.

This section describes how to generate the public key with the help of `ManagedCryptoAPI`. That is, this section covers step 2 of the "Overview of Steps for Sharing a Session Key" section. To export a public key through the `ManagedCryptoAPI`, follow these steps:

1. Acquire an instance of the `ManagedCryptoAPI` class.
2. Acquire and retain a handle to a key store in a CSP by calling `ManagedCryptoAPI.AcquireNamedContext` or `ManagedCryptoAPI.AcquireDefaultContext`.

3. Acquire the bytes of the public key exchange key by calling `ManagedCryptoAPI.ExportPublicKey`. Pass in the handle to the key store and a reference to an `Int32`. `ExportPublicKey` returns a byte array holding the bytes of the public key exchange key and sets the `Int32` to the number of bytes in the returned byte array that actually hold bytes of the public key exchange key.

Listing 14.11, derived from the ExportPublicKeyDemo sample application, acquires the bytes of the public key exchange key and dumps them into a file named `PublicExchangeKey.blob`:

LISTING 14.11 Code will acquire bytes of the public key exchange key and dump them into a file named `PublicExchangeKey.blob`

```
C#
ManagedCryptoAPI l_Crypto = new ManagedCryptoAPI();

IntPtr l_hProvider = IntPtr.Zero;
Int32  l_PubKeySize = 0;

l_hProvider = l_Crypto.AcquireNamedContext("KICKSTART");

byte[] l_PublicExchangeKeyBytes = l_Crypto.ExportPublicKey(l_hProvider,
   ref l_PubKeySize);

System.IO.BinaryWriter l_Writer = new BinaryWriter(new
   System.IO.FileStream("\\PublicExchangeKey.blob",
   System.IO.FileMode.CreateNew));

l_Writer.Write(l_PublicExchangeKeyBytes, 0, (int)l_PubKeySize);

l_Writer.Close();

VB
Dim l_Crypto As ManagedCryptoAPI = New ManagedCryptoAPI
Dim l_hProvider As IntPtr = IntPtr.Zero
Dim l_PubKeySize As Int32 = 0

l_hProvider = l_Crypto.AcquireNamedContext("KICKSTART")

Dim l_PublicExchangeKeyBytes() As Byte = l_Crypto.ExportPublicKey(l_hProvider,
        l_PubKeySize)

If (System.IO.File.Exists("\PublicExchangeKey.blob")) Then
   System.IO.File.Delete("\PublicExchangeKey.blob")
```

```
End If

Dim l_Writer As System.IO.BinaryWriter = New System.IO.BinaryWriter(New
        System.IO.FileStream("\PublicExchangeKey.blob",
        System.IO.FileMode.CreateNew))

l_Writer.Write(l_PublicExchangeKeyBytes, 0, Convert.ToInt32(l_PubKeySize))
l_Writer.Close()
```

Exporting a Public Key in a Sample Application

The ExportPublicKeyDemo sample application is located in the folder `\SampleApplications\Chapter14`. There are C# and Visual Basic versions. To use the application, click the button labeled Export Public Key Exchange Key. The program uses the `ManagedCryptoAPI` to acquire a byte array holding the public key exchange key, and it writes the bytes to a file called `PublicExchangeKey.blob`. This file can be copied to another device. The other device uses it to encrypt a session key in such a way that only the first device can use it the session key. The act of exporting a session key in this way is step 4 in the "Overview of Steps for Sharing a Session Key" subsection, and it is covered in the next subsection, "Importing a Public Key Exchange Key."

Importing a Public Key Exchange Key

In the scenario where two devices, Device A and Device B, must share a session key, assume that Device A has generated a public key exchange key, that the bytes are stored in a file, and that the file has been copied to Device B. Device B must import the public key exchange key and use it to *export* a session key. When Device B exports the session key in this way, the session key is encrypted so that only Device A can use it. The exported session key is copied back to Device A and imported as outlined in the section titled "Importing a Session Key from Another Device."

Device B also saves a copy of the same session key that it can use itself, as outlined in the section titled "Storing a Session Key for Use Only by the Device That Created It."

The `ManagedCryptoAPI` class makes it easy to import a public key exchange key and use it to export a session key. To do so, follow these steps:

1. Acquire an instance of the `ManagedCryptoAPI` class.
2. Acquire and retain a handle to a key store in a CSP by calling `ManagedCryptoAPI.AcquireNamedContext` or `ManagedCryptoAPI.AcquireDefaultContext`.
3. Generate the session key that will be exported.

4. Acquire the bytes of the public exchange key that was generated on another device as a byte array.

5. Import the bytes of the public key exchange key to acquire a handle to the imported key. The easiest way to do this is to call `ManagedCryptoAPI.ImportPublicKey`.

6. Export the session key to another array of bytes. The easiest way to do this is to call `ManagedCryptoAPI.ExportSessionKey`. Pass in the handle to the public key, acquired in step 5 and the handle to the session key to export, acquired in step 3. Also pass in a reference to an `Int32`. `ExportSessionKey` sets the `Int32` to the number of bytes in the returned array that hold the bytes of the exported session key.

ALSO STORE A SESSION KEY LOCALLY WHEN YOU EXPORT IT

Make sure to save this session key locally, as well, as depicted in the section titled "Storing a Session Key for Use Only by the Device That Created It," so that both devices will access the same key. Retain a handle to the session key.

7. Copy the bytes of the exported session key to the remote device through any convenient means, such as saving to a file or even transmitting over a socket connection. It is safe to do this because the session key is encrypted with the remote device's public key exchange key.

8. The remote device now has a session key. It can import the key as depicted in the section "Importing a Session Key from Another Device."

Listing 14.12 illustrates steps 1 through 7.

LISTING 14.12 Code that demonstrates steps 1 through 7 of importing a public key exchange key

```
C#
private void DropExportedSessionKey(IntPtr in_hProvider, IntPtr
  in_hSessionKey, ManagedCryptoAPI in_Crypto)
{
   IntPtr l_PublicExchangeKey = IntPtr.Zero;
   Int32 l_SessionKeySize = 0;

   // Get the bytes of the public exchange key which we already have
   // gotten from another device.
   System.IO.BinaryReader l_Reader = new BinaryReader(new
      System.IO.FileStream("\\PublicExchangeKey.blob",
      System.IO.FileMode.Open));

   byte[] l_PublicExchangeKeyBytes = l_Reader.ReadBytes(1000);
```

```
    l_Reader.Close();

    l_PublicExchangeKey = in_Crypto.ImportPublicKey(in_hProvider,
       l_PublicExchangeKeyBytes);

    byte[] l_ExportedSessionKeyBytes =
       in_Crypto.ExportSessionKey(l_PublicExchangeKey, in_hSessionKey,
       ref l_SessionKeySize);

    // The session key is encrypted using the public exchange key from
    // the other device.

    // We store the bytes of the session key on disk
    if (File.Exists("\\ExportedSessionKey.blob"))
    {
       File.Delete("\\ExportedSessionKey.blob");
    }

    System.IO.BinaryWriter l_Writer = new BinaryWriter(new
       System.IO.FileStream("\\ExportedSessionKey.blob",
       System.IO.FileMode.CreateNew));

    l_Writer.Write(l_ExportedSessionKeyBytes, 0, (int)l_SessionKeySize);
    l_Writer.Close();
}

// This code will drop an exportable session key
ManagedCryptoAPI l_Crypto = new ManagedCryptoAPI();

IntPtr l_hProvider = IntPtr.Zero;
IntPtr l_hSessionKey = IntPtr.Zero;

l_hProvider = l_Crypto.AcquireNamedContext("KICKSTART");
l_hSessionKey = l_Crypto.GenerateSessionKey(l_hProvider);

DropExportedSessionKey(l_hProvider, l_hSessionKey, l_Crypto);
```

VB

```
Private Sub DropExportedSessionKey(ByVal in_hProvider As IntPtr,
        ByVal in_hSessionKey As IntPtr, ByVal in_Crypto As ManagedCryptoAPI)
   Dim l_PublicExchangeKey As IntPtr = IntPtr.Zero
   Dim l_SessionKeySize As Int32 = 0
```

```
    ' Get the bytes of the public exchange key which we already have gotten
    ' from another device.
    Dim l_Reader As System.IO.BinaryReader = New System.IO.BinaryReader(New
            System.IO.FileStream("\PublicExchangeKey.blob",
            System.IO.FileMode.Open))
    Dim l_PublicExchangeKeyBytes() As Byte = l_Reader.ReadBytes(1000)
    l_Reader.Close()

    l_PublicExchangeKey = in_Crypto.ImportPublicKey(in_hProvider,
            l_PublicExchangeKeyBytes)

    Dim l_ExportedSessionKeyBytes() As Byte = in_Crypto.ExportSessionKey
             (l_PublicExchangeKey, in_hSessionKey, l_SessionKeySize)

    ' The session key is encrypted using the public exchange key from the other
    ' device. We store the bytes of the session key on disk
    If (System.IO.File.Exists("\ExportedSessionKey.blob")) Then
       System.IO.File.Delete("\ExportedSessionKey.blob")
    End If
    Dim l_Writer As System.IO.BinaryWriter = New System.IO.BinaryWriter(New
            System.IO.FileStream("\ExportedSessionKey.blob",
            System.IO.FileMode.CreateNew))
    l_Writer.Write(l_ExportedSessionKeyBytes, 0, Convert.ToInt32
             (l_SessionKeySize))
    l_Writer.Close()
End Sub

' This code will drop an exportable session key
Dim l_Crypto As ManagedCryptoAPI = New ManagedCryptoAPI

Dim l_hProvider As IntPtr = IntPtr.Zero
Dim l_hSessionKey As IntPtr = IntPtr.Zero

l_hProvider = l_Crypto.AcquireNamedContext("KICKSTART")
l_hSessionKey = l_Crypto.GenerateSessionKey(l_hProvider)

DropExportedSessionKey(l_hProvider, l_hSessionKey, l_Crypto)
```

Exporting a Session Key in a Sample Application

The ExportSessionKey sample application is located in the folder `\SampleApplications\Chapter14`, with C# and Visual Basic versions in separate folders. This application can export session keys for remote devices. It can also export session keys that the same device can use later by following the steps outlined in the section titled "Importing a Previously Saved Session into the Same Device That Created It." The application first imports a public key exchange key before it exports the session key for use by another device.

To make the application export a session key by using the public key exchange key from another device, follow these steps:

1. Run the ExportPublicKeyDemo application in another device to create a public exchange key that is stored in a file named `PublicExchangeKey.blob`.
2. Copy the `PublicExchangeKey.blob` file to the root folder of the device that will run ExportSessionKey.
3. Run ExportSessionKey and click the Drop Exported Session Key button. The program responds by importing the public key exchange key from the `PublicExchangeKey.blob` file and exporting a session key to a file named `ExportedSessionKey.blob`.

Importing a Session Key from Another Device

Before starting this section, it is assumed that the following has occurred between Device A and Device B:

- Device A has created a public key exchange key and transferred it to Device B, as described in the "Generating the Public Key" section.
- Device B has imported the public key exchange key and used it to export a session key, as described in the section titled "Importing a Public Key Exchange Key." The session key has been copied back to Device A.

> **SIMULTANEOUSLY DROPPING A SESSION KEY FOR LOCAL AND EXPORTED USE**
>
> You can press the Drop Pair of Session Keys button to drop both an exported session key called `ExportedSessionKey.blob` and a session key for use by the same device called `LocalSessionKey.blob`.

Device A is now ready to import the session key that has been copied to it. The act of importing the session key means that the CryptoAPI has read the bytes of the key and returned a handle to the session key. That handle can be used to perform encryption and decryption, as described in the section titled "Encrypting and Decrypting by Using a Handle to a Session Key."

`ManagedCryptoAPI` provides a simple means of importing the session key and acquiring a handle to the key that can be used for encryption and decryption. To do this, follow these steps:

1. Acquire an instance of the `ManagedCryptoAPI` class.
2. Acquire and retain a handle to a key store in a CSP by calling `ManagedCryptoAPI.AcquireNamedContext` or `ManagedCryptoAPI.AcquireDefaultContext`.

3. Acquire the bytes of the session key to import into a byte array.
4. Call `ManagedCryptoAPI.LoadSharedSessionKey`, passing in a handle to the key store and the byte array holding the session key to import. `LoadSharedSessionKey` returns a handle to the session key that can be used for encryption and decryption, as shown in the next section, "Encrypting and Decrypting by Using a Handle to a Session Key."

The following sample code is taken from the sample application ImportLocalSessionKey. It loads the bytes of the session key to import and gets a handle to it.

```
C#
l_Reader = new BinaryReader(new System.IO.FileStream
        ("\\ExportedSessionKey.blob", System.IO.FileMode.Open));

byte[] l_SharedKeyBytes = l_Reader.ReadBytes(1000);
l_Reader.Close();
l_hPrivateSessionKey = l_Crypto.LoadSharedSessionKey(l_hProvider,
        l_SharedKeyBytes);

// Use l_hPrivateSessionKey for encryption and decryption

VB
l_Reader = New System.io.BinaryReader(New System.IO.FileStream
        ("\ExportedSessionKey.blob", System.IO.FileMode.Open))

Dim l_SharedKeyBytes() As Byte = l_Reader.ReadBytes(1000)
l_Reader.Close()
l_hPrivateSessionKey = l_Crypto.LoadSharedSessionKey(l_hProvider,
        l_SharedKeyBytes)

' Use l_hPrivateSessionKey for encryption and decryption
```

Encrypting and Decrypting by Using a Handle to a Session Key

The `ManagedCryptoAPI` class provides a way to encrypt and decrypt data using a handle to a session key. You can get the session key in three ways:

- Import a session key that was exported by your own device, as described in the section titled "Importing a Previously Saved Session into the Same Device That Created It."
- Import a session key that was exported by another device, as described in the "Importing a Session Key from Another Device" section.

- Call ManagedCryptoAPI.GenerateSessionKey directly. You can see an example of calling GenerateSessionKey by looking at the section titled "Encrypting and Decrypting by Using a Handle to a Session Key."

No matter how you got a handle to a session key, you use it in the same way to encrypt and decrypt data. To encrypt data, follow these steps:

1. Acquire an instance of ManagedCryptoAPI.
2. Acquire a handle to a session key by using one of the three methods described earlier.
3. Acquire the data to encrypt as a byte array.
4. Call ManagedCryptoAPI.KeyEncrypt, passing in the data to encrypt and the session key handle. KeyEncrypt returns an array of bytes with the encrypted data.

The following code demonstrates these steps:

```
C#
// Get the bytes to encrypt from the user interface
byte[] l_PlainTextBytes = System.Text.Encoding.ASCII.GetBytes
        (this.txtInitialText.Text);

// l_hPrivateSessionKey holds the handle to the session key.
byte[] l_Encrypted = l_Crypto.KeyEncrypt(l_hPrivateSessionKey,
        l_PlainTextBytes);
```

```
VB
' Get the bytes to encrypt from the user interface
Dim l_PlainTextBytes() As Byte = System.Text.Encoding.ASCII.GetBytes
        (Me.txtInitialText.Text)

' l_hPrivateSessionKey holds the handle to the session key.
Dim l_Encrypted() As Byte = l_Crypto.KeyEncrypt(l_hPrivateSessionKey,
        l_PlainTextBytes)
```

To decrypt data, follow these steps:

1. Acquire an instance of ManagedCryptoAPI.
2. Acquire a handle to a session key using one of the three methods described earlier.
3. Acquire the data to decrypt as a byte array.
4. Call ManagedCryptoAPI.KeyDecrypt, passing in an array of bytes holding the data to decrypt, the session key handle, and a reference to an Int32 that holds the number of bytes in the array that was passed in and holds encrypted data. KeyDecrypt returns an

array of bytes and sets the `Int32` to the number of bytes in the array that holds decrypted data.

The following code demonstrates these steps:

```
C#
// l_Encrypted holds the encrypted data
// l_hPrivateSessionKey holds the handle to the session key.
Int32 l_TotalDecrypted = l_Encrypted.Length;
byte[] l_Decrypted = l_Crypto.KeyDecrypt(l_hPrivateSessionKey, l_Encrypted,
        ref l_TotalDecrypted);
```

```
VB
' l_Encrypted holds the encrypted data
' l_hPrivateSessionKey holds the handle to the session key.
Dim l_TotalDecrypted As Int32 = l_Encrypted.Length
Dim l_Decrypted() As Byte = l_Crypto.KeyDecrypt(l_hPrivateSessionKey,
        l_Encrypted, l_TotalDecrypted)
```

Importing a Session Key with a Sample Application

The C# and Visual Basic versions of the ImportSessionKey sample application are located in `SampleApplications\Chapter14`. This application demonstrates how to import session keys created by the same device and how to import session keys created by another device.

If you want to use a session key that was created by the same device running ImportSessionKey, then run the ExportSessionKey application to generate a file holding the bytes of the session key. Make sure that the file is named `LocalSessionKey.blob` on the root directory of the device. Make sure that the checkbox labeled "Use shared session key" is not checked, and then click the Go button. The sample application imports the locally created session key and uses it to encrypt the data in the textbox labeled Initial Text. Then the encrypted data is painted into the Encrypted Bytes textbox. Next it is decrypted and painted into the Decrypted Text textbox.

If you want to use a session key that was created by another device, follow these steps to use this sample application:

1. It is assumed that Device A will run ImportSessionKey and it will import a session key from Device B.
2. Run the ExportPublicKeyDemo sample application on Device A. This program drops a public key named `PublicExchangeKey.blob` for sharing session keys, as discussed in the section titled "Generating the Public Key." Copy this file to Device B.
3. On Device B, run ExportSessionKey and click the button labeled Drop Exported Session Key. This causes Device B to export a session key encrypted with the public key from Device A. Thus, only Device A can use this encrypted version of the session key. The encrypted session key is stored in a file called `ExportedSessionKey.blob`. Copy this file back to Device A.

4. Run ImportSessionKey on Device A and be sure that the checkbox labeled "Use shared session key" is checked. Click the Go button. The sample application imports the session key from Device B and uses it to encrypt the data in the textbox labeled Initial Text. Then the encrypted data is painted into the Encrypted Bytes textbox. Next it is decrypted and painted into the Decrypted Text textbox.

Determining CryptoAPI Constants with the ConstFinder Sample Application

Writing the `ManagedCryptoAPI` class required knowing a lot of the constant values that are defined as part of CryptoAPI. When writing native code, using these constants is easy because developers simply need to include the correct header files to access them.

The ConstFinder is an application for Embedded Visual C++ 3.0 that was used to determine the actual constant values needed to write `ManagedCryptoAPI`. This project will also load under Embedded Visual C++ 4.0. It is included in the folder `\SampleApplications\Chapter14\Windows_ConstFinder`.

ConstFinder simply sets local variables to the values of the constants, which makes it easy to learn their values by using a debugger.

If you don't have Embedded Visual C++ or are simply not interested in knowing how the various constant values in `ManagedCryptoAPI` were discovered, you don't need to worry about the ConstFinder. It is included only for your convenience.

In Brief

- The native CryptoAPI is a brittle programming interface that is capable of most common cryptography related functions. Users of the .NET Compact Framework must target the CryptoAPI to perform cryptography in their applications.
- Due to the brittleness and large number of pitfalls associated with targeting the CryptoAPI class directory from managed code, we provide a simpler-to-use wrapper class called `ManagedCryptoAPI`.
- `ManagedCryptoAPI` uses central concepts related to CryptoAPI. Specifically, developers must understand the notion of a handle and a CryptoAPI context.
- CryptoAPI is extensible. Developers can plug in their own "provider" to perform cryptography inside CryptoAPI. Because of its ubiquitous availability, the `ManagedCryptoAPI` uses the `PROV_RSA_FULL` provider.

- `ManagedCryptoAPI` can aquire a CryptoAPI context by calling either `ManagedCryptoAPI.AcquireDefaultContext` or `ManagedCryptoAPI.AcquireNamedContext`.
- There are many pitfalls to avoid when acquiring or creating a context.
- The `ManagedCryptoAPI.ManagedComputeHash` method computes a hash on data.
- The `ManagedCryptoAPI.PasswordEncrypt` method encrypts data based on a password.
- The `ManagedCryptoAPI.PasswordDecrypt` method decrypts data based on a password.
- The `ManagedCryptoAPI` class lets developers share a session key for encrypting and decrypting data on two or more devices.
- `ManagedCryptoAPI.GenerateSessionKey` creates a session key and returns a handle to it.
- `ManagedCryptoAPI.LocalSaveSessionKey` exports a session key into a byte array in such a way that only the device that created the byte array can use it again.
- `ManagedCryptoAPI.LoadPrivateSessionKey` loads the bytes of a previously saved session key. It can only load bytes of a session key created earlier on the same device. It returns a handle to an encryption key with which you can encrypt or decrypt data.
- `ManagedCryptoAPI.ExportPublicKey` exports a public key as a byte array. The public key can be used by other devices to encrypt a session key, which the original device can also use.
- `ManagedCryptoAPI.ImportPublicKey` can be used by a device to import the public key of a remote device. Then the device can call `ManagedCryptoAPI.ExportSessionKey` to export a session key as a set of bytes that the remote device can use. The bytes are encrypted so that only the remote device is capable of deciphering them. Thus, the bytes can be safely transmitted over an insecure network.
- A device can load a session key that a remote device wants to share by passing the session key bytes into `ManagedCryptoAPI.LoadSharedSessionKey`. The session key bytes must be encrypted with the local device's public key.
- `ManagedCryptoAPI.KeyEncrypt` encrypts data using a handle to a session key that is passed into it. The session key can be derived in a variety of ways, such as by the same device or imported from another device.
- `ManagedCryptoAPI.KeyDecrypt` decrypts data using a handle to a session key that is passed into it. It is the complementary method of `KeyEncrypt`.

Measuring the Performance of a .NET Compact Framework Application

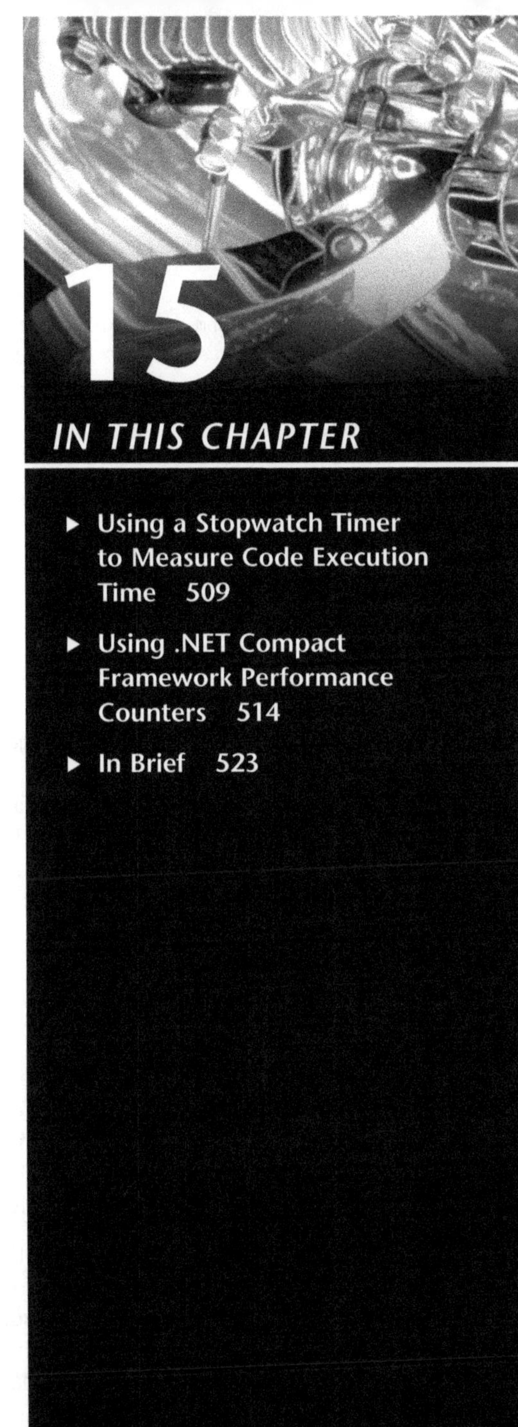

IN THIS CHAPTER

Using a Stopwatch Timer to Measure Code Execution Time

If you are working on a production application that will ship on the .NET Compact Framework platform, chances are you will be doing performance analysis on the application code. As this chapter is being written, the performance profiling tools on the market for the .NET Compact Framework are few and far between. Despite this lack of performance tools, there are some simple techniques that can be used to profile an application's performance traits.

First, it is important to have a reliable way to measure the execution time of a block of code. The .NET Compact Framework provides the `TickCount` property on the `Environment` class in the `System` namespace. This property can be used to calculate the execution time of a block of code. For example, the code in Listing 15.1 measures the execution time of the `XmlDocument.Load` method.

LISTING 15.1

```
C#
XmlDocument dom = new XmlDocument();

int startTime = System.Environment.TickCount;
dom.Load("test.xml");
int executionTime = System.Environment.TickCount - startTime;

VB
Dim startTime As Int32
Dim executionTime As Int32
Dim dom As New XmlDocument

startTime = System.Environment.TickCount
dom.Load("test.xml")
executionTime = System.Environment.TickCount - startTime
```

The `executionTime` variable contains the number of milliseconds the `XmlDocument.Load` method call took to execute. Although this is a very simple way to measure execution time, the `TickCount` property can return inaccurate results. The `TickCount` property is based on the OS implementation of the `GetTickCount` function. The `GetTickCount` function retrieves the number of milliseconds that have elapsed since the system was started. The resolution of the return value is limited to the system timer, which does not usually have a high resolution. Also, the `GetTickCount` counter will wrap around to zero if the system is running continuously for 49.7 days.

A high-resolution timer, like `QueryPerformanceCounter`, will give more accurate and reliable results. If a high-resolution performance counter does exist on the system, the `QueryPerformanceCounter` method can be used to retrieve the frequency of the counter in counts per second. The frequency of the counter is processor dependent. On some processors the frequency might be the cycle rate of the processor clock.

The `QueryPerformanceCounter` method retrieves the current value of the high-resolution performance counter. By calling this function at the beginning and end of a section of code, the counter is used as a high-resolution timer. Suppose that `QueryPerformanceFrequency` indicates that the frequency of the high-resolution performance counter is 50,000 counts per second. If the application calls `QueryPerformanceCounter` immediately before and immediately after the section of code to be timed, the counter values might be 3000 and 6500, respectively. These values would indicate that .07 seconds (3500 counts) elapsed while the code executed.

The .NET Compact Framework does not provide an API to access these functions. Despite this, we can use Platform Invocation, P/Invoke, to get access to this high-resolution counter. The `QueryPerformanceCounter` and `QueryPerformanceFrequency` functions are located in coredll.dll. The DllImport attributes can provide access to these functions. It is important to note that QueryPerformanceCounter and QueryPerformanceFrequency provide access only to a high-resolution timer. If the original equipment manufacturer, OEM, did not provide an implementation of a high-resolution counter, then these APIs will fail. If this is the case, your timer code should fall back to the Environment.TickCount property or some other timer implementation. Listing 15.2 contains an example of a stopwatch-like timer using a high-resolution counter.

LISTING 15.2

```
C#
public sealed class StopWatch {
  // Native Methods
  [DllImport("coredll.dll")]
  public static extern bool QueryPerformanceCounter(out long c);
  [DllImport("coredll.dll")]
  public static extern bool QueryPerformanceFrequency(out long c);

  // Static data
  static long frequency;
  static bool perfCounterSupported;

  // Instance Data
  long startTime = 0;

  // Initialize the counter frequency static
  static StopWatch() {
    perfCounterSupported = QueryPerformanceFrequency (out frequency);
  }

  // Start the StopWatch
  public void Start() {
    startTime = TickCount;
  }

  // Stop the StopWatch
  public long Stop() {
    if(perfCounterSupported)
      return (long)((float)(TickCount-startTime)/frequency)*1000;
    else
      return (TickCount - startTime);
  }
```

```
    // Get the current tick count
    public static long TickCount {
      get {
        long val;
        bool perfCounterSupported = QueryPerformanceCounter(out val);

        if (perfCounterSupported)
          return val;
        else
          return (long)System.Environment.TickCount;
      }
    }
  }
```

```
VB
Public Class StopWatch
    ' Native Methods
    Declare Function QueryPerformanceCounter Lib _
    "coredll.dll" (ByRef c As Long) As Boolean
    Declare Function QueryPerformanceFrequency Lib _
    "coredll.dll" (ByRef c As Long) As Boolean

    ' Static Data
    Shared frequency As Long
    Shared perfCounterSupported As Boolean

    ' Instance Data
    Dim startTime As Long

    ' Initialize the counter frequency
    Shared Sub New()
        perfCounterSupported = QueryPerformanceFrequency(frequency)
    End Sub

    ' Initialize the instance data
    Sub New()
        startTime = 0
    End Sub

    ' Start the StopWatch
    Public Sub Start()
        startTime = TickCount()
```

```
    End Sub

    ' Stop the StopWatch
    Public Function [Stop]() As Long
        If perfCounterSupported Then
            Return CLng(CDbl(TickCount() - startTime) / frequency * 1000)
        Else
            Return (TickCount() - startTime)
        End If
    End Function

    ' Get the current tick count
    Public Shared ReadOnly Property TickCount() As Long
        Get
            Dim val As Long
            perfCounterSupported = QueryPerformanceCounter(val)

            If perfCounterSupported Then
                Return val
            Else
                Return CLng(System.Environment.TickCount)
            End If
        End Get
    End Property
End Class
```

This code provides a general-purpose, high-resolution stopwatch timer that can measure the execution time of an operation. The `Stop` method returns the number of milliseconds that have elapsed since calling the `Start` method. Listing 15.3 contains an example of how one could use the `StopWatch` class.

LISTING 15.3

```
C#
StopWatch timer = new StopWatch();

timer.Start();
Thread.Sleep(1000);
long elapseTime = timer.Stop();

MessageBox.Show("The stop watch recorded " + elapseTime + "ms");
```

```
VB
Dim timer As New StopWatch
Dim elapseTime As Int64

timer.Start()
Thread.Sleep(1000)
elapseTime = timer.Stop()

MessageBox.Show("The stop watch recorded " & elapseTime & "ms")
```

Using .NET Compact Framework Performance Counters

Performance testing and tuning are not confined only to measuring the execution time of the code. Performance also depends on the working set of the application. This includes the amount of memory allocated and de-allocated while the application executes. The presence of the .NET Compact Framework garbage collector makes it difficult to calculate how much memory is used and recycled as the application executes. The .NET Compact Framework's execution engine provides a small window into an application's working set.

The .NET Compact Framework's execution engine has the ability to keep track of several performance statistics as your application runs. These statistics include how much memory is allocated, how much memory was garbage collected, the number of P/Invokes, and so on. By default, this information is not collected or displayed to the user, but with some registry tweaking these performance counters can be turned on.

Turn On the .NET Compact Framework Performance Counters

The .NET Compact Framework performance counters are controlled by a registry key setting. The registry key is

```
HKEY_LOCAL_MACHINE\SOFTWARE\Microsoft\.NETCompactFramework\PerfMonitor
```

The DWORD registry value is "`Counters`." The `Counters` value can be either zero or one. Setting the value to zero turns the performance counters off, while setting the value to one turns them on. If this registry key and value are not present, then the performance counters are off. This is the default state of the .NET Compact Framework when it is installed on the device.

When the execution engine is first invoked, it checks the value of this registry key. If the value is one, the performance counters are collected. The registry key is checked at startup time only. Thus, changing the registry value after the execution engine is started will not change its behavior until it is shut down and restarted.

Viewing the Performance Counters

To view the performance counters, set the registry key to one and then kick off any .NET Compact Framework application. When your application is done executing, the execution engine will output a file named `mscoree.stat` to the root directory of the device. This is a plain text file that contains the values of the performance counters collected during the run of the application. See Figure 15.1 for an example of the `mscoree.stat` file.

Understanding the Performance Counters

mscoree.stat - WordPad

File Edit View Insert Format Help

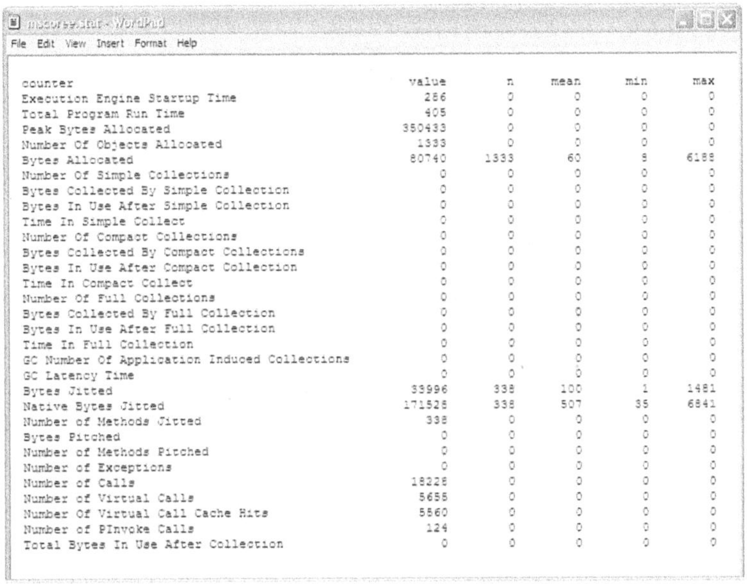

counter	value	n	mean	min	max
Execution Engine Startup Time	286	0	0	0	0
Total Program Run Time	405	0	0	0	0
Peak Bytes Allocated	350433	0	0	0	0
Number Of Objects Allocated	1333	0	0	0	0
Bytes Allocated	80740	1333	60	8	6188
Number Of Simple Collections	0	0	0	0	0
Bytes Collected By Simple Collection	0	0	0	0	0
Bytes In Use After Simple Collection	0	0	0	0	0
Time In Simple Collect	0	0	0	0	0
Number Of Compact Collections	0	0	0	0	0
Bytes Collected By Compact Collections	0	0	0	0	0
Bytes In Use After Compact Collection	0	0	0	0	0
Time In Compact Collect	0	0	0	0	0
Number Of Full Collections	0	0	0	0	0
Bytes Collected By Full Collection	0	0	0	0	0
Bytes In Use After Full Collection	0	0	0	0	0
Time In Full Collection	0	0	0	0	0
GC Number Of Application Induced Collections	0	0	0	0	0
GC Latency Time	0	0	0	0	0
Bytes Jitted	33996	338	100	1	1481
Native Bytes Jitted	171528	338	507	35	6841
Number of Methods Jitted	338	0	0	0	0
Bytes Pitched	0	0	0	0	0
Number of Methods Pitched	0	0	0	0	0
Number of Exceptions	0	0	0	0	0
Number of Calls	18228	0	0	0	0
Number of Virtual Calls	5655	0	0	0	0
Number Of Virtual Call Cache Hits	5560	0	0	0	0
Number of PInvoke Calls	124	0	0	0	0
Total Bytes In Use After Collection	0	0	0	0	0

FIGURE 15.1 An `mscoree.stat` file is generated by the execution engine.

There are thirty performance counters, and, despite their descriptive names, their meanings are not all self-evident. The performance counters can be divided into five groups using the information they collect as the division criteria. Table 15.1 describes the five counter groups and the information they collect.

TABLE 15.1
The .NET Compact Framework Performance Counter Groups

COUNTER GROUP	INFORMATION COLLECTED
Execution Time	The execution time of the application and execution engine startup
Allocation	Characteristics of memory allocation
Collection	Characteristics of the memory collected by the garbage collector
JIT Compilation	Characteristics of the Just-In-Time (JIT) compiler and the code it generates
Method Call	Characteristics of the methods called while the application ran

Measuring Application Execution Time Characteristics

The counters in the Execution Time group measure how long the application and execution engine executed. This includes total application runtime and execution engine startup time. The following list describes the counters in the Execution Time group:

Execution Engine Startup Time This counter measures the time from when the execution engine first gets control, after the OS has loaded the EXEs and DLLs, to when the execution engine is about to call the main method of the application. This time, which does not include creating and initializing the application's static objects, is measured in milliseconds. The values *n,* mean, min, and max are not collected for this counter.

Total Program Run Time The time from when the execution engine first gets control to when the execution engine terminates. The time is measured in milliseconds. The values for *n,* mean, min, and max are not collected for this counter.

Measuring Allocation Characteristics

The counters in the Allocation group measure memory allocation characteristics. These counters measure memory in terms of byte as well as objects, but the counters do not provide information on what types of objects were allocated.

Peak Bytes Allocated This counter measures the largest number of bytes allocated at once. The values *n,* mean, min, and max are not collected for this counter.

Number Of Objects Allocated This is the total number of objects allocated by the execution engine. This includes every object that was created while the program was executing. The values *n,* mean, min, and max are not collected for this counter. This counter is measured in objects.

Bytes Allocated The number of bytes allocated for all of the objects allocated in the system. The value *n* is the number of objects allocated. The mean value is the mean number of bytes allocated per object. The min value is the size of the smallest object. The max value is the size of the largest object allocated. All values except *n* are measured in bytes. The *n* value is measured in objects and should be equal to the value of the Number of Objects Allocated counter.

Using the Counters in the Collection Group

The Collection counters collect information about the operations performed by the garbage collector (GC). This is the largest group of counters. The GC does three types of collections, and there are four counters collected for each type of collection. See the "Understanding the .NET Compact Framework's Garbage Collector" section for a description of each type of collection. The total amount of time spent in the GC is also tracked. The following list describes the many counters in the Collection group:

Number Of Simple Collections This counter measures the number of simple collections performed by the garbage collector during the execution of the application. See the section titled "Understanding the .NET Compact Framework's Garbage Collector" for a description of a simple collection. The values *n,* mean, min, and max are not collected for this counter.

Bytes Collected By Simple Collection This counter measures the total number of bytes freed during all of the simple collections that occurred during the program run. The *n* value is the number of simple collections that occurred and should be equal to the value of the Number of Simple Collections counter. The mean, min, and max values are self-explanatory.

Bytes In Use After Simple Collection This counter measures the sum of all the bytes that were still actively being referenced after each simple collection. The *n* value is the number of simple collections that occurred and should be equal to the value of the `Number of Simple Collections` counter. The mean, min, and max values are self-explanatory.

Time In Simple Collect This counter measures the total amount of time spent by the garbage collector doing simple collections. This counter is measured in milliseconds, and the *n,* mean, min, and max values are not collected.

Number Of Compact Collections Measures the total number of compact collections that occurred during application execution. See the section titled "Understanding the .NET Compact Framework's Garbage Collector" for a description of a compact collection. The values *n,* mean, min, and max are not collected for this counter.

Bytes Collected By Compact Collections This one measures the total number of bytes freed during all of the compact collections that occurred during application execution. The *n* value is the total number of compact collections that occurred and should be equal to the value of the Number of Compact Collections counter. The mean, min, and max values are self-explanatory.

Bytes In Use After Compact Collection This counter measures the sum of all the bytes that were still actively being referenced after each compact collection occurred. The *n* value is the number of compact collections that occurred and should be equal to the value of the `Number of Compact Collections` counter. The mean, min, and max values are self-explanatory.

Time In Compact Collect This counter measures the total amount of time that the garbage collector spent making compact collections. This counter is measured in milliseconds. The values *n,* mean, min, and max are not collected for this counter.

Number Of Full Collections This is the total number of full collections carried out by the garbage collector during application execution. See the section titled "Understanding the .NET Compact Framework's Garbage Collector" for a description of a full collection. The values *n,* mean, min, and max are not collected for this counter.

Bytes Collected By Full Collection This counter measures the total number of bytes freed during all of the full collections that occurred during application execution. The *n* value is the total number of full collections that occurred and should be equal to the value of the Number of Full Collections counter. The mean, min, and max values are self-explanatory.

Bytes In Use After Full Collection This counter measures the sum of all the bytes that were still actively being referenced after each full collection occurred. The *n* value is the number of full collections that occurred and should be equal to the value of the Number of Full Collections counter. The mean, min, and max values are self-explanatory.

Time In Full Collection This tool measures the total amount of time that the garbage collector spent executing full collections. This counter is measured in milliseconds. The values *n,* mean, min, and max are not collected for this counter.

GC Number Of Application Induced Collections This counter measures the number of collections that were manually invoked by the application by calling `GC.Collect`. The values *n,* mean, min, and max are not collected for this counter.

GC Latency Time This counter measures the total amount of time spent by the garbage collector making all collections during the application execution. The values *n,* mean, min, and max are not collected for this counter.

Total Bytes In Use After Collection The peak memory, managed and unmanaged, in use by the application as measured immediately following each garbage collection. The *n* value is the number of garbage collections that occurred during application execution. The mean, min, and max values are self-explanatory.

Discovering JIT Compiler Performance Characteristics

The execution engine also collects information about the JIT compiler. This includes information about the code it compiled as well as the code it pitched. The term *pitching* means to remove JIT-compiled code from the code cache, an in-memory cache for JIT-compiled code. This provides fast access for methods called more than once because they need to be compiled on only the first method call. Subsequent calls will execute the JIT-compiled code in the code cache. That said, code pitching will actually slow your application down because cached methods that were pitched will need to be recompiled the next time they are called. The following list describes the counters in the JIT Compilation group:

Bytes Jitted This counter measures the total amount of MSIL code that was Just-In-Time compiled. The *n* value is the number of methods jitted and should be equal to the Number of Methods Jitted counter. The mean value is the mean number of bytes jitted per method. The min value is the smallest number of bytes jitted in one method, and the max value is the largest number of bytes jitted in one method.

Native Bytes Jitted This tool measures the total amount of native instructions created while jitting the MSIL of the application. The *n* value is the number of methods jitted and should be equal to the Number of Methods Jitted counter. The mean value is the mean number of native instructions, in bytes, created per method. The min value is the smallest number of native instructions, in bytes, created by jitting a method. The max value is the largest number of native instructions, in bytes, created by jitting a method.

Number of Methods Jitted This counter measures the total number of methods jitted during the program run. The values *n,* mean, min, and max are not collected for this counter.

Bytes Pitched This one counts the total number of native instructions, measured in bytes, released from memory after being Just-In-Time compiled. The values of *n,* mean, min, and max are not collected for this counter.

Number of Methods Pitched This counter measures the total number of methods released from memory after having been Just-In-Time compiled. The values *n,* mean, min, and max are not collected for this counter.

Measure Method Call Characteristics

The execution engine can also collect information about the method calls and exceptions thrown. The counters in the Method Call group retain this information, as described in this list:

Number of Exceptions This counter measures the number of exceptions that were thrown during application execution. Values for *n,* mean, min, and max are not collected for this counter.

Number of Calls Measures the number of calls that the JIT could make directly. This means the JIT could generate an instruction(s) resulting in a direct call to the method. Values for *n,* mean, min, and max are not collected for this counter.

CALCULATING THE VIRTUAL CALL CACHE HIT RATE

Using the Number of Virtual Call Cache Hits counter and the `Number of Virtual Calls` counter, you can calculate the virtual call cache hit rate. This hit rate is the percentage of virtual calls that were resolved in the virtual call cache. The following formula can be used to find the virtual call cache hit rate:

(Number of Virtual Call Cache Hits / Number of Virtual Calls) × 100%

For example, if there were 5000 virtual calls and the virtual call cache was hit 4900 times, then the hit rate would equal 98%:

(4900/5000) × 100% = 98%

The higher the hit rate, the better, because virtual calls that were resolved in the cache have a smaller overall execution time.

Number of Virtual Calls This one measures the number of calls that the JIT could not make directly. This means that in order to call the method, the JIT first needed to compile the method and then determine how to call the method.

Number Of Virtual Call Cache Hits This counter measures how many virtual calls were resolved by the virtual call lookup cache. Values for *n,* mean, min, and max are not collected for this counter.

Number of PInvoke Calls This counter measures the total number of P/Invoke calls that were made during application execution. The values for *n,* mean, min, and max are not collected for this counter.

Total Bytes in Use After Collection The final counter measures the peak memory, managed and unmanaged, in use by the application as measured immediately following each garbage collection. The *n* value is the number of garbage collections that occurred during application execution. The mean, min, and max values are self-explanatory.

Understanding the .NET Compact Framework's Garbage Collector

The .NET Compact Framework garbage collector uses a completely different algorithm from the desktop garbage collector when freeing memory. Whereas the desktop garbage collector uses an algorithm based on memory lifetimes and generations, the .NET Compact Framework has three different types of garbage collection that it will execute when memory is needed.

The first type of garbage collection is simple collection. This is the fastest algorithm the garbage collector uses to free memory. This algorithm is a basic mark-and-sweep algorithm. It passes through the memory, marking any memory that is being used and freeing that which is not.

The second type of garbage collection is compact collection. First a compact collection performs a simple collection to free any unused memory. Then the compact collection attempts to rearrange allocated memory into one contiguous block of memory. Since a compact collection includes a simple collection, the compact collection is a slower collection.

The third and slowest garbage collection algorithm is the full collection. In addition to a compact collection, the full collection will pitch code until there is enough free memory. Any code that is pitched will have to be re-jitted when it is needed again.

The garbage collector does not do all three collections every time memory is needed. The garbage collector will first run a simple collection and test whether enough memory has been released. If not, a compact collection is executed. If there still is not enough memory released, the garbage collector will finally run the full collection algorithm. If there is still not enough memory after a full collection, a `System.OutOfMemoryException` will be thrown.

The NetCF Performance Counter Explorer

In the support media of this book, in the directory for this chapter, you will find a Visual Studio .NET project file for the .NET Compact Framework Performance Counter Explorer. This is a desktop application that helps analyze the performance counters in the `mscoree.stat` file. To use this application, first create a `mscoree.stat` file and put it on a desktop computer. Next, start the .NET Compact Framework Performance Counter Explorer. Then load an `mscoree.stat` file by selecting the Open... menu option from the File menu. The open file dialog will appear. Find the `mscoree.stat` file and click the Open button. This will load the file into the performance counter table (see Figure 15.2). The table now contains all of the performance counters from the `mscoree.stat` files as well as all of their values.

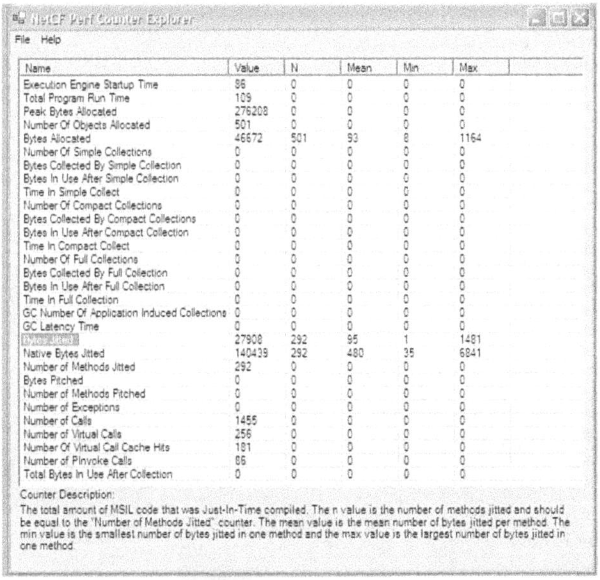

NetCF Perf Counter Explorer

File Help

Name	Value	N	Mean	Min	Max
Execution Engine Startup Time	86	0	0	0	0
Total Program Run Time	109	0	0	0	0
Peak Bytes Allocated	276208	0	0	0	0
Number Of Objects Allocated	501	0	0	0	0
Bytes Allocated	46672	501	93	8	1164
Number Of Simple Collections	0	0	0	0	0
Bytes Collected By Simple Collection	0	0	0	0	0
Bytes In Use After Simple Collection	0	0	0	0	0
Time In Simple Collect	0	0	0	0	0
Number Of Compact Collections	0	0	0	0	0
Bytes Collected By Compact Collections	0	0	0	0	0
Bytes In Use After Compact Collection	0	0	0	0	0
Time In Compact Collect	0	0	0	0	0
Number Of Full Collections	0	0	0	0	0
Bytes Collected By Full Collection	0	0	0	0	0
Bytes In Use After Full Collection	0	0	0	0	0
Time In Full Collection	0	0	0	0	0
GC Number Of Application Induced Collections	0	0	0	0	0
GC Latency Time	0	0	0	0	0
Bytes Jitted	27908	292	95	1	1481
Native Bytes Jitted	140439	292	480	35	6841
Number of Methods Jitted	292	0	0	0	0
Bytes Pitched	0	0	0	0	0
Number of Methods Pitched	0	0	0	0	0
Number of Exceptions	0	0	0	0	0
Number of Calls	1455	0	0	0	0
Number of Virtual Calls	256	0	0	0	0
Number Of Virtual Call Cache Hits	181	0	0	0	0
Number of PInvoke Calls	86	0	0	0	0
Total Bytes In Use After Collection	0	0	0	0	0

Counter Description:

The total amount of MSIL code that was Just-In-Time compiled. The n value is the number of methods jitted and should be equal to the "Number of Methods Jitted" counter. The mean value is the mean number of bytes jitted per method. The min value is the smallest number of bytes jitted in one method and the max value is the largest number of bytes jitted in one method.

FIGURE 15.2 A `mscoree.stat` file is loaded into the NetCF Perf Counter Explorer.

Now click one of the performance counters. The description of the counter is displayed in the Counter Description area at the bottom of the application.

The second feature of the .NET Compact Framework Performance Counter Explorer is its ability to compare two `mscoree.stat` files. This comes in handy when attempting to measure the performance impact of a source code change. First, create a `mscoree.stat` file without the source code change and put it on the desktop computer. Next, apply the source code changes to your application and create another `mscoree.stat` file. Put this `mscoree.stat` file on the desktop computer as well. Then select the Compare... option from the File menu. The open file dialog will appear. Select the first `mscoree.stat` file to compare and click Open. Another open file dialog will appear. Now select the second `mscoree.stat` file to compare. Data will again be loaded into the Perf Counter table. The performance counter names will be the same, but the values have a different meaning. The values are now the differences between the first and second `mscoree.stat` files. This provides an easy way to view the differences between two different `mscoree.stat` files (see Figure 15.3).

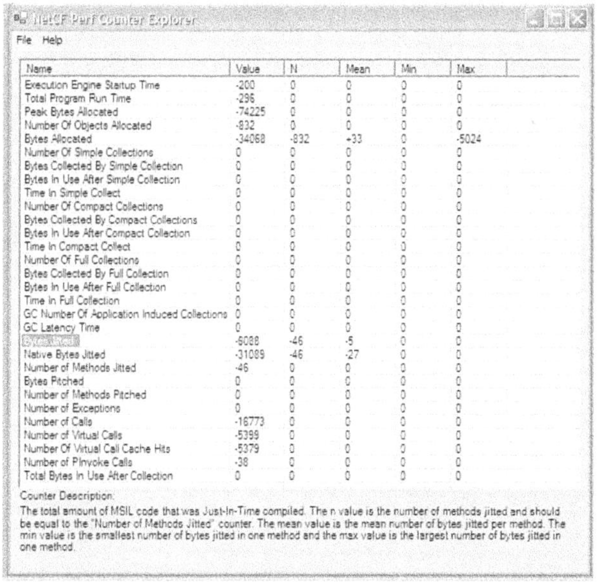

NetCF Perf Counter Explorer

File Help

Name	Value	N	Mean	Min	Max
Execution Engine Startup Time	-200	0	0	0	0
Total Program Run Time	-296	0	0	0	0
Peak Bytes Allocated	-74225	0	0	0	0
Number Of Objects Allocated	-832	0	0	0	0
Bytes Allocated	-34068	-832	+33	0	-5024
Number Of Simple Collections	0	0	0	0	0
Bytes Collected By Simple Collection	0	0	0	0	0
Bytes In Use After Simple Collection	0	0	0	0	0
Time In Simple Collect	0	0	0	0	0
Number Of Compact Collections	0	0	0	0	0
Bytes Collected By Compact Collections	0	0	0	0	0
Bytes In Use After Compact Collection	0	0	0	0	0
Time In Compact Collect	0	0	0	0	0
Number Of Full Collections	0	0	0	0	0
Bytes Collected By Full Collection	0	0	0	0	0
Bytes In Use After Full Collection	0	0	0	0	0
Time In Full Collection	0	0	0	0	0
GC Number Of Application Induced Collections	0	0	0	0	0
GC Latency Time	0	0	0	0	0
[illegible]	-6088	-46	-5	0	0
Native Bytes Jitted	-31089	-46	-27	0	0
Number of Methods Jitted	-46	0	0	0	0
Bytes Pitched	0	0	0	0	0
Number of Methods Pitched	0	0	0	0	0
Number of Exceptions	0	0	0	0	0
Number of Calls	-16773	0	0	0	0
Number of Virtual Calls	-5399	0	0	0	0
Number Of Virtual Call Cache Hits	-5379	0	0	0	0
Number of PInvoke Calls	-38	0	0	0	0
Total Bytes In Use After Collection	0	0	0	0	0

Counter Description:

The total amount of MSIL code that was Just-In-Time compiled. The n value is the number of methods jitted and should be equal to the "Number of Methods Jitted" counter. The mean value is the mean number of bytes jitted per method. The min value is the smallest number of bytes jitted in one method and the max value is the largest number of bytes jitted in one method.

FIGURE 15.3 The NetCF Perf Counter Explorer is now loaded to compare two `mscoree.stat` files.

Selecting a performance counter will again display its description in the Counter Description area. The Perf Counter table displays only the differences between the values. You can get to the original values by double-clicking a performance counter or selecting the performance counter name and pressing Enter. This will display a window that contains the Value, N, Mean, Max, and Min values of the performance counter from both File One and File Two (see Figure 15.4).

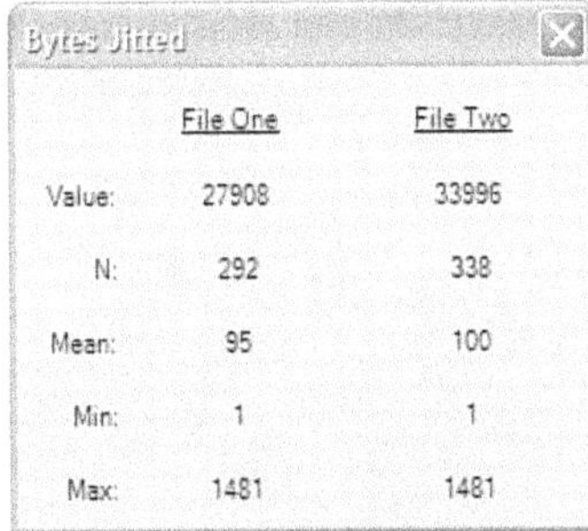

FIGURE 15.4 The counter comparison dialog in the NetCF Perf Counter Explorer lets you directly compare the values from two files.

In Brief

- Building an accurate stopwatch timer to measure the execution time of an application
- Turning on and viewing the .NET Compact Framework performance counters
- Viewing and comparing the .NET Compact Framework performance counters with the .NET Compact Framework Performance Counter Explorer

Deploying a .NET Application

IN THIS CHAPTER

Packaging a .NET Compact Framework Application

Once you finish building, testing, and debugging your application, you will need to distribute it to your eager customers in a simple, reliable manner. In the .NET Compact Framework, this is done in two steps. First your application must be packaged into a self-extracting cabinet file. These files usually have the `.cab` file extension and will here be referred to as CAB files. One CAB file contains all the files required for your application, which reduces the chances of partial installation.

Using Visual Studio .NET to Package an Application

You can use Visual Studio .NET to create a CAB file for your application. This is definitely the quickest and easiest way to build a CAB file for your application. It is also the least flexible way to build a CAB file. Later in this chapter we will discuss a more flexible way to create a CAB file, but for now let's discuss CAB file generation with Visual Studio .NET.

Visual Studio .NET provides a simple UI-driven way to build a CAB for your application. You can simply select the Build CAB File option on the Build menu, which will build a CAB file in the `<project_name>/cab/<build_type>` directory, "`<project name>`" being the name of your project and

"<build_type>" being either the debug or the release, depending on the current build mode settings.

There are three other files that are created in the `<project_name>/obj/<build_type>` directory: a batch file, a configuration file, and a dependencies file. The batch file is named `BuildCab.bat` and can be used to rebuild the CAB file. This is particularly useful if you have made changes to the configuration file and you do not want the configuration file to be regenerated by Visual Studio .NET by choosing the Build CAB File option again. The configuration file's name depends on the name and target platform of the project. For example, the project named CalcDotNet target for the Pocket PC platform would be named `CalcDotNet_PPC.inf`. The configuration file is described in detail in the "Using the CAB Wizard to Package an Application" section. The dependencies file is named `Dependencies_PPC.inf`. This file is used to record the dependencies for the CAB file. When the CAB is being installed, the installer will check that the dependencies in this file have already been installed on the device.

That is all there is to it. Now you have a CAB file that can be installed on a device.

Using CAB Wizard to Package an Application

Although using Visual Studio .NET to create a CAB file is insanely easy, it is very inflexible. Using the CAB Wizard takes more time and understanding but provides great flexibility for building the CAB file.

USING VISUAL STUDIO .NET AND THE CAB WIZARD TOGETHER TO CREATE A CAB FILE

One great way to build a CAB file for your application is to use both Visual Studio .NET and the CAB Wizard. You can first use Visual Studio .NET to create the configuration file. This file is output into the `<project_name>/obj/<build_type>` directory. See the "Using Visual Studio .NET to Package an Application" section for details. Then you can edit the configuration file to customize your CAB file. Finally, you can use the `BuildCab.bat` file, also output in the `<project_name>/obj/<build_type>` directory, which uses the CAB Wizard to rebuild your CAB file.

Before you can use the CAB Wizard, you must have a configuration file that tells the wizard how to build the CAB file. The configuration file has the `.inf` file extension and consists of a number of sections that describe the target locations of the files and shortcuts as well as registry settings that will be contained with the CAB file.

Throughout the `.inf` file you will need to specify filepaths for different installation points. The CAB Wizard has a set of macro strings that represent certain directories on the Windows CE file system. Table 16.1 lists these macros and their corresponding Windows CE directories.

TABLE 16.1
Windows CE Directory Macros

DIRECTORY MACRO	WINDOWS CE DIRECTORY
%CE1%	\Program Files
%CE2%	\Windows
%CE4%	\Windows\StartUp
%CE5%	\My Documents
%CE8%	\Program Files\Games
%CE11%	\Windows\Start Menu\Programs
%CE14%	\Windows\Start Menu\Programs\Games
%CE15%	\Windows\Fonts
%CE17%	\Windows\Start Menu

The configuration file consists of several different sections that describe the files that will be placed in the CAB file. Each section is now described.

Specifying the Application Version in the `Version` Section

```
[Version]
Signature = "signature_name"
Provider = "provider_name"
CESignature = "$Windows CE$"
```

The `Version` section is used to specify the creator of the CAB file. The `Version` section must always be present in the configuration file.

`Signature` Must be either "$Windows NT$" or "$Windows 95$". It does not matter which string you choose, making this key almost useless.

`Provider` It is recommended that this be the name of the company that is creating the application. This is recommended because the value will be displayed during installation of the CAB.

`CESignature` Must be "$Windows CE$". This key seems almost as useless as the `Signature` key, but it may have some use in later versions of CAB Wizard.

Understanding the `CEStrings` Section

```
[CEStrings]
AppName = "Calc.NET"
InstallDir = %CE1%\%AppName%
```

The CEStrings section provides substitution strings for the application name and default installation directory. The CEStrings section is required.

AppName The name of your application. Any further occurrence of %AppName% will be resolved to AppName's value.

InstallDir The name of the default installation directory. Any further occurrence of %InstallDir% will be resolved to InstallDir's value.

In the earlier example this would be \Program Files\CalcDotNet.

Specifying Custom Substitution Strings in the Strings Section

```
[Strings]
db_file = db.mdb
config_file = config.xml
```

The Strings section is used to define one or more custom substitution strings. Think of this section as a dictionary of custom substitution strings that can be used throughout the configuration file. This section can contain one or more entries. In the earlier example any occurrence of %db_file% or %config_file% will be replaced with db.mdb or config.xml, respectively.

Understanding the DefaultInstall Section

```
 [DefaultInstall]
CopyFiles = Files.Common
CEShortcuts = Shortcuts

[DefaultInstall.ARM]
CopyFiles=Files.ARM
CESetupDLL=vsd_setup.dll

[Shortcuts]
Calc.NET, 0, CalDotNet.exe

[Files.Common]
CalcDotNet.exe,,,0

[Files.ARM]
vsd_setup.dll,,,0
```

The `DefaultInstall` section specifies the application files that will be installed. This section is required. Additional `DefaultInstallation` sections can be specified for the various processor types supported. In the earlier example the `DefaultInstall.ARM` section would define the default installation files for the ARM processor. The `DefaultInstallation` section can contain five different keys, but only one, `CopyFiles`, is required.

The `CopyFiles` key defines the files that will be copied to the device when the application installs. The value is the name of a custom section that is declared in the configuration file. The custom section contains the list of files to be installed. In the example the `File.Common` section and the `Files.ARM` section define custom `CopyFiles` sections. The `File.Common` section tells the installer to copy the `CalcDotNet.exe` file, whereas the `Files.Arm` copies `vsd_setup.dll`. For more information on the `CopyFiles` section, see the Windows CE SDK documentation.

The `AddReg` key adds entries into the registry when the application is installed. This section is optional. For more information, see the Windows CE SDK documentation.

Specifying Application Shortcuts with the `CEShortcuts` Key

`CEShortcuts` defines the shortcuts that the installer creates on the device. The value of the `CEShortcuts` key is the name of a custom section that lists the shortcuts to create. In the previous example the `Shortcuts` section is this custom section. The Shortcuts section contains the list of shortcuts that will be created when the application is installed. A shortcut list entry takes the following form, as expanded upon in the list that follows:
`filename,type,target,destination`.

- `filename` is the name of the shortcut.
- `type` is a numeric value. Zero represents a shortcut to a file, whereas any nonzero value represents a shortcut to a directory.
- `target` defines the destination file or directory for the shortcut. If a file is specified, it must also be specified in the `CopyFiles` section. Use a `file_list_section` name as defined in the `DestinationDirs` section or the `%InstallDir%` substitution string.
- `destination` defines the destination path for the shortcut file. Only a directory identifier from Table 16.1 or `%InstallDir%` can be used. This is optional, and if it is not specified, the shortcut file will be placed in the default directory destination, which is defined in the `DestinationDir` section.

In the earlier example, the shortcut named `Calc.NET`, which targets `CalcDotNet.exe`, is created in the default directory destination.

Specifying the Desktop Location of the Application in `SourceDisksNames` Section

```
[SourceDisksNames]
1 = ,"Common files",,C:\myApp\MyCommonFiles

[SourceDisksNames.ARM]
2 = ,"ARM Files",,C:\MyApp\ARMFiles
```

The `SourceDisksNames` section specifies the desktop paths to where the application files can be retrieved when the CAB file is being built. The SourceDisksNames section is required. The keys in this section take the following form: `disk_id = ,comment, ,path`

- `disk_id` is the source identifier used to specify the source directory. You will see this identifier again in the `SourceDisksFiles` section.
- `comment` is a friendly description of the source directory.
- `path` is the desktop file system path to where the application's files live.

Additional sections that are processor specific can be specified by appending the suffix `.ProcessorName` to `SourceDisksNames`. In the previous example a `SourceDisksNames` section for the ARM processor is defined.

The `SourceDisksFiles` Section

```
[SourceDisksFiles]
intro.wav=1,"\sounds"

[SourceDisksFiles.ARM]
sample_arm.exe=2, "\samples"
```

The `SourceDisksFiles` section describes the name and path of the files for the application. The `SourceDisksFiles` is also required. The keys in this section take the following format: `filename=disk_id[, subdir]`

- `filename` is the source filename.
- `disk_id` is the source identifier defined in `SourceDisksNames` section.
- `subdir` is the subdirectory under the `disk_id` directory that contains the file. The parameter is optional.

Additional sections that are processor specific can be specified by appending the suffix `.ProcessorName` to `SourceDisksFiles`. In the example a `SourceDisksFiles` section for the ARM processor is defined.

Specifying the Final Application Destination Directory in the DestinationDirs Section

```
[DestinationDirs]
Shortcuts = 0, %CE17%
Files.Common = 0, %InstallDir%
Files.ARM = 0, %InstallDir%
```

The DestinationDirs section is a required section that specifies the paths of the destination directories on the target device. The format of the entries in this section is as follows:

```
file_list_section = 0,subdir
```

> file_list_section The section that contains the list of files that will be installed in the location specified by the subdir parameter.
>
> subdir The destination directory. The subdir key can be an absolute filepath, a directory macro, or the %InstallDir% macro. In the previous example, the files in the Shortcuts section are installed in the \Windows\Start Menu directory. The files from the Files.Common section and the Files.ARM section are installed in %InstallDir%.

Listing 16.1 is a complete configuration file for the application Calc.NET.

LISTING 16.1

```
[Version]
Signature="$Windows NT$"
Provider="Calc-Net-Company"
CESignature="$Windows CE$"

[CEStrings]
AppName="Calc.NET"
InstallDir=%CE1%\%AppName%

[CEDevice]
VersionMin=3.00
VersionMax=3.99

[DefaultInstall]
CEShortcuts=Shortcuts
CopyFiles=Files.Common

[DefaultInstall.ARM]
CopyFiles=Files.ARM
CESetupDLL=vsd_setup.dll
```

```
[DefaultInstall.SH3]
CopyFiles=Files.SH3
CESetupDLL=vsd_setup.dll

[DefaultInstall.MIPS]
CopyFiles=Files.MIPS
CESetupDLL=vsd_setup.dll

[DefaultInstall.X86]
CopyFiles=Files.X86
CESetupDLL=vsd_setup.dll

[SourceDisksNames]
1=,"Common1",,"E:\Projects\Calc.NET\obj\Debug\"

[SourceDisksNames.ARM]
2=,"ARM2",,"E:\Projects\Calc.NET\obj\Debug\"
3=,"ARM_Setup",,"E:\Program Files\Microsoft Visual Studio .NET
2003\CompactFrameworkSDK\v1.0.5000\Windows CE\wce300\ARM\"

[SourceDisksNames.SH3]
4=,"SH34",,"E:\Projects\Calc.NET\obj\Debug\"
5=,"SH3_Setup",,"E:\Program Files\Microsoft Visual Studio .NET
2003\CompactFrameworkSDK\v1.0.5000\Windows CE\wce300\SH3\"

[SourceDisksNames.MIPS]
6=,"MIPS6",,"E:\Projects\Calc.NET\obj\Debug\"
7=,"MIPS_Setup",,"E:\Program Files\Microsoft Visual Studio .NET
2003\CompactFrameworkSDK\v1.0.5000\Windows CE\wce300\MIPS\"

[SourceDisksNames.X86]
8=,"X868",,"E:\Projects\Calc.NET\obj\Debug\"
9=,"X86_Setup",,"E:\Program Files\Microsoft Visual Studio .NET
2003\CompactFrameworkSDK\v1.0.5000\Windows CE\wce300\X86\"

[SourceDisksFiles]
Calc.NET.exe=1

[SourceDisksFiles.ARM]
vsd_config.txt.ARM=2
vsd_setup.dll=3
```

```
[SourceDisksFiles.SH3]
vsd_config.txt.SH3=4
vsd_setup.dll=5

[SourceDisksFiles.MIPS]
vsd_config.txt.MIPS=6
vsd_setup.dll=7

[SourceDisksFiles.X86]
vsd_config.txt.X86=8
vsd_setup.dll=9

[DestinationDirs]
Files.Common=0,%InstallDir%
Shortcuts=0,%CE2%\Start Menu
Files.ARM=0,%InstallDir%
Files.SH3=0,%InstallDir%
Files.MIPS=0,%InstallDir%
Files.X86=0,%InstallDir%

[Files.Common]
Calc.NET.exe,,,0

[Files.ARM]
vsd_config.txt,vsd_config.txt.ARM,,0
vsd_setup.dll,,,0

[Files.SH3]
vsd_config.txt,vsd_config.txt.SH3,,0
vsd_setup.dll,,,0

[Files.MIPS]
vsd_config.txt,vsd_config.txt.MIPS,,0
vsd_setup.dll,,,0

[Files.X86]
vsd_config.txt,vsd_config.txt.X86,,0
vsd_setup.dll,,,0

[Shortcuts]
Calc.NET,0,Calc.NET.exe
```

Running the CAB Wizard

Now that you can create and customize a configuration file, we can look at using the CAB Wizard, a command line tool with the following command syntax: `cabwiz.exe "inf_file" [/dest dest_dir] [/err err_file] [/cpu platform_label [platform_label]]`

`inf_file` The full path and filename to the configuration file described in the previous sections.

`dest_dir` The directory where the CAB files will be placed. This parameter is optional, and if it is not supplied, the wizard will place the CAB file in the directory in which the configuration file was found.

`error_file` The name and absolute path of the file that the CAB Wizard will use to write information about errors that occur while creating the CAB file. This parameter is also optional, and if not specified, the Wizard will display the errors in message boxes. The CAB Wizard runs without UI if this parameter is specified.

`platform_label` If one or more of these parameters are specified, then a CAB file is created for each platform_label that you specify. The `/cpu` option must be the last option used on the command line.

The following example creates a CAB file for the ARM processor: `Cabwiz.exe "c:\myApp\myconfig.inf" /err errors.txt /cpu arm`.

Distributing a .NET Compact Framework Application

Now that you have packaged your application in a self-extracting CAB file, you will need to make this CAB file available to your customers. Before we move on to discussing ways to distribute your application, let us examine what other files, if any, you need to distribute with your application.

It is not guaranteed that the target device will have the .NET Compact Framework installed. You may need to include the .NET Compact Framework CAB file in your distribution. Also, most commercial programs rely on components, managed or native, that may not be included in the CAB file for your application. You may need to package these other components in their own CAB file.

Now we can discuss the different ways to distribute your application. Any Windows CE application, not just a .NET Compact Framework application, can be distributed through these four different sources:

- From a Web site
- From a file share
- From a memory card
- Through ActiveSync

In addition to these four distribution strategies, you can distribute your application from one device to another over an infrared connection. This is not a serious distribution strategy and will not be discussed in this chapter.

Distributing Your Application from a Web Site

Distributing your CAB files via a Web site is an easy and efficient solution. Your customers never have to go to the store or order from a catalog. They simply browse your Web site through Pocket Internet Explorer and download the CAB files from your Web server. Once the CAB files are on the device, the user simply clicks the CAB files, and installation begins. Also, you can make updates, service packs, patches, and new releases available on the Web site at any time.

Distributing Your Application from a File Share

Distributing your CAB files via a file share is just as easy and efficient as distributing them from a Web site. This is not a sufficient solution for worldwide deployment of your application, as a Web site would be, but this deployment scenario is very attractive if you are an enterprise developer who is creating a solution for the employees in your company. The employees would have access to some internal file shares from which they could access and download product CAB files.

Distributing Your Application on a Memory Card

The Pocket PC OS provides a special feature that allows applications to be installed and uninstalled when a memory card is inserted and removed from the card slot. You could distribute your application to Pocket PC users on a memory card. When they insert the card into the Pocket PC, the OS will search the root directory for a subdirectory that has the same name as the device's processor. If that subdirectory is found, the OS will search it for an `Autorun.exe` file. If this file is found, it is copied to the `\Windows` directory. Next the OS attempts to execute the `Autorun.exe` file with the *install* parameter. When the memory card is removed, the OS will execute `Autorun.exe` with the *uninstall* parameter. Fortunately, the Windows Platform SDK for Pocket PC provides an example `Autorun.exe` file. You can use this to install and uninstall your application when you distribute it to your Pocket PC users on a memory card.

Distributing Your Application Through ActiveSync

You can also use the ActiveSync connection between the user's desktop and device to distribute your application. ActiveSync provides a desktop component called the Application Manager to manage the installation of applications on a device.

The Application Manager uses an initialization (`.ini`) file to manage the installation of the application. The Application Manager also requires a desktop setup application that must check whether a current version of Application Manager is installed. If the Application Manager is installed, then the Application Manager is called, passing the `.ini` file.

Creating an Application Manager Initialization File

Let's examine the initialization file. The `.ini` file has the following format:

```
[CEAppManager]
Version      = 1.0
Component    = component_name

[component_name]
Description  = descriptive_name
[Uninstall   = uninstall_name]
[IconFile    = icon_filename]
[IconIndex   = icon_index]
[DeviceFile  = device_filename]
CabFiles     = cab_filename [,cab_filename]
```

The key values are now described:

`component_name` This is the name of the section that contains the application initalization keys.

`descriptive_name` This is the name that will appear in the `Description` field of the Application Manager when a user chooses the application.

`uninstall_name` Identifying the application's Windows Uninstall registry key name, this name must match the key name found in the `HKLM\Software\Microsoft\Windows\CurrentVersion\Uninstall` registry key. This allows the Application Manager to remove the application from the desktop computer and the device automatically.

`icon_filename` This identifies the desktop icon file. This string is used to display the `device_filename` when the filename is viewed in ActiveSync.

`icon_index` This numeric index in `icon_filename` is used to display the `device_filename` when it is viewed in ActiveSync. If this key is nonexistent, the first icon in `icon_filename` is used.

`device_filename` This filename on the device will display the icon specified by `icon_filename` and `icon_index` when the `device_filename` is viewed in ActiveSync.

`cab_filename` This is the filename of the available CAB files, relative to `install_directory`. Use commas to separate multiple `cab_filenames`.

Creating an Application Manager Desktop Setup File

The Application Manager also requires a desktop setup application. This setup application must check to make sure that Application Manager is installed and then calls Application Manager, passing the `.ini` file. Listing 16.2 presents part of a managed Application Manager desktop setup application written in C#:

LISTING 16.2

```
static string APPMAN_REGKEY_NAME =
  "software\\Microsoft\\Windows\\" +
  "CurrentVersion\\App Paths\\CEAppMgr.exe";


static bool GetAppManPath(out string appManPath) {
  appManPath = null;
  RegistryKey appManRegKey =
    Registry.LocalMachine.OpenSubKey(APPMAN_REGKEY_NAME);

  if(appManRegKey != null) {
    MessageBox.Show("Application Manager not found");
      return false;
  }

  appManPath = (string)appManRegKey.GetValue(null);
  if(appManPath == null) {
    MessageBox.Show("Application Manager not found");
    return false;
  }

  return true;
}
```

The `GetAppManPath` is used to get the path and filename to the Application Manager. This is done by opening registry key `software\Microsoft\Windows\CurrentVersion\App Paths\CEAppMgr.exe`, which is located in the Local Machine registry hive. If this registry key does not exist, the Application Manager is not installed, and the application displays an error message.

Next the function gets the value of the default key under software\Microsoft\Windows\CurrentVersion\App Paths\CEAppMgr.exe. This value is the path to the Application Manager. If this value does not exist, something must have gone wrong with the Application Manager installation, and we cannot run the Application Manager. In this case an error message is displayed to the user. If the path is found, then it is returned in the appManPath out parameter. Listing 16.3 describes how to launch Application Manager.

LISTING 16.3

```
static void LaunchAppMan(string appManPath, string iniFilePath) {
  string quotedIniFilePath= "\" + iniFilePath + "\"";

  ProcessStartInfo processInfo =
    new ProcessStartInfo( appManPath,quotedIniFilePath)
  processInfo.CreateNoWindow = true;
  Process.Start( processInfo );
}
```

The LaunchAppMan function is a process that launches the Application Manager with the .ini filename as the sole parameter. The function assumes that the filename is completely qualified. Then, it launches the Application Manager with the fully qualified .ini filename. This function will end after launching the Application Manager without waiting for the Application Manager to end. Listing 16.4 contains the driver function for the Application Manager executable.

LISTING 16.4

```
static int Main(string[] args)
{
  string appManPath = null;

  if(!AppManInstaller.GetAppManPath(out appManPath)
     || appManPath == null) {
     return 1;
  }

  if(args.Length > 0)
    MessageBox.Show("No initialization file was specified");

  LaunchAppMan(appManPath, string.Empty);

  return 0;
}
```

The `Main` function glues the `GetAppManPath` method and `LaunchAppMan` method together. First the path to Application Manager is found using the `GetAppManPath` method. If the path cannot be found, then an error code of 1 is returned. Then we pull the setup `.ini` filename from the command line arguments and use it along with the path to Application Manager to call the `LaunchAppMan` method.

You now have a desktop setup application that you can use to distribute and install your application through ActiveSync.

In Brief

- All .NET Compact Framework applications should be packaged and distributed in a CAB file.
- You can use Visual Studio .NET to build your CAB file.
- You can also use the CAB Wizard to build your CAB file.
- The CAB Wizard requires a configuration file, which you can edit, to build the CAB file.
- You can distribute your application four different ways:

  From a Web site

  From a file share

  From a memory card

  Through ActiveSync
- Distributing your application through ActiveSync requires an initialization (`.ini`) file and a desktop setup application.

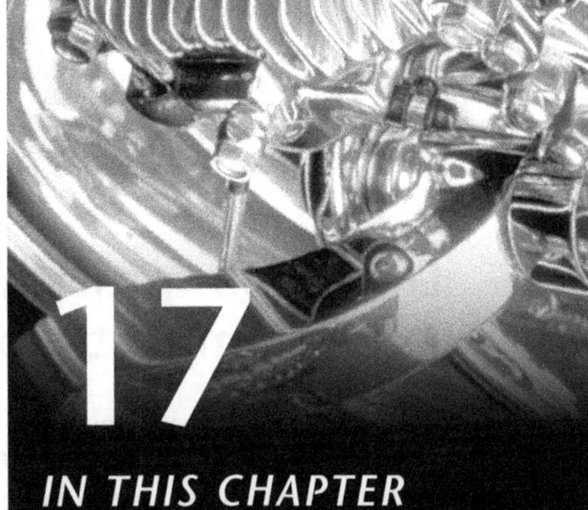

Developing for the SmartPhone

IN THIS CHAPTER

Introducing the SmartPhone

The SmartPhone is essentially a cell phone that runs the Pocket PC operating system. To facilitate the small form factor of a cellular telephone, the SmartPhone deviates from a standard Pocket PC device in two important ways:

- The screen size for the SmartPhone is smaller than those found on standard Pocket PC devices. The SmartPhone screen resolution is 176 × 220, compared with 240 × 320 for the standard Pocket PC form factor.
- The SmartPhone screen is not touch-sensitive. This radically changes the paradigm by which users enter information to applications. Users interact with applications by pressing physical buttons on the phone.

These two differences make developing for SmartPhones a different endeavor from developing for a Pocket PC or a Windows CE device. The most obvious difference is the smaller screen size, which requires developers to think about screen real estate more carefully. But the lack of a touch-sensitive screen and keyboard means that a whole swath of user interface objects, including the venerable button, stop making sense.

This chapter explores how to develop for the SmartPhone by using the .NET Compact Framework while working around the SmartPhone's inherent limitations.

Developing for the SmartPhone by Using the .NET Compact Framework

To develop for SmartPhone, you must install the SmartPhone support package for Visual Studio, which is available from Microsoft. By default, Smart Device Extensions for Visual Studio lets you create projects for the Pocket PC or the Windows CE platform. By adding the SmartPhone support package, the SmartPhone platform is added as an available project type.

Because the SmartPhone add-on is simply an extension for Smart Device Extensions, you get the same development experience when developing for the SmartPhone platform as when developing for Pocket PC or Windows CE. The major differences are that there are new emulators to deploy to and the controls not supported by the SmartPhone are ghosted out in the ToolBox. These controls are disabled:

- `Button`
- `RadioButton`
- `ListBox`
- `TabControl`
- `DomainUpDown`
- `NumericUpDown`
- `TrackBar`
- `ContextMenu`
- `ToolBar`
- `StatusBar`
- `OpenFileDialog`
- `SaveFileDialog`
- `InputPanel`

The major challenge in working with the SmartPhone is designing an effective user interface by using the controls available to you in the ToolBox. The power of the rest of the .NET Compact Framework is still available to you.

Miscellaneous SmartPhone OS Issues

It is important to be aware of differences in the SmartPhone file system compared with regular Pocket PC and Windows CE platforms. On the SmartPhone only the `\Storage` directory is writable. Thus, all managed applications are deployed to the `\Storage` directory, and you can write files only in the `\Storage` directory.

Writing an Application for SmartPhone—XMLDataSetViewer

We'll use a tutorial approach to build a simple XML `DataSet` viewer. The XML `DataSet` viewer shows the first table in a `DataSet`. The tutorial provides developers with an existing SmartPhone project to experiment with. It also demonstrates how the .NET Compact Framework development experience is nearly unchanged when working with SmartPhone.

> **MANAGING FILES ON THE SMARTPHONE**
>
> Managing files on the SmartPhone OS can be cumbersome because there is no touch screen or File Explorer program. With physical SmartPhone devices, you at least have the option of accessing the SmartPhone's file system via ActiveSync. But this problem gets especially annoying when working with the emulator, because by default, you cannot access the emulator's file system through ActiveSync. Keep a lookout at `http://www.gotdotnet.com` in case Microsoft releases any utilities to work around this limitation.

Before starting the tutorial, we assume that the SmartPhone add-on has been installed. The steps that follow demonstrate how to create the XMLDataSetViewer utility from scratch. Finished versions of the XMLDataSetViewer are available in the directory `\SampleApplications\Chapter17`. There are C# and Visual Basic versions of the XMLDataSetViewer.

Building the DataSetViewer from Scratch—Tutorial Steps

1. Launch Visual Studio .NET and create a new project. You can choose a Smart Device Application project for either C# or Visual Basic.
2. Follow the same steps as you normally would for creating a new Smart Device Application, except when asked to choose a platform to target, choose SmartPhone instead of Pocket PC or Windows CE. This step is shown in Figure 17.1.

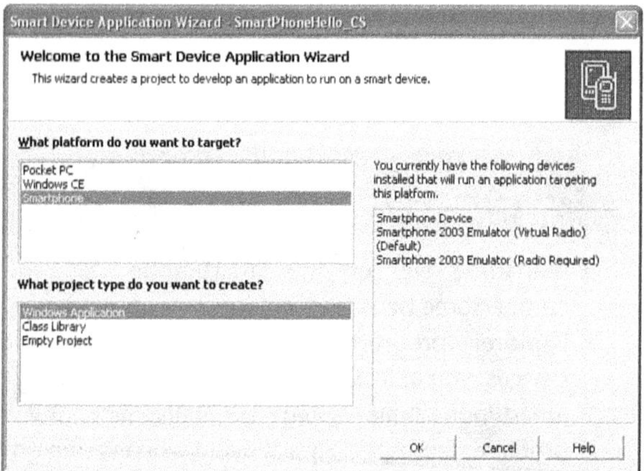

FIGURE 17.1 When creating a new project, choose the SmartPhone platform.

3. When you have finished setting up the new application, you will see the form editor and the Toolbox, as shown in Figure 17.2. This screen looks almost identical to that for a Pocket PC project, except that the form is smaller to reflect the smaller screen size on SmartPhones. Also, some of the controls in the Toolbox are ghosted out.

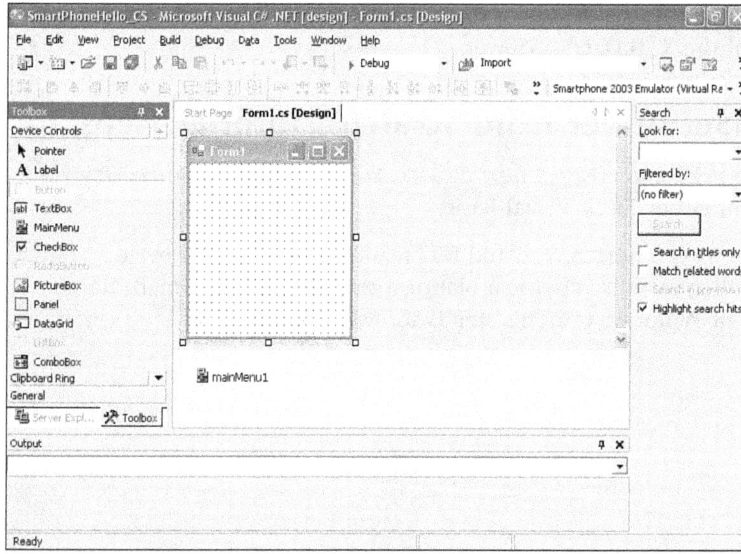

FIGURE 17.2 The form editor and Toolbox for a SmartPhone project contain some ghosted-out items.

4. Drag a DataGrid and a TextBox onto the control. Name the DataGrid `dgDataSet` and the TextBox `txtXmlToLoad`. For convenience, use this text as the default for the TextBox: `\Storage\Program Files\XMLDataSetViewer_CS\SampleDataSet.xml`.

5. Add a reference to the DataGrid to your project. To do this, right-click the solution name (for example, `XmlDataSetView_CS`) in the Solution Explorer. Then, select Add Reference. You will see a dialog window in which you can choose from a variety of DLLs. Choose the Browse button and move to the directory in which you installed Visual Studio. This is typically `C:\Program Files\Microsoft Visual Studio .NET 2003`. In this directory select the file `CompactFrameworkSDK\v1.0.5000\Windows CE\System.Windows.Forms.DataGrid.dll`.

6. Add menu items by clicking the `MainMenu1` icon that appears below the form editor in the IDE. You can add menu items by clicking new menu slots and typing the text for the menu item. For this project we chose two items: `Exit` and `Load XML`.

7. Add code to the `Exit` menu item by double-clicking it. The IDE brings up the method that gets called when the `Exit` menu item is clicked. Add this code:

```
C#
Application.Exit();
```

```
VB
Application.Exit()
```

8. Add a declaration for a `DataSet`, called `m_DataSet`, at the top of the `Form1` class. For example, the class member declarations for the project would look like this:

```
C#
public class Form1 : System.Windows.Forms.Form
{
   private System.Windows.Forms.DataGrid dgDataSet;
   private System.Windows.Forms.MenuItem menuItem1;
   private System.Windows.Forms.MenuItem menuItem2;
   private System.Windows.Forms.TextBox txtXmlToLoad;
   private System.Windows.Forms.MainMenu mainMenu1;

   private DataSet m_DataSet;
// Rest of class Form1 not shown here...
```

```
VB
Public Class Form1
   Inherits System.Windows.Forms.Form
   Friend WithEvents dgDataSet As System.Windows.Forms.DataGrid
   Friend WithEvents MainMenu1 As System.Windows.Forms.MainMenu

   Private m_DataSet As DataSet
   'Rest of class Form1 not shown here...
```

9. Add this code to the `Load XML` menu item by double-clicking the menu item and inserting the code in the code editor:

```
C#
if (this.m_DataSet == null)
{
   this.m_DataSet = new DataSet();
}
this.m_DataSet.Clear();
try
{
   m_DataSet.ReadXml(this.txtXmlToLoad.Text);

   // Set up a DataView
   DataView l_DataView = new DataView(m_DataSet.Tables[0]);
   this.dgDataSet.DataSource = l_DataView;
}
catch (Exception ex)
{
   MessageBox.Show(ex.ToString());
}
```

```
VB
If (Me.m_DataSet Is Nothing) Then
   Me.m_DataSet = New DataSet
   End If
Me.m_DataSet.Clear()
Try
   m_DataSet.ReadXml(Me.txtXmlToLoad.Text)

   ' Set up a DataView
   Dim l_DataView As DataView = New DataView(m_DataSet.Tables(0))
   Me.dgDataSet.DataSource = l_DataView
```

```
Catch ex As Exception
   MessageBox.Show(ex.ToString())
End Try
```

10. Add the default XML file, `SampleDataSet.xml`, to your application. To do this, view the Solution Explorer by pressing `Ctrl+Alt+L` and hover the mouse over the solution name (for example, `XMLDataSetViewer_CS`). Right-click and select `Add`, `Add Existing Item`, and then select the file called `SampleDataSet.xml`. You can find this file included with the completed XMLDataSetViewer project, either the C# or Visual Basic version.

11. Build and deploy your application! If you don't have any SmartPhone hardware, deploy to the emulator with the Virtual Radio.

Using the XMLDataSetViewer

Using SmartPhone applications is different from using Pocket PC applications because there is no touch screen. Instead, there are two menu buttons, a directional button, and a keypad for navigating an application.

To quit the application, click the left menu button, which triggers the Exit menu item.

To select the XML file to load in this simple application, you must insert the full text path of the XML file to load into the textbox. To do this, first make the textbox the active control in the application. Once the textbox is the active control, you can move the cursor with the cursor keys and insert text by using the numeric keypad.

To load the selected XML file, click the right menu button, which triggers the `Load XML` menu item.

The DataGrid is very small on a SmartPhone screen. To navigate through the data, first select the DataGrid as the active control by pressing the down button on the directional keypad. Once the DataGrid is the active control, you can scroll up and down by pressing the up and down buttons on the directional keypad. If the data table being displayed is too wide to fit in the DataGrid, you can navigate left and right by using the left and right buttons on the directional keypad.

In Brief

- The SmartPhone platform is essentially a cell phone that runs a modified version of the Pocket PC operating system.

- The SmartPhone screen size is smaller than a standard Pocket PC, and it is not touch-sensitive. This changes the means by which SmartPhone applications interact with users when compared to the Pocket PC.
- To develop for SmartPhone by using the Visual Studio Smart Device Extensions and the .NET Compact Framework, you must install the SmartPhone support package from Microsoft.
- Developing .NET Compact Framework applications that target the SmartPhone is nearly the same experience as developing a Pocket PC application, except that many UI controls that only make sense with a large touch-sensitive screen are missing.
- Another major difference between the SmartPhone OS and the Pocket PC OS is the file system. On the SmartPhone, only the `\Storage` directory is writable.

Index

A

E

H

I

J

K

L

M

N

O

P

Q

R

S

T

U

V

X

www.ingramcontent.com/pod-product-compliance
Lightning Source LLC
Chambersburg PA
CBHW082117210326
41599CB00031B/5788

* 9 7 8 0 6 7 2 3 2 5 7 0 0 *